AF527896

Diseases of Small Ruminant

Diseases of Small Ruminant

by :

Prof. (Dr) V.S.Vihan

B.V.Sc., & A.H., M.V.Sc., Ph.D., JICA Fellow (JAPAN)
Principal Scientist & Head, Division of Animal Health
Central Institute for Research on Goats, Makhdoom, Farah, Mathura (U.P.)

2010

SATISH SERIAL PUBLISHING HOUSE

403, Express Tower, Commercial Complex, Azadpur, Delhi-110033 (India)
Phone : 011-27672852, Fax : 91-11-27672046
E-mail :info@satishserial.com, hkjain1975@yahoo.com
Website : www.satishserial.com

Published by :

SATISH SERIAL PUBLISHING HOUSE
403, Express Tower, Commercial Complex, Azadpur, Delhi-110033 (INDIA)
Phone : 011-27672852 Fax : 91-11-27672046
E-mail : info@satishserial.com, hkjain1975@yahoo.com

© Publisher, 2010

ISBN 81-89304-85-2

© 2010. All rights reserved, no part of this publication may be reproduced, stored in a retrieval system or transmitted in any form or by any means, electronic, mechanical, photocopying, recording or otherwise without the prior written permission of the publisher/author.

This book contains information obtained from authentic and highly regarded sources. Reasonable efforts have been made to publish reliable data and information, but the authors/editor(s)/contributors and publisher cannot assume responsibility for the validity of all materials or the consequences of their use. The authors/editor(s)/contributors and publisher have attempted to trace and acknowledge the copyright holders of all material reproduced in this publication and apologize to copyright holders if permission and acknowledgements to publish in this form have not been given. If any copyright material has not been acknowledged please write and let us know so we may rectify it.

Typeset at: **Harminder** *for* **Laxmi Art Creations** # 98 11 48 23 28
Printed at: Salasar Imaging Systems, Delhi-35

Preface

The goat and sheep, the two small ruminants, are thought to have been the earliest domesticated ruminants. The two species are widely distributed in all the agro-ecological zones of India. They support a large population of the rural landless and marginal farmers and have an important role to play in any programme aimed to alleviate rural poverty and livelihood security particularly in arid, semiarid and hilly regions of the country. Small ruminant greatly contribute to the food and nutrition security of the marginal and small farmers and landless rural people through out the country. Presently, the country possesses over 124 million goats and 61 million sheep which annually contribute over Rs.13,500 million to the national economy. Presently, the gross output form sheep and goat sector alone has been estimated to be more than Rs. 100 billion through meat, milk, wool, fiber and other products. The socio-economic importance of small ruminants in India is evident by the sharp increase in their numbers and contributions. The two species substantially contribute milk, meat, fiber, skin and manure to the subsistence of small holders and landless rural poor. They also play an important role in income and employment generation and house hold nutrition. Importance of goats and sheep lies in the fact that human population is increasing and economy is also growing rapidly creating increasing demands for animal protein foods. On the other hand feed resources for maintaining large ruminants are decreasing due to shrinkage of grazing lands. This demand can, therefore, be met by increasing productivity of goats and sheep. The productivity of Indian sheep and goats is, however, adversely affected due to incidence of several diseases.

The incidence of diseases and parasitic infestations in small ruminant is one of the major constraints in the development of this enterprise and is reported to cause huge loss to the poor farmers and consequently to the Indian economy. Diseases in small ruminant result into very high mortality which ranges from 5 to 25%. In addition, diseases also cause morbidity losses in terms of low productivity of animals. Much effort in this country has been made to diagnose diseases, develop vaccines and immunize them. In order to maximize the productivity of the two species, in addition to the components of breeding, nutrition, management, optimum health protection plays an important role. There are heavy losses of good germ plasm due to diseases. Several diseases of the two ruminants are emerging and re-emerging due to the shift in the management system. Disease is indeed a major cause of economic wastage and inefficient production in small ruminants. Control of animal diseases will reduce the yield gap between the actual and potential production leading to a substantial increase in the outputs.

Over the years a number of animal diseases have been introduced through import of livestock and livestock products and many of them have established themselves and are causing huge economic losses every year. There is always a risk of introduction of new diseases/strains of pathogens. Trans-boundary animal diseases and emerging, re-emerging and exotic animal diseases are assuming a new dimension to the livestock production systems in several countries. The key intrinsic factors for making these diseases important at a global scale are mainly the mode of survival and spread of the causal agents, variability and the availability of hosts, alternate hosts and vectors. Infectious diseases do not respect National boundaries. Human epidemic of plague in 1994 in India, alarming panic created by SARS and bird-flu epidemics in 2003-06 resulted in declaring a **Global Alert** for the first time in the history of WHO. Animal health interventions have to some extent helped in reducing losses due to mortality and morbidity. But a lot more is required to be done for providing improved diagnostics and vaccines needed at affordable costs. The prevalence of animal diseases may preclude entry of a country in the lucrative international markets of meat, hides and skins. The success in increasing export of livestock products would depend on making our livestock free from diseases to meet the SPS requirements laid down by WTO. Launching of systematic disease control and eradication programs for OIE listed diseases along with effective disease surveillance program and strict enforcement of SPS measures in processing livestock products are critical for international acceptance of our products.

The availability of reliable data especially fȯr OIE listed 'A' and 'B' diseases viz., PPR, Foot and Mouth, Blue-Tongue and Brucellosis is a pre requisite for their control program. PPR is a dreadful disease of small ruminants. With the eradication of Rinderpest from the country, the prevalence of PPR has gained significance. The endo-parasites are ever-present health hazard in small ruminant production. The Gastro Intestinal Nematode parasitic control for the past thirty years has rested firmly on the shoulders of anthelmintics. Emergence of anthelmintics resistance against major anthelmintic groups is a threat to future parasitic control program. Investigations carried out so far indicate that more than 80% of sheep flocks have resistance strains of *Haemonchus contortus* to Benzimidazole group of anthelmintic and the situation with other group of anthelmintics is equally discouraging. A detailed study of the toxic plants responsible for malicious and inadvertent poisoning in animals during grazing and mapping of the toxicological plants in various zones of the country has greatly helped in developing feeding systems. During the past few years, the world has witnessed a global emergence and re-emergence of several infectious sheep and goat diseases that have had, and are still having, significant impact on both animal and human health. It has been worked out that economic gain from small ruminants could be easily doubled by giving due care to the prevention and control of goat and sheep diseases. The author of the book **"Diseases of Small Ruminant"** have made an attempt to cover most of the information on managemental practices, clinical examination, collection, preservation and dispatch of materials, important bacterial, viral, chlamydial, mycoplasmal, rickettsial, fungal, parasitic, emerging and exotic diseases, production and metabolic diseases and toxicities which have been included in this book. The author would like to thanks and duly acknowledge different scientists/ researchers whose references, photographs and informations have been used in completing the manuscript of this book.

The encouragement and blessings poured by friends and well wishers are gratefully acknowledged. I feel pleasure in extending thanks to various scientists who has provided the information on various topics included in this book. It is a matter of pleasure in extending thanks to all our colleagues of different divisions of the Institute for their help at different stages in preparing the manuscript.

I am sure this book will serve as a reference book to all those engaged in small ruminant production and control of small ruminant diseases in the country.

V.S. Vihan

Contents

1

Introduction

The rapidly changing patterns of demand for livestock and livestock products point to livestock production being an increasing component of the agricultural economies of India. The extent to which the rural poor will benefit from these changes depends on how livestock can be integrated into developing markets and whether cheaper livestock products benefit the rural poor as consumers as well as producers. There is scope for goats to play an important role for smallholder farmers in accessing these new markets. Their significance, which is now being exploited in several countries, is that they are small livestock in high demand and can thrive on low inputs and local resources. Poor men's cow - the Goat is thought to have been the earliest domesticated ruminant and of all the species of domesticated animals except dog, has the widest ecological range. Originating in Asia, goats have spread over all the continents and inhabit almost all-climatic zones from arctic-circle to the equator. Goats have served the mankind earlier and longer than cattle and sheep. They are often termed as the poor man's cow. They have distinct social, economical, managerial and biological advantages over other livestock species. They significantly contribute to the agrarian economy and play a very vital role in the livelihood security of the small and marginal farmers and landless labourers especially in arid, semi-arid and mountainous regions of the country. Goats have several advantages over other animals. They are widely adapted. They thrive well and reproduce in tropical, cold, humid as well as dry regions. Their small size compared to cattle and buffaloes permit them to be maintained on a limited area. The breeding animals are inexpensive. Goats consume a wide variety of grasses, weeds, forbs, bushes, shrubs, tree leaves and crop residues that would otherwise go waste and cause pollution. They are gentle and easy to control. Their small size makes them suitable for home slaughter and the meat can be consumed before it spoils in warm climate where no refrigeration facilities are available. The goats can be milked any time of the day and are therefore named as the moving refrigerators. Goat milk is prescribed for children, old and sick as it is easily digestible and has medicinal value. Goat meat (chevon) is preferred over other meats because it is leaner and there are no religious taboos against its consumption.

1.1 Goat Population and Production

World's current goat population is around 790 million. Asian region has the highest and over 63.57% of the world population of goats. India possesses 120.0 million making 23.89% of Asia and 15.19% of the world goat population. The developing countries of the world have about 94.5% of the goats and the developed ones only the rest 5.5%. While the goat population has increased in Asian and African regions, it decreased in America, Europe and Russia during the last over three decades. The over all world growth rate, however, has been almost uniform during the last about 25 years. The growth rate in India varied from 0.94 to 5.13% with an average of 3.05 % during 1951-2004. The goats around the world contributed 13827.1 TMT of milk, 4953.0 TMT of meat and 910.5 TMT of fresh skins annually.

The Asian region contributed 56.71% of the milk, 74.86% of the meat and 78.80% of the fresh skins of the world production of goats. India produced 27.40% of the milk, 9.60% of the meat and 14.23% of the fresh skins of the world and 48.33% of the milk, 12.80% of the meat and 18.06% of the fresh skins of the Asian goat production. Around 89% of the goats in the world are reared primarily for meat. In India about 40% of the total population of goats is slaughtered for meat every year. Goats around the world contributed only 2.12% of the total meat and 1.60% of the total milk production from all species.

1.2 Ecology

The livestock in general and the goats in particular have been held responsible by the foresters and environmentalists for causing ecological degradation and desertification in Asian continent. There has been a lot of controversy over the role of goats in ecological degradation and in desertification. Two conflicting views prevail on the goat's role in land use (1) The goat is the major cause of deforestation, rangeland destruction and soil erosion, and as such, its propagation should be checked, and (2) Goat acts as regenerator of vegetation through dispersal of seeds in its droppings and vegetative propagation through browsing. It is undoubtedly true that an uncontrolled population of goats roaming in open can be extremely harmful to the growth and development of vegetation. But it is just as true that any over grazing by any other ruminant will have similar consequence. The high pressure and shrinking of the grazing lands, especially in view of their low carrying capacity results in over use of the land and depletion of natural vegetation. The man to land ratio is fast declining. The livestock population is increasing due to low farm incomes from marginalizing of land holdings and increased dependence on livestock, leading to over exploitation of the shrinking grazing lands. In addition, indiscriminate felling of trees and deforestation, removal of shrubs, bushes and roots for fuel purposes, use of mechanical powers for cultivation and continuous neglect of the grazing lands are the real reasons causing damage to the ecology. Sheep because of their bifid upper lip are able to graze closer to the ground and in doing so in dry weather and on loose soils, frequently uproot the small grass species and damage the soil. The cattle with their large size hooves and heavier body weights cause more loosening of the soil rendering it susceptible to water and wind erosion. When the sheep move towards the water points or night shelters, they follow each other in a row because of the flocking instinct and generate tracks, which become free of vegetation and subject to erosion. Agriculture operations in general and on the arid and semi arid lands in particular cause more soil erosion than by the livestock grazing. Goats are mainly browsers and would seldom go for grown up trees. They dislike the leaves of timber trees. They also do not prevent the establishment and spread of grasses that are so essential for soil conservation. It was found that the goats spend more than 90 % time on browsing and hardly graze for 10 % time on surface vegetation. Goats have been found to defoliate smallest branches of trees without damaging the twigs in behavioural studies. Their browsing habit tends to reclaim saline soils by consuming salt-laden leaves of the range plants and contribute fertility to soil by even distribution of essential manure on the lands they graze. Cattle exploit 5-6 times more of biomass than goats. Goats also help in dispersal of grass, bush and tree seeds. They consume grass and bush seeds and fodder tree pods while grazing and defecate hard coat undigested seeds of pod-bearing and xerophytic trees after providing acid treatment while passing through digestive system and fortifying with nutrients in the form of the faecal pellets and spread uniformly all over the grazing land. These seeds germinate in large number as soon as soil moisture conditions are favourable. The goat saliva left on the bitten foliage adds nitrogen directly to the plant cells inducing quick regrowth. Saliva is also said to detoxify the high tannin available in many of the tree leaves found in the desert areas. The biting tender leaves and twigs by goats also induce tillering and faster luxuriant regrowth. Goat keeping has proved economical in the drier areas not suitable for water spreading or dry land cropping. Goats are less prone to toxic effects of noxious shrubs. The National Commission on Agriculture (1976) recommended reduction in goat numbers from then existing 67 to 40 million. The population has since then doubled. A Task Force to evaluate the impact of sheep and goat rearing in ecologically fragile zones was constituted by the Government of India in 1987 under the Chairmanship of Professor K. Hanumantha Rao, one of the leading economists and then a Member of Planning Commission, and with a large number of foresters and environmentalists as members. The task force observed that there was no definite evidence to prove that goats pose a threat to the ecology, as was generally believed. In any case, their negative effect, if any, on ecology has been vastly exaggerated. It further observed that within the desired grazing pressure, goats are more economical and less harmful than large ruminants. It was, therefore, recommended that goats should not be categorized as animals responsible for the destruction of ecology.

1.3 Socio-Economic Gains

The socio-economic importance of goats in India is evident by the sharp increase in their numbers and contributions during the last about 30 years. Goats contribute milk, meat, fibre, skins and manure to the subsistence of small holders and landless rural poor. They play an important role in capital storage, income and employment generation and house hold nutrition. Importance of goats lies in the fact that human population is increasing very rapidly creating increasing demands for animal protein foods on the one hand and the feed resources for increasing large ruminants are decreasing due to shrinkage of grazing lands on the other. This demand can, therefore, be met with by increasing population of goats. It is easier to increase the population of goats because the capital investment is relatively low, land requirements are small, reproductive rates are higher both due to shorter breeding interval and high prolificacy and they can be managed by spare family labour and do not require any serious housing facilities and management skills. There is much less risk in goat farming in drought prone areas where large mortality occurs due to frequent droughts. The goats act as an insurance against disaster under pastoral and agriculture subsistence system. Goats have religious and ritualistic importance in India. They are offered as sacrificial animals both by Muslims on Id and by Hindus especially the worshippers of Goddess Kali. They are worshipped for their creative and generative powers and sexual virility. There are no religious taboos against consumption of goat meat. Goat meat is preferred over other meats as it is leaner. Goat milk is easily digestible because of small sized fat globules. It has much less allergic problems than the milk of other livestock species. It also has medicinal value and can ward off many diseases as the goats browse on variety of plants including medicinal ones. Goats can be milked as often as required and are, therefore, called moving refrigerator. The goat skins are highly valued and have large export potential. The value addition in goat skin processing is much larger than other products. The bones of slaughtered and dead goats are also utilized for bone meal manufacture. A goat produces about 150 kg of dry manure per year for use in crop production and gardening. Goat browsing accelerates growth of trees, shrubs and surface vegetation. They also act as seeding machines. They are much less prone to the adverse effects of toxic plants. They have higher dry matter and fibre digestibility and can subsist on poor woody vegetation. Goats are able to obtain more nutrients from the given environment in all seasons than other livestock species. Goats make important economic contributions in India. They are so vital to a large human population that their contribution to national economy cannot be over looked. Goats require relatively much lower investments and facilities in terms of housing, feed, labour and health care. There is quick pay off due to fast multiplication and early maturity. Further, the risk involved in goat farming is much lower when compared to other livestock and crop production. Goats are reported to be more economical than cattle and sheep under natural grazing/browsing. The indigenous goats were found 2.5 times more economical than indigenous sheep when maintained on a free range grazing. Significantly more meat and milk production per unit live weight per year from goat than buffalo, camel and sheep have been recorded. The cost of production of goat milk worked out to be less than half than for cow milk while milk from buffaloes was intermediate. The results of a socio-economic survey in Rajasthan conducted have revealed that the number of goats increased 3 times between 1951- 1983 and goats accounted for 28-31% of the value of the livestock assets and for 16-19% of the gross receipts from crops and livestock. Some studies have revealed that the goats contributed up to 50-55% to the total cash income of a farm family in the hot arid region of the country. There are growing scientific evidences to suggest that goat keeping is more economical in drier regions that are not suitable for water spreading and dry land grain or fodder cropping. It is thus evident that goats make very sizeable economic contributions to the economy of India.

1.4 Goat Genetic Resources

Approximately 300 breeds of goats are reported to be existing world over of which 22 are in India. Majority of these are primarily meat breeds and found in the tropics with only a few tropical dairy breeds. Goat breeds belong to dairy, meat, fibre and pelt types or their various combinations and meat being an important component in economics of production of most of the goat breeds except highly specialized dairy goat breeds. In addition to the important descript breeds, there are a large number of non-descript populations; India is endowed with large and biologically diverse population of goats. They are widely distributed through out the country. There is serious need to describe and evaluate the existing genetic resources in terms of their population, population trends, flock size and structure, ecology, feed resources, management practices, body size and weight, confirmation and production characteristics using modern biological and statistical methods for accurate description and determining genetic differences. An Expert Panel on Animal Genetic Resource Conservation constituted by FAO has prepared descriptors for various livestock species so that uniform system of describing available genetic resources in different countries is utilized for comparison among different breeds within and across different countries. Some breeds have been identified as improver breeds and utilized within the country or in the region. Jamunapari, Barbari and Beetal have been utilized as improver breeds for milk and meat within India and in a number of South East Asian countries. Specialized dairy breeds viz. Saanen, Alpine, Toggenburg and AngloNubian have been used all over the world for improving milk production. Most of the goat breeds in India are found very well adapted to harsh climate, long migration, tropical diseases, poor nutrition and shortage of drinking water.

Indian Breeds

a. **Northern-Western Region:** (Punjab, Haryana, Rajasthan, Gujarat and U.P.)
Beetal, Barbari, Jamunapari, Sirohi, Marwari, Jhakrana, Kutchi, Gohilwadi, Zalawadi, Mehsana and Surti.

b. **Temperate Himalayan Region:** (Uttrakhand, H. P. and J. & K.)
Gaddi, Chegu and Changthangi

c. **Eastern Region:** (Eastern U.P., Bihar, Bengal, Orissa and Assam)
Bengal, Ganjam and Assam Hill

d. **Southern Peninsular Region:** (Maharashtra, Andhra Pradesh, Karnataka,Tamilnadu and Kerala)
Osmanabadi, Sangamneri, Malabari and Kannaiadu

Threatened Breeds of Goats and Their Conservation

The International Union for Conservation of Nature (IUCN) has classified the animals into seven categories on the basis of their viability viz. a. Extinct, b. Endangered, c. Vulnerable, d. Rare, e. Intermediate, f. In-sufficiently known and g. Out of danger. The goat breeds available in India are generally placed under Endangered and Vulnerable categories. A breed is considered endangered when its effective number is too small to prevent genetic loss of the population. As per FAO, a breed with a population below 5000 breeding females may be declared as Endangered. Some of the recognized goat breeds in India are threatened due to lack of breeding policy, non- availability of breeding males and indiscriminate cross breeding and intermixing with local populations. The other problems are spread of home tract of a goat breed in two or more districts/states, non availability of breed wise figures in the Indian Livestock Census Reports, migration of flocks, high slaughter rate of fast growing goats and the increase of ratio of non- descript to descript animals in different parts of home tracts of recognized breeds. A breed is vulnerable, when the population is rapidly declining numerically or its security is under threat. The purebred Jamunapari population has declined seriously and it is less than 8,500 in its habitat due to different reasons. Beetal is drawing attention for conservation as the number is declining due to changing agricultural patterns in Punjab. Barbari is facing dilution in breed characteristics due to cross breeding with Sirohi and Jamunpari breeds in its home tract. Due to decrease in natural browse and geographical restriction, Surti of Gujarat and Jakhrana of Rajasthan are decreasing in number. The Changthangi breed of the ladakh region is the subject for special concern as it is declining both in number and performance. More weightage on improvement of some breeds and extensive use of breeds such as Beetal, Barbari and Jamunapari may result in elimination of other distinct breeds due to cross breeding. Means for conservation, multiplication and improvement of these breeds using Open Nucleus Breeding Scheme (ONBS) have been suggested.

It is apparent from the current population size, loss of habitat and introduction of other breeds in the area that Jamunapari, Beetal, Jakhrana and Surti in North Western Region, Sangamneri, Osmanabadi and Malabari in Southern Region, Khasi and Ganjam in Eastern Region and Chegu and Changthangi breeds of goats in the Temperate Himalayan Region are threatened. The position of several other breeds is also not very comfortable due to acute shortage of elite breeding males except Marwari, Sirohi and Black Bengal which presently seem to be in safe position with respect to population and size of their breeding tracts. The conservation of goat genetic resources is essential and is of paramount importance owing to their widespread destruction, over exploitation and degradation all over India.

Generally two approaches viz. *ex situ* and *in situ* methods of conservation may be adopted for conservation of the goat breeds in India. Since the generation and loss of alleles is a dynamic process, it should be maintained at close equilibrium through sound management system. The maintenance of animals in natural habitat or in their adaptive environments as close as possible is called *in situ* conservation. Under *ex-situ* conservation the sperms and embryos of goats like other farm animal species can be frozen and subsequently used to produce normal off-springs. Embryos are considered to be the best material for *ex-situ* conservation because they store all genetic material in a single entity, which can give rise to new off-springs. Cryogenic storage of DNA is another method for preservation of genetic material. It has several advantages over live cells viz. avoiding the spread of diseases, requirement of less space and more economical. DNA can also be conserved in the form of transgenic animals. The other important aspects of conservation of goat genetic resources may involve establishment of National Goat Data Bank, National Gene Bank, National Goat Conservation Board, Regional Goat Gene-cum Data Bank and Global Data Bank etc.

1.5 Breeding Strategy

The breeding strategy involves decision on objective and approach in bringing genetic improvement. This also depends on socio-economic and ecological consideration, existing genetic resources, their productivity, possibility of their improvement through selection within a breed, grading with indigenous improver breeds, or replacing an indigenous breed within exotic breed if it is arising from similar ecological region and thus will be adaptable, or cross-breeding for evolving new breed combining the adaptation and hardiness of local breed and higher productivity of the exotic breed.

Breeding Objectives

Major breeding objectives may be for milk, meat and fibre or their combinations. For milk production, major consideration will be given to lactation yield, lactation length and milk composition. In meat production, live-weight gain, efficiency of feed conversion, carcass yield and quality, reproductive performance, survival and milk yield of the dam are important characters. In fibre production the greasy fleece weight, scoured fleece yield, fibre diameter and fibre length are important characters.

Breeding Approaches

Breeding approach may involve (i) selection within local breeds (ii) grading a local breed with superior indigenous breed (iii) crossbreeding with exotic breed for evolving new breeds combining adaptation and hardiness of the indigenous and higher productivity of the exotic and (iv) replacement of an indigenous breed with higher productivity exotic breed. The breeding approach will have to be decided on the basis of the performance of local breed, the size of the genetic variability in the characters of interest, physical environment and feed resources available to sustain higher production. Decision will also depend upon the demand for the products in markets close to the centre of production and distant markets, and handling, processing and marketing infrastructure available. There is likelihood that increased production or improvement in quality without assessing market demand and developing handling and marketing system may lead to serious economic deprivation to the producers.

Selection Criteria

Selection criteria will depend upon the phenotypic and genetic variances in and co-variances among important characters, their relative economic values and sources of information. The age at first kidding and kidding interval have moderate to high heritability and litter size and pre-weaning survival have low heritability. Body weight, growth rate, carcass yield and quality have reasonably high heritability. Milk yield and milk composition have medium heritability. Most of these characters except reproduction and survival have large phenotypic variability allowing superior animals to be identified easily both through subjective and objective assessment. Attempts have been made to develop selection criteria for improving production for various economic traits. For improving milk production in goats, selection based on age at first kidding and the first lactation milk yield combined into an index will provide maximum gains in reproduction and milk production. For improving meat production, selection on 6-month body weight may be most feasible and bring reasonable genetic progress in meat production through improving reproduction, body weight gain and carcass yield.

Performance Recording

Institution of selection would require identification of animals and recording of their performance. On station recording is more convenient, but the population size is generally small to provide sufficient intensity of selection and even accuracy of selection. Further it does not allow production of required number of superior males. The subjective assessment done by the farmers themselves also results in some improvement. Objective assessment both in terms of quantity and quality would be more desirable. In addition to the production, reproduction and survival are also important traits and will need recording. Once a month AM-PM milk recording is sufficient to determine lactation yield. Milk recording may be done after separating the kids from their mothers on the day of recording. In case of meat production, recording of body weight at weaning and 6 months of age may be sufficient. Where there are possibilities of studying individual feed lot performance, the fortnightly body weight gain along with fortnightly feed consumption, efficiency of feed conversion, carcass yield and quality could also be recorded and involved in selection of males. For improving prolificacy in meat breeds, selection of males from larger litters and on their body weight at 6-months can help in improving mothering ability and reproductive performance in addition to body weight gains.

Selection in Farmer Flocks

Through recording of performance of a large number of farmers' flocks preferably through objective assessment, flocks which are superior than average can be identified and used as flocks for producing genetically superior breeding males which could subsequently be distributed to the flocks which are inferior to the average of the population. This will require identification of superior males and females based on the characters of importance and mating them to produce male progeny, getting them reared in the farmers' flocks and/ or in rearing centres and making them available to the flocks to be improved on male to male replacement basis.

Selection in Institutional Flocks

Institutional flocks may be established through selection of superior males and females from farmers' flocks. This is the usual system of genetic improvement in majority of developing countries where the government has responsibility for providing superior breeding males to the flock owners. Such flocks are properly recorded and performance data are utilized in selecting the breeding males and females. The best males are retained in the institutional flock for breeding with elite females for producing young males and the rest are made available to the flock owners. In majority of cases the flock size is small allowing a few males to be used for breeding and thus the performance, may not be better than in the private flocks and males produced on such farms are not thus acceptable to the farmers. The number of males available are also rather small than those required. In case of selection for milk, the number of daughters per sire is usually small and considering the reproductive performance, the age by which the progeny performance becomes available, the male is too old to breed. A large number of superior males can also be produced utilizing the technique of multiple ovulation embryo transfer (MOET). This can be done both in the farmers' flocks and in institutional farms.

Selection in Open Nucleus Flocks

It may be desirable to organize the farmers into co-operative group breeding schemes. Open nucleus flocks can be established through selection of better performing animals on the basis of objective assessment from the farmers' flock and utilized for producing breeding males through intensive selection and their distribution to the cooperating farmers. Keeping nucleus flock open would help in reducing in breeding and increasing selection intensity through continuous introduction of superior animals from the farmers' flocks. After distribution of the males to the cooperating flocks, rest could be distributed to other flocks to be improved. Subsequently, flocks of cooperating farmers could also be utilized in breeding males using the selected males from the nucleus (multiplier) flocks. The use of MOET can help in improving genetic gains through selection. Such a selection scheme will bring faster genetic progress specially through reducing the generation interval. These selected breeding males could be more extensively utilized through artificial insemination with frozen semen.

Grading with Superior Indigenous Breeds

Grading with superior indigenous breeds will require similar approach as described earlier for producing superior males and their distribution to the flocks to be improved. This could be done with the farmers' flocks or in institutional farms. There are a number of superior breeds of goats in India viz. Jamunapari, Beetal, and Barbari that have been utilized with success for improving body size, milk yield and prolificacy in medium and small size breeds.

Crossbreeding with Exotic Breeds

A number of specialized dairy breeds viz. Alpine, Toggenburg, Saanen and Anglo-Nubian from Europe and North America have been utilized for improving milk production in indigenous goat breeds. The results indicate that there is substantial improvement of crossbreeds in milk production and reproductive performance specially age at first kidding although there is a slight decline in prolificacy. Half-breds have the best performance and there is little to gain by increasing exotic inheritance beyond 50%.

Utilization of MOET and AI

It will be essential that similar approach as in dairy cattle both in selection as well as in extensive utilization of selected breeders be followed for bringing quicker genetic improvement for milk. Utilization of embryo technology and artificial insemination with frozen semen will be necessary for this purpose. Extensive research and development effort has been made in developing method for long- term preservation of semen in goats with variable success. This is an area that would need serious attention in case quicker genetic improvement in goats is desired. Similarly, for making genetic progress through selection in the indigenous breeds or their crossbreds with exotic breeds where performance recording in farmers' flocks on continuous basis is not currently being practised and is difficult, the utilization of open nucleus breeding flocks with MOET can help in generating large superior male germ plasm which can be extensively utilized through artificial insemination with frozen semen.

1.6 Commercial Chevon Production

The goat meat available in the Indian markets comes either from old and culled adults or from male kids slaughtered any time between 6 months to 1 year of age. The quantity and quality of this meat is poor due to poor market weights, lower dressing percentage and narrow bone: meat ratio as these kids are maintained on scrub vegetation like their dams and hence hardly attain a body weight of 15-16 kg when they are normally marketed. The dressing percentage varies from 35 to 40 and bone: meat ratio from 1: 3 to 1: 4. The studies conducted in the country have revealed that a marked improvement can be achieved both in the finishing weights and carcass quality through intensive feeding of the male kids. Based on these findings two packages of improved practices have been developed and recommended for adoption by the goat farmers and the entrepreneurs for commercial goat meat production. A brief account of these packages is given below.

1.6.1 Package for Progressive Farmers

Goat Farmer Cooperative Societies may be formed in the goat rearing areas. These Societies in association with the village Panchayats should develop improved silvipastures on the available community grazing lands with financial assistance from the financial Institutions and subsidies being provided by the Central and State Governments.

The registered goat flocks may be allowed to graze on these pastures judiciously for 6 to 8 hrs daily. In addition to grazing, the pregnant does during last 30 days of pregnancy and the lactating does during first 60 days of lactation be supplemented with 300g/h/d concentrate mixture containing 12 % DCP and 65 % TDN to ensure 2.5 to 3 kg birth weights and 14 to 15 kg weaning weights in male kids. To ensure a weaning weight of 14 to 15 kg, these male kids are provided *ad lib* suckling, creep ration and green/dry leguminous fodders during pre weaning period and completely weaned at 60 days of age. These weaners may then be fed *ad lib* on complete feeds comprised of 50 % concentrate and 50 % roughage under stalls till they attain 25 to 30 kg finishing weight at around 6 to 7 months of age. Alternatively the kids be allowed to graze on available pastures and supplemented with concentrate mixture @ 1.5 to 2% of the body weight till they reach the desired finishing weights. The finisher kids should then be sold by the farmers or their Cooperative Societies for slaughter to the consumers or the traders directly avoiding involvement of middlemen.

1.6.2 Package for Entrepreneurs

The male kids produced and reared by the farmers in general and the progressive goat breeders following the recommended package of practices in particular may be purchased by the entrepreneurs at around 60 days of age and transferred from the villages to the Meat Production Complexes established near the cities. This will help in reducing the grazing pressure on shrinking pasture lands, early rebreeding and easy management of the flocks. These Complexes may be equipped with a Feed Compounding Plant capable of incorporating higher proportion of cheaper low grade roughages and agro industrial by-products to manufacture economic complete feeds, Feedlot Animal Houses for intensive feeding of the kids, Modern Slaughter House, Meat Processing, Product Manufacturing and By- Products Handling Machineries. The kids procured from the villages may be intensively fed on complete 50:50 or 60:40 concentrate: roughage feeds under feedlot from 2 months to 6 - 7 months age when they attain finishing weight of 25-30 kg. These intensively fed kids may then be sold as live animals in the National or International markets or slaughtered for selling fresh meat or their meat may be processed and converted into different meat products for export. Similarly the slaughterhouse by-products may be converted into commercial products for sale.

1.7 Economic Losses in Goat Production Due to Diseases

Goats occupy a premier place in the livestock sector and contribute largely to the economic and social welfare of the rural poor. Infact, there has been a dramatic increase in the worldwide interest in goats during the past decade and new initiatives are being undertaken for expansion of the goat industry in our Country. However, diseases pose a serious threat to the development of this sector. Thus serious efforts are being undertaken to facilitate improvement in disease recognition and diagnosis in order to increase opportunities for international trade in goats. In this regard PPR, Goat pox, FMD, Bluetongue and Contagious ecthyma are important diseases in small ruminants which inflict substantial losses in terms of reduced productivity in addition to lowering the quality of wool and leather obtained from them. They pose a major obstacle to the intensive rearing of goats and considerably hamper international trade. An indepth understanding of these diseases would be useful in the diagnosis, prevention and control as well as the management of these diseases in an effective manner. Although the diseases can be easily identified from the clinical signs and species affected, laboratory tests are essential for confirmation. A number of diagnostic tests are available for identification of these diseases; however, the best way to control these diseases is through prophylactic immunization of susceptible animals with a potent and efficacious vaccine, especially in areas where these diseases are endemic in nature.

Monetary Losses Due to Diseases and Parasites

The Royal Commission on Agriculture (1928) aptly, remarked "in no sphere has scientific research conferred greater benefit on agriculture than by providing a means of controlling livestock diseases". In 1966, the Chief, Animal Production and Health, FAO stated that "All the extra food required to meet the current world shortfall could certainly be provided if the annual losses due to animal diseases and parasitism are effectively controlled". This is more obvious under Indian conditions where mortality rate in goats have been reported to be as high as 77.3% in Jamunapari, 64.4% in Black Bengal, 62.6% in Barbari, 56.2% in Sirohi and 47.7% in Assam local goats by various research workers. Unfortunately, there is no readily available compiled information in the country indicating the monetary losses due to goat diseases. The incidence of goat diseases was surveyed by many workers. It was observed that in most of the states of India, information on the incidence of goat diseases is scanty. Small and marginal farmers and nomads who mainly keep goats in the country rarely go to veterinary hospitals for treatment of their animals and thus losses caused due to various diseases remain un-recorded. Many times, it has been observed that sick goats are slaughtered before death. However, it is a well known fact that losses in goat production due to various diseases as well as endo and ecto-parasites are quite high being highest in young kids and breeding goats. However, the economic losses due to diseases and parasitic infestation in goats may broadly be put into the two main classes (1) Mortality losses (2) Morbidity losses.

(1) Mortality losses

Death causes a direct financial loss to the goat farmers. In addition, it also causes loss of improved genetic material and expenditure on the treatment of animal is another cause of economic loss to the farmers.

(a) Kid mortality

Much has been written about specific causative agents and the disease they cause during their first six months. Various conditions responsible for mortality could be infectious and non-infectious. The important infectious conditions are colisepticemia and colibacillosis, various manifestations of chlamydia, contagious ecthyma, clostridial infections, FMD, PPR and pneumonia. The non-infectious conditions include feed oriented toxicity, metabolic diseases, deficiency of specific nutrient. Parasitic diseases include gastro-intestinal parasites, blood parasites and ecto-parasites. One study reported that 10 % kid mortality rate resulted in 38 % reduction in net profit. Thus, according to the calculation profitability from goat farming is lowered because of high kid mortality. The cost of rearing kids upto 0-3 months of age is Rs.187.37. The economic returns from dead kids could be added with the cost of rearing of kids upto the time of death to determine the exact cost of the dead kids. This will indicate the economic loss due to death upto weaning. But to make it more realistic, every kid should be assigned some value at the time of birth, depending upon the breed and this value should also be added to the figure of loss estimated for every kid. There is also a possibility of expenditures in terms of medical care on such diseased kids and such expenses should also be taken into account while working out losses due to kid mortality.

(b) Adult mortality

Under normal husbandry practice 5.0 % adult mortality has been accepted as normal. However, it can be improved further to save the huge economic losses. Goats suffer from a variety of bacterial, viral and parasitic diseases. The most important single cause of death in adult goats was pneumonia which accounted for 39.5 % deaths. The cause of pneumonia was attributed to pleuropneumonia like organisms (PPLO), chlamydia and parainfluenza-3 virus. Majority of adult goat mortality was found in does as they become weak after 4-5 kidding, while male goats are generally slaughtered before they get old and become unproductive. Secondly, under intensive or farm management condition adult mortality is significantly higher than when goats are kept in small flocks. Similarly, mortality has been reported to be high in migratory goats when they are kept under sedentary management conditions. In humid region incidence of parasitic infestation causes greater monetary losses, while in cold climate, incidence of pneumonia related deaths has been reported to be more.

(c) Abortion and still-births

Losses in goat production due to abortion and still birth are primarily because of decreased milk production by animals which have suffered abortion, increased rate of infertility, inter-kidding period and also loss of foetus. Some death of adult animals occurs as a result of acute metritis following retention of placenta. The causative agents of abortion are multifold, some are specific while others are non-specific. Organisms which constitute normal flora of the genital tract and are non specific, under favorable conditions, may become invasive. These include *E.coli, Mycoplasma* and *S. pyogenes*. The other organisms causing abortions in goats are *Brucella melitensis, Leptospira, Listeria,* Viruses, Chlamydia and Fungi. Ingestion of excessive amounts of preformed estrogenic substances in the diet may cause abortion. Some plants such as *Astragalus* contain estrogenic substances leading to abortions. Economic losses due to abortion and still-birth could be up to 10-15 % of the total loss due to disease incidence in goats.

Expected Economic Loss Due to Mortality

There is a considerable economic loss especially to the goat producers and to the country, in view of the mortality in kids and adult goats. As per livestock census 2007, there were about 124 million goats in India, of which nearly 50-60% are breedable females.

Assuming, that 90 % of the goats will be kidded once a year, then the total number of kids produced in a year will be 62-70 million. Again, as per reported figure if 20 % of these kids and 16.0 % of adult goats die, the total mortality would be around 15-20 million kids and 8-10 million adult females. Assuming average 4.0 % abortion and still births and value of reduced milk, meat and fibre production along with the expenses incurred on health care during the disease period then the estimates of losses will further move up. If all these losses could be checked by proper preventive health care measures, meat production would increase. The estimated losses due to mortality, abortions and still birth would be about Rs.10000 million. 20-25 % of such losses could easily be checked by taking preventive health care of goats in the rural areas.

Morbidity Losses

It is generally believed that goats are resistant to various diseases. But it has been observed that right from birth, many diseases may affect the goats leading to losses in their productive potential. The major economic losses due to morbidity may be grouped into four areas.

(1) *Production losses*: Direct losses through reduction of milk yield, retarded growth rate and alopecia.

(2) *Reproductive losses*: Reduction of reproductive efficiency, delayed sexual maturity, low twin-ability, increased inter-kidding period, infertility, decreased libido in bucks and poor quality of semen.

(3) Lowered efficiency in feed utilization.

(4) Cost for the treatment of individual animal, production of different vaccines in private and public laboratories, deworming and dipping of goats and extra labour involvement.

(5) Carrier of different diseases and parasites.

Economic Losses Due to Morbidity

It is very difficult to quantify the economic loss due to various types of production losses due to morbidity. Under feed lot experiment, production potential of different Indian goat breeds has been studied. Under normal farm management condition, they can produce only upto 50% of their production potential. The average carcass weight of Indian goats has been considered to be only 10.0 kg while in

Sri Lanka and Pakistan it is reported to be 20 and 21 kg, respectively. Moreover, when some of them have a case history of pneumonia, coccidiosis, parasitic infestation, Johne's disease, diarrhoea, nutritional toxicity as well as specific nutrient deficiency, their growth rate and milk yield is further reduced. Thus, the present economic returns could be significantly increased. Average carcass weight could be increased from the present value of 10 kg to 15 to 18 kg by checking production losses giving additional returns of about Rs. 4000-6000 million per annum.

□□□

2

Managemental Practices

2.1 Feeding Management

India is a group of States characterized by mixed farming systems. Livestock are common to all systems. They are an essential part of existing systems and offer opportunities for high-value production. Although a large number of goat feeding management systems are in practice and vary from region to region within the country. These can essentially be described under three systems viz. extensive, semi-intensive and intensive. The ultimate aim of adopting a particular feeding system should be to bring improvement in production, reproduction and survival rates of the goat flock. Extensive studies on evaluation of common grazing lands, developed pastures, supplementary feeding with fodders and concentrates under semi-intensive and intensive feeding under feed lot vis-a-vis performance and production levels in goats have been conducted.

Extensive System

Goats are generally reared under extensive system and play only a secondary role to crop as well as other livestock production. This system includes migratory, transhumance, free range, pasture and range grazing management of goats. The small size of goats has distinct advantages over other livestock species to contribute significantly to the sustainability of small farm situations that predominate in this country. The goats usually owned by small farmers and landless labourers are grazed together and tend to be herded over long distances in search of feed and water. The flock sizes are larger and goats belonging to several owners are run together. A low level of unpaid family labour represents the main input. The system is principally one of low resource use and a low level of productivity emerges from poor nutritional availability. There is marked fluctuation in feed availability and its nutritive value among different regions and among years and seasons. Goats require feed all the year round. They produce all the time if it is always plentiful and of the right kind. If, on the other hand, it is short or of the wrong kind only for a month or two there is less milk, less kids and less meat. The greatest limitation in our rangelands and natural pastures is on the availability of adequate energy throughout the year and adequate amounts of protein for more than half the year. During the period from January to June the problem of energy is very acute and the intake is much less than even the maintenance requirements. Although the grazing on the rangelands is considered to be the cheapest method for goat production, the migration of the animals and over grazing of available lands is causing serious problems of soil erosion and degradation leading to desertification. Due to long and dry summers, erratic rains and light textured soils deficient in organic content having tendency to salination, the natural potential of these lands falls short of the nutritional requirements of goats. The adult body weights, reproduction rates and milk production are lower and morbidity and mortality rates are higher. The birth, weaning and slaughter weights and dressing percentage are poor and bone: meat ratio is narrow in kids maintained solely on grazing on the degraded lands. The common grazing/ rangelands can hardly carry one goat per hectare per year when unprotected and two when protected from biotic factors. There is great prospect of

Fig.1 Extensive grazing system of goats

improving these natural rangelands. Reseeding with *Cenchrus ciliaris* and *Lasiurus sindicus* perennial grasses increased the grass yield from 0.5 ton to 1.5- 2.0 ton per hectare per year. Intercropping of cowpea in *Cenchrus ciliaris* pasture increased the dry matter yield by three times. Introduction of *Dolichos lablab* legume improved the nutritive value and increased the dry matter yield from 1-1.5 ton to 2.5-3.0 ton per hectare. Such a grass-legume pasture could carry five goats per hectare for five years. Plantation of 50 fodder trees of *Prosopis cineraria, Ailanthus excelsa* and *Leucaena leucocephala* had no adverse effect on the pasture growth and provided an additional yield of 1 ton dry matter when fully grown and lopped twice a year. Areas in arid and semi-arid regions that can't support cattle and buffaloes may be identified and utilized for small ruminant production with low investment. Rural common grazing lands should be improved through reseeding with nutritious, perennial and high yielding grasses and legumes. Fodder tree plantation should be taken up on pasturelands, wastelands, riverbanks and roadsides and bunds of ponds, canals and agriculture fields on large scale.

Semi-Intensive System

A kind of compromise between extensive and intensive systems is referred to as semi-intensive system of goat management. It is a combination of limited free range grazing and feeding in stalls. Integration of goat rearing with arable cropping and tree cropping is also included wherein either the goats are tethered or cut and carry system of available fodder is employed. Animals belonging to several owners may be combined for grazing which is mostly done morning and evening for 4 to 6 hours. The animals are supplemented with kitchen wastes, concentrate mixtures, crop residues, green and dry fodders and tree leaves etc. as per the availability. Thus goats utilize all available feed resources including natural grasses, shrubs, bushes, tree leaves, crop residues, stubbles, weeds, cultivated fodders and concentrates etc. under this system. The level of nutrition is just optimum or low but surely better than that under extensive system. Production performance depends on the quantity and quality of grazing/browsing availability and supplementary feeding provided. Studies have shown that the natural range/grazing lands cannot support growth of kids for commercial meat production and additional supplementation with concentrate mixture and/or leguminous fodders and tree leaves is recommended. Supplementation of additional energy and protein through concentrates has been advantageous in increasing the growth rate and attaining the desired slaughter weights at relatively younger age. It has been found that the kids fed on 50 concentrate: 50 roughage ration @ 4% of body weight attained 25 kg live weight at about 6 months of age. If the roughage portion of the diet can be made available from grazing, the remaining 50 % of the diet has to be supplemented through concentrate mixture. It is seen that although the goats are primarily browsing animals but a marked improvement in kidding rates, milk yield, live weight gains and quantity and quality of meat can be achieved by grazing/browsing the goats on developed two-tier pastures and supplementing with concentrate mixtures and/ or cultivated leguminous fodders and tree leaves at appropriate levels.

Intensive System

The intensive system of goat feeding management includes grazing on developed pastures and/or feeding completely in stalls on cultivated fresh or conserved fodders, crop residues and concentrates. Although goats prefer to browse as compared to grazing, they are quite capable of making efficient use of cultivated pastures for meat and milk production. Stocking rates of 16 to 60 goats per hectare are feasible depending on the type of grass, level of fertilization and the presence and absence of legumes and fodder trees. This system requires high labour and capital investment and is suitable for only intensive meat production from goats. In addition to providing better milk, growth and carcass quality, it removes pressure from the grazing lands. Several studies have been conducted on fattening of indigenous and crossbred kids under *ad libidum* feeding on concentrates including barley grain, maize grain, oil cakes, wheat and rice brans, molasses etc. and roughages including tree leaves, cultivated grasses and legumes in the ratio of 40:60, 50:50 and 60:40. The male kids weaned at 2-3 month age and fed under feedlot system achieved slaughter weights of 25 kg at 5 to 6 months age with a dressing % of 48 to 51. The kids maintained under semi-intensive system reached the target live weight of 25 kg earlier than those under intensive system. The kids under semi-intensive required less feed for a kg of gain than those under intensive feeding. The dressing % was superior under intensive. The bone and lean meat % was higher under semi-intensive and fat % under intensive system. The daily gains, milk intake, meat quantity and quality and feed efficiency were found superior in the kids maintained under semi-intensive as compared to those maintained under intensive or extensive system. The milk yield during 150 days of lactation was higher under intensive system than that under semi-intensive and extensive systems. The total confinement and complete feeding of goats in stalls has, however, proved to be detrimental some times. Integration of goats with tree cropping especially with reseeding area among rows of trees with shade loving nutritious perennial grasses and legumes has considerable potential. This system has the advantage of increased fertility of the soil via the return of dung and urine, control of weeds and thus reducing use of weedicides and fertilizers in addition to easier management of the crop. The value of goats is also very significant in the wider context of massive agro-forestry programmes taken up in our country. Since liberal use of crop residues and agro- industrial by-products has to be made for compounding feedlot rations, efforts to identify and utilize them should be intensified. Production and consumption of balanced goat feed mixtures as pellets and blocks based on maximum use of by-products should be developed in a coordinated manner. Cereal straws are an important roughage resource and their nutritive value can be improved with nitrogen-supplementation, ammoniation, urea solution and chemical treatments. Studies have shown that the low-grade roughages like straws and stovers added with little oil cakes, brans and mineral supplements may be very suitably compounded into Complete Feed Blocks and Pellets for goats. The technology of compounding Complete Feed Blocks is advantageous in several ways. Feed Blocks require less than one-third storage space as compared to the raw material. Transportation of Feed Blocks is easier and cheaper. Selectivity of ingredients by the animals from the Feed Blocks is avoided. Palatability and nutritive value of the ration is improved. Live weight gains and milk production are also improved.

2.2 Shelter Management

The role of shelter is of great significance for improvement in the productivity of goats. Unfortunately, due to socio-economic conditions of the farmers the housing facilities available for goats in our country are sub-optimal. Scientific efforts have been made to find out building material, the orientation of building, floor space and floor type, ventilation rate, height and shape of the roofs required for housing of goats.

Houses without Covered Space

This kind of houses are commonly practiced in Rajasthan where flock size is large. In this system, the area is fenced with thorny material possibly under the tree. Risk of exposure of animals to inclement weather and chances of theft and predation are always high in this kind of shelters.

Table 1: Production performance of goats under different systems of feeding management

Particulars	Natural rangeland (Extensive)	Developed Pasture (Semi-extensive)	Developed Pasture + Cone. Suppl (Semi-intensive)	Intensive
Kidding Rate (%)	68.0	87.0	113.0	108.0
Birth Weight (kg)	2.7	2.9	3.4	3.2
3m Weight (kg)	9.0	11.0	16.0	14.2
6m Weight (kg)	14.0	17.5	27.5	25.5
9m Weight (kg)	18.4	22.5	32.8	29.7
Dressing (%)	40.5	44.3	50.5	52.0
150 days milk yield (kg)	65.0	86.0	110.0	94.0
Adult Mortality (%)	10.0	7.5	2.5	5.0
Kid Mortality (%)	20.0	15.0	5.0	10.0

Fig. 2 : Goat Housing

Houses with Provision of Covered and Open Space

This kind of housing is provided where flock size is small or sufficient capital is available for construction of such shelters. The houses of this kind are suitable to provide complete protection from inclement weather and safety from theft and predation.

Slatted Floor Houses

Houses with slatted floor are suitable for the regions where humidity and rainfall period is of long duration. Slatted floor is prepared 1 meter above the ground level allowing sufficient space for cleaning. Wooden strips of about 2.5 cm thickness and 5 cm breadth are placed parallel to each to prepare the slat. A gap of 1 cm between two strips is kept so that fecal pellets and urine are easily passed though it and collected on the ground.

Underground Houses

Under ground houses are used in the temperate regions to protect the animals from sever cold. Underground sheds are capable to maintain warmth and comfortable microenvironment within the house. Provision of sufficient ventilation should be ensured. The floor of the pens should be dry and soft. Their frequent cleaning and disinfection is must. Otherwise, they may become an important source of infection. A 10 cm thick bedding of soft, dry grasses may be used in kidding/ lambing pens to prevent the newborns from possible infection through soil. Bedding is also a cheap source to prevent kids from exposure to cold.

Building Materials

Mud or bricks can be used to prepare walls. Chain link or welded mash may be utilized to fence the open paddocks as well as open space on the long axis walls. Thatch panels plastered with non-erodable mud mortar can also be utilized for preparing building structure.

Roofing Materials

Corrugated asbestos or iron sheets, mud tiles, traditional thatch, improved thatch are commonly used for roofing materials. The asbestos or iron sheets and mud tiles are long lasting materials. Iron sheets are not suitable as the temperature within such sheds during summer is observed higher than the outside temperature and during winters it is lower than the outside. Traditional thatch provide relatively much comfortable microenvironment than any other roofing materials. The life of such roofs is short. 'A' shape roof is most suitable to maintain even atmospheric conditions throughout the shed. Slope of the roof may be decided taking into consideration the intensity of rainfall in the area. Roof height of 3.5 to 4 meter may be kept in the centre and about 3 meter at the eaves.

Optimum Floor Space Requirement of Goats

The animal sheds should have a provision of open as well covered space. The area of open space is normally double of the covered area. The large area available in the open space provides an opportunity to the animals to move around freely. The covered area is utilized mainly to provide the shelter to the animals to protect them from inclement weather. The floor space requirement of different categories of goats under Indian conditions has been determined at Central Institute for Research on Goats (CIRG), Makhdoom Mathura, India.

Table 2 : Floor space requirements

Age/Category of animals	Floor space requirement per goat (mts)	
	Covered space	Open space
1. Up to 3 months age	0.20 - 0.30	0.40 - 0.60
2. 3 months to 9 months age	0.60 - 0.75	1.20 - 1.50
3. 9 - 12 months age	0.75 - 1.00	1.50 - 2.00
4. Adult goats	1.00	2.00
5. Pregnant and Lactating does	1.50	3.00
5. Breeding buck	1.50 - 2.00	3.00 - 4.00

Ventilation and Orientation of Shelters

The provision of proper ventilation in the goat sheds is essential to permit continuous movement of fresh air to meet the oxygen requirement of the animals and to remove the moisture and gases generating as a result of respiration, urination, defecation etc. This is possible when 30-50 per cent indoor wind velocity is maintained with the provision of sufficient ventilation. The work carried out at CIRG, Makhdoom suggested providing more than 100% ventilation during summer and rainy season. During winter season, when temperature reaches below 10°c, the ventilation at the rate of 2-5 per cent should be provided on the long axis walls of the shed. Rest of the open area may be covered using thatch panels of 10 cm thickness. Use of flexible ventilation during winter season by way of removing the thatch panels during day time will allow maximum air flow through the sheds and availability of sunlight to help in removal of moisture from the sheds at the faster rate.

Feeding and Watering Devices

Use of faulty feeding and watering devices may be the reason for contamination of feed material with soil, urine and faeces of the animal itself. It is observed that in traditional feeders, goats tend to contaminate the feed by voiding the faeces and urine, putting their legs, standing or sitting inside the feeder. Contaminated feed is the main reason for refusal and wastage of feed in higher quantities and for occurrence of many of the common diseases. Therefore, use of feeding and watering devices capable of preventing contamination to feed is of paramount significance. Feeders are of both rectangular and Hexagonal shape. On hexagonal feeders the feeder space requirement is reduced to about 40%. Their height can be adjusted as per height of the animal. The height of the feeding trough equivalent to 80% of the shoulder point height is found more comfortable to the goats.

Fig. 3 : Rectangular feeders

Fig. 4 : Hexagonal feeders

2.3 Management of Newborn to Weaning

Disease Problems

At birth, the major obstacles to survival are hypoxia, hypothermia, hypoglycemia and susceptibility to infectious disease. The latter is exacerbated by delayed or inadequate intake of colostrum rich in maternal immunoglobulins. Failure of passive transfer of maternal antibodies to newborn kids leads to increased morbidity and mortality throughout the postnatal period, particularly from septicemia in the first week of life and later, from pneumonia as kids reach weaning age and diminishing concentrations of maternal antibodies have not been fully replaced by immunoglobulins actively produced by the kid.

Failure to disinfect the navel after birth and maintaining kids in a dirty environment will also lead to increased incidence of omphalophlebitis, septicemia, polyarthritis, and pneumonia. Diarrheal diseases are common in this period. Numerous etiological agents can cause neonatal diarrhea over the first 2 weeks of life. After 2 weeks of age, coccidiosis is an increasing threat. A pneumoenteritis complex is sometimes seen in young kids, suggesting poor husbandry, heavy environmental pathogen loads and inadequate colostral intake. Contagious ecthyma also may appear clinically in this age group. Newborn kids should be inspected carefully for congenital defects such as cleft palate, rectovaginal fistula, atresia ani, hydrocephalus, umbilical hernia, and ventricular septal defect. The intersex condition is especially common in goats.

Managemental Practices

Kidding should be a well-anticipated event, not an unexpected surprise. Being prepared for routine processing of kids at birth and being ready to respond to emergencies will reduce neonatal losses. A clean, warm, dry, well-ventilated and well-bedded kidding area should be designated and maintained. Appropriate supplies should be on hand such as towels for drying off kids, feeding tubes for kids that do not suckle and tincture of iodine for dipping navels. A colostrum management program should be worked out before the onset of the kidding season. Proper identification and record keeping are signs of a well-managed farm. Ear tattooing of kids is recommended for permanent identification and should be performed after 3 month of age. Kids can be housed in group pens but should not be overcrowded. Pen design should allow easy and thorough cleaning. The spread of diarrheal agents is exacerbated by poor sanitation and overstocking. Early culling or marketing of unwanted male kids will reduce kid populations and the attendant risk of increased environmental contamination and the spread of infectious disease.

Castration is best carried out between 4 and 14 days of age. The reason for castration must be clearly defined. For meat kids, the intended market and the age at marketing will determine the need for castration. In some ethnic markets, uncastrated goats are preferred. If goats are kept longer for finishing, then they need to be castrated when the market will not tolerate a buck odor to the meat.

Vaccination and the timing of vaccination of neonates depend a good deal on the overall heard vaccination program. If does have been vaccinated during the dry period and kids receive adequate colostrum from vaccinated does, then kids can be considered to be protected by maternal immunity for those specific diseases through the first 8 weeks of life. Because the waning of maternal antibody varies with initial concentration, vaccination of kids should begin at 3 months with booster inoculations given at intervals if required. Kids are immunocompetent at birth. If kids are born to unvaccinated does, vaccination should begin during the fourth week of life and be boostered according to vaccine directions. Prophylactic antisera may be used during the first week concurrently with bacterin-toxoids to protect kids while active antibody is forming. Tetanus antitoxin and *Clostridium perfringens* type C and type D vaccines are the products most commonly used in kids. The vaccinations required for kids will vary with geography. However a universal recommendation is the vaccination of kids for enterotoxemia due to *Clostridium perfringens* type C and type D, Foot and Mouth, PPR, Goat pox and Pasteurellosis vaccine are mandated or recommended as part of National disease control programs. Still other vaccines such as anthrax, blackleg, contagious ecthyma vaccines may be used on the basis of knowledge that the disease occurs commonly in a specific region. Indiscriminate use of vaccines should be avoided, especially the use of live vaccines not specifically designed for goats. Where soils are selenium-deficient, nutritional muscular dystrophy in young kids is a real concern and parenteral administration of a combined vitamin E and sodium selenite injection to kids at birth is indicated.

Weaning may occur as early as 6 weeks or as late as 12 weeks of age, depending on the feeding system used and other management considerations. Regardless of the milk feeding regimen, it is important that kids be exposed to hay and grains early in life to promote proper rumen development. Abrupt cessation of milk feeding without adequate adaptation to solid feeds can predispose to indigestion, bloat and possibly polioencephalomalacia.

2.4 Management of Kids from Weaning to Breeding

Disease Problems

Coccidiosis and pneumonia are the dominant problems in this age group, particularly under conditions of confinement housing. Regarding coccidiosis, the transition to a diet of solid feed encourages more ingestion of coccidial oocysts while feeding. With regard to pneumonia, the progressive decline of maternal antibodies and the stresses of weaning appear to predispose kids to respiratory infections after 8 weeks of age. When kids are turned out to pasture, they are very susceptible to gastrointestinal parasitism and lungworms due largely to the absence of a developed local immunity in the alimentary and respiratory tracts. The severity of clinical parasitism will depend on the concentration of infective larvae on pasture, the stocking rate, weather conditions and other factors. Other disease problems also may occur in this group. There is a recurrent risk of enterotoxemia when concentrates are fed.

Managemental Practices

For housed kids, it is essential to maintain housing and feeding systems that minimize the incidence of pneumonia and coccidiosis however, coccidiosis can occur even on the most conscientiously managed farms. Therefore, in addition to good sanitary practices, it may be necessary to use a coccidiostat in the feed or water. Currently, there are no vaccines that can be confidently recommended for the control of respiratory disease in young goats. The role of viruses usually implicated in respiratory disease is unclear, and the available *Pasteurella* bacterins are of unproven efficacy. In countries where vaccination for caprine brucellosis is carried out, kids are usually vaccinated between 3 and 8 months of age.

Regarding pasture parasites, young goats should be turned out to pastures that have not been in use for at least year and should be kept on separate pastures from adults if possible. A broad spectrum anthelmintic should be administered before turnout and parasite loads monitored throughout the season by microscopic examination of composite fecal samples collected at pasture. Tactical anthelmintic treatments can be given accordingly as needed. Male and female kids should be separated by 3 months of age to avoid unwanted breeding among sexually precocious individuals. External genitalia of all kids should be examined before breeding to identify any intersexes present. In temperate regions, goats are seasonal breeders induced to begin estrous cycles by decreasing day length. There is much interest in manipulating the hormonal signals for estrus to permit goats to cycle year round. Year round breeding is of particular importance to large scale commercial goat farming who must provide a consistent volume of product on a year round basis. The two main techniques for bringing does and doelings into heat during the normally anestrous period of the year are the use of progesterone supplements, and the manipulation of day length through confinement and exposure to artificial light.

2.5 Management of Bred Does/Ewes

Managemental Practices

Does in lactation are usually given a dry period coincident with the last 2 to 3 months of gestation. Bred doelings that have not yet milked are also nonlactating in advanced pregnancy. The key objective for these animals is proper nutritional management to avoid obesity in early gestation (late lactation) and to provide adequate nutrients to support the rapid growth of fetuses in the last trimester of pregnancy. This is particularly important regarding pregnancy toxemia. The disease can be avoided by curtailing grain in late lactation (early gestation) when milk production no longer justifies it. These goats can be carried through mid gestation on high quality legume hay and little or no grain. Grain feeding should be resumed at the beginning of the last trimester gradually increased until term. In flocks already experiencing a pregnancy toxemia problem, periodic checking of urine samples from does and doelings for ketones with commercially available powders or dipsticks will help to detect incipient cases.

There are many potential infectious and non-infectious causes of late abortion in goats. To protect pregnant does from infectious abortion, no new animals should introduce in the flock when pregnant animals are present. This includes breeding bucks. Any animal that aborts should be isolated immediately and the area where the abortion occurred quarantined and disinfected. Attempts should be made to make a definitive diagnosis of the cause of abortion. All fetal tissues and placentas should be submitted for diagnosis of the cause of abortion. All fetal tissues and placentas not submitted for diagnosis should be buried or incinerated.

The dry period is the appropriate time to vaccinate pregnant does in order to increase the concentration of specific antibodies in the colostrum and thereby enhance passive immunity in the kids. Vaccines should be given to dry does 3 to 5 weeks before kidding. Pregnant does with nematode gastrointestinal parasites may experience a peri parturient egg rise with increased shedding of eggs around the time of kidding. This is a survival strategy of the parasite to enhance the likelihood of infecting a new generation of hosts. Therefore, pregnant does should be given a broad spectrum anthelmintic 2 to 3 weeks before kidding. Though no anthelmintic at proper dose has been definitively shown to cause abortion or teratogenic effects in goats, levamisole and albendazole have been suspected to do so, thus inhibiting their veterinary use in pregnant does.

2.6 Management of Milking Animals

Disease Problems

Not unexpectedly, mastitis is the major disease challenge in lactating does. A variety of pathogens has been identified in milking goats. *Mycoplasma* mastitis can be epidemic in some flocks. Mastitis due to *Staphylococcus aureus* can also be costly and difficult to control if aggressive culling of affected goats is not practiced. Caseous lymphadenitis can occur in goats of all ages but is particularly problematic in milking does. Abscessed lymph nodes most commonly occur on the head and neck. The restraint of milking does in head catches for milking increases the risk of abscess rupture and transmission of the disease. Some weight loss is normal in early lactation, particularly in high producing animals. The stresses of kidding and subsequent lactation however may trigger clinical manifestations of weight loss in carrier animals subclinically infected with *Mycobacterium paratuberculosis.* Animals that continue to lose weight during lactation should be tested for paratuberculosis.

When milking does are pastured, gastrointestinal parasitism must be anticipated as a potential problem. Though goats do develop an age and exposure related resistance to parasites, it is not absolute. Even if adult milkers do not develop overt clinical parasitism, the subclinical effects of parasites on milk production may be significant.

Managemental Practices

The major goal in managing lactating does is to maximize their milk output in a cost efficient manner. To accomplish this, both general health and udder health must be maintained. The feeding program must optimize production and be tailored to suit the needs of individual goats depending on their stage and level of lactation. In addition, a breeding and culling program that selects for superior milk production must be implemented. In support of these goals, health and production records, though they need not be elaborate, must be conscientiously maintained.

Preventive medicine procedures for milking does include vaccinations, foot care, and when conditions dictate, deworming. Proper foot care in the form of regular hoof trimming is extremely important. It reduces the likelihood of footrot, teat injuries, abnormal gait, and general sore footedness leading to decreased feed intake. The frequency of trimming depends to a large extent on the amount of exercise goats get. Natural wearing or trimming occurs through browsing and grazing, especially on rough or rocky surfaces.

Goats flock left out in pasture should be checked periodically for nematode parasites by fecal examination. Milking does can be dewormed tactically as required but there are two concerns.

Anthelmintics should be those approved for lactating animals and appropriate milk discard times observed. Good udder health depends on selection for does with good udder conformation and attachment, the provision of clean, well-bedded living quarters for milking does, properly functioning milking equipment, and an established protocol of udder hygiene during milking. Intramammary infusions of antibiotics at the beginning of the dry period can be used in flocks with an active mastitis problem. Clipping the hair on the udder and hindquarters will improve hygienic condition during milking. Milk quality is an important issue. Too often in the past, milk odors and taints have dampened consumer acceptance of goat milk as a wholesome, desirable product. Buck contact with milking does should be kept to a minimum and bucks should not be maintained near milking parlors or milk storage areas because buck odor will permeate the milk. Milking does should not be allowed to forage indiscriminately in unimproved pastures or woodlands as many plants may lead to off flavors or odors in milk. Proper handling of milk after extraction from the doe is extremely important. Excessive agitation and delays in pasteurization promote lipolysis of the fat in goat milk adding to the so called "goaty" flavor of the milk. No matter what is fed, there must be adequate trough or feeder space for each and every doe in the milking flock so that submissive does are not deprived of feed by dominant does. In addition, dominant does may be so busy bullying other goats that they may not eat adequately themselves and may need to be culled. Another point related to feeding is that the feeding of silage is associated, though not consistently, with the development of Listeriosis in goats.

2.7 Management of Bucks/Rams

This reflects the typical position of the bucks in the flock when it comes to the implementation of herd health practices. The buck by virtue of his offensive odor is too often left off the program. This omission should be conscientiously avoided because healthy bucks play a key role in reproductive efficiency and overall improvement of herd productivity.

Disease Problems

Disease conditions unique to bucks are limited. Obstructive urolithiasis is probably the most important. Though more common in castrated males, urinary calculi can also cause obstructive disease in intact males and may result in their destruction as breeding animals.

Bucks do become quite aggressive in the breeding season and can injure each other while fighting. They can also injure people, and workers should always exercise caution around breeding bucks. A common problem associated with fighting is the breaking off of a horn scur with profuse bleeding from the poll. The profuse bleeding, though rarely serious, is a frequent cause of alarm. Proper disbudding prevents this problem. In the breeding season, bucks acquire the habit of urinating on themselves, particularly on the face and the back of the forelimbs. This can cause severe urine scald and secondary bacterial dermatitis. Petroleum jelly applied to the back of the forelimbs will inhibit the development of serious problems.

Managemental Practices

To obtain maximum performance from bucks, they should be "tuned up" before the onset of breeding season. There is a definite tendency for bucks to become thin as the breeding season progresses. This appears to be physiologic rather than pathologic. Nevertheless, it underscores the need for bucks to be in excellent body condition before breeding begins. This requires regular deworming and proper feeding. Of equal importance is proper foot care. Mounting of does for breeding puts considerable strain on the back legs and poorly trimmed feet can produce sufficient pain to inhibit the buck from performing properly, particularly if the buck is used heavily. Proper foot care is too often overlooked because workers do not want to handle smelly bucks. Similarly bucks should not be left out of herd wide vaccinations simply because they are a nuisance to handle.

A complete reproductive examination should be performed on breeding bucks prior to use. This should include a general physical examination, genital examination, semen evaluation, and assessment of libido. This should not be done too long before the onset of the breeding season as semen quality may be diminished during the season of deep anestrous. If bucks are housed in groups year round, they will get used to each other and even at the onset of the breeding season are unlikely to do serious damage to each other by fighting. However, introduction of a new buck, especially a younger or smaller one, to the group during the breeding season could prove fatal to the newcomer. Clipping the beards and hair of bucks, especially on the forequarters, may reduce the generally offensive odor of the buck that results from urine retention on the hair. This should be done before the onset of breeding activity. The presence of a buck near the does is quite helpful in detecting estrus. Fence line contact is adequate and the buck need not commingle with the does to elicit estrus behavior.

2.8 Flock Health Management

In general commercial goat farming should be managed as closed herds to minimize disease problems. Taking goats to shows and fairs should be discouraged. If showing is necessary for the financial well-being of the farm through the sale of breeding stock, then isolation and quarantine procedures should be followed for animals returning from outside. A separation period of 3 weeks from the herd is warranted from the disease control point of view, artificial insemination is preferable to introduction of bucks or trucking of does to other farms for breeding.

Goats are very clever about undoing fasteners and gate closings, so gates and doors must be very secure to avoid unwanted escape and accidental access to feed and chemical storage areas which could result in engorgement toxemia, bloat, or poisoning. Goats are also notorious for chewing on the materials from which their enclosures are constructed. Therefore old sheds with lead based paints should be avoided as goat housing. When new housing is constructed, timber treated with potentially toxic chemicals such as pentachlorophenol used as wood preservatives should be avoided.

Goats are also very hard on fences because of their propensity for climbing and improper selection of fencing materials can be costly to owners. Fences with horizontal runs or slats encourage climbing and destruction of fences. Chain link fences predispose goats to catching their limbs in the fence with possible fractures resulting. If vertical slats or bars are used on gates or fencing, the spacing should be carefully evaluated to be sure that goats cannot become trapped and strangled, especially horned goats.

Goats, like sheep, are susceptible to predation due to their small size and lack of potent defenses. Kids are particularly defenseless and may fall prey to raptors and ground-dwelling predators. As the main defense of small ruminants is fleeing, the practice of tethering goats can be deadly where predators are common. Goats, like other livestock, can experience suboptimal performance or over disease from mineral deficiencies in feedstuffs related to soil composition in different parts of the world. The veterinarian and producer need to be aware of the mineral deficiencies occurring locally. Deficiencies well documented to cause financial loss in goats are selenium, iodine, cobalt, and copper. Fertilization, top dressing of pastures, supplemental feeding, and provision of trace mineral salt licks or parenteral administration of needed substances can be used to address these problems. Another potential local problem for goats is consumption of noxious or toxic plants.

The occurrence of one or the other disease causes great loss to the goat farmers. Death of animals results in direct loss whereas morbid condition causes indirect loss by affecting the production adversely. Sudden change in the ration, poor nutrition, un-hygienic conditions in the animal sheds, contaminated feed and fodder are some of the important reasons causing morbidity and mortality in goats. Therefore, it is important to watch the animals closely for their health status on regular basis. Detection of sick goats is relatively difficult. A sick goat may lag behind while grazing. Occurrence of dullness, reduced feed intake, dry and rough coat, abnormal physiological responses and decrease in body weight may be considered as the abnormal behavior of the animal. Animal showing abnormal behavior should be immediately separated out from the flock for detailed clinical examination. Knowledge of symptoms of various diseases will help in deciding appropriate line of treatment. The common diseases of goats on the basis of their etiology can be divided in to Viral, Bacterial and Parasitic diseases. The important viral diseases of goats are Pox, Contagious ecthyma, Foot and Mouth Diseases (F.M.D), Blue Tongue, Peste-des-petits ruminants (P.P.R.) and Pneumonia. Important bacterial diseases are Haemorrhagic septecaemia (H.S.), Johne's disease (J.D.), Enterotoxaemia (ET.) and Brucellosis. Parasitic diseases are Coccidiosis, Haemonchosis and Flukes. The non-specific cases of goats include Scours, Navel ill and Bloat.

❑❑❑

3

Clinical Examination

A complete clinical examination consists of three major elements: history taking, physical examination and inspection of the environment. Because many diseases seen in individual goats are likely to represent potential herd problems, prompt diagnosis of clinical cases is essential so that, in addition to therapy, appropriate preventive measures can be introduced into the overall management program. In many caprine diseases, subclinical cases often exist in addition to the obvious clinical ones, and additional diagnostic testing may be required to identify them. The existence of sub-clinical infections and carrier states is a troublesome one for performing prepurchase health examinations or writing health certificates for exportation or interstate travel.

3.1 Examination of Animals

Diagnosis and treatment of debilitating conditions or diseases are complex tasks. An observed condition may have more than one cause. Thus, it is difficult to arrive at a diagnosis solely based on symptoms. However, the range of diseases to be considered for diagnosis is narrowed through observation of symptoms. Carefully conducted systematic clinical examination is a must in any diagnostic procedure. The examination has to be conducted always in the same order and all observations have to be noted down. In order to document each case properly an examination checklist that includes the history of the case and the proposed treatment should be used. A systematic collection and storage of such data is useful for epidemiological studies and surveillance programmes at a later stage. Before conducting clinical examination of the animal, question the goat keeper thoroughly about the development of the disease. All the goat keeper's observations that can be verified during the examination should be taken into consideration.

Example

Pose clearly formulated question to owners in a friendly and respectful way.

How long has the animal been sick?

What symptoms did the farmer observe first? What was the subsequent development?

Is this the first time the animal has shown these symptoms?

Did the farmers give any treatment?

How many animals, including other species or humans are sick?

Has the flock had any contact with other animals?

What sort of fodder and how much is the animal getting?

A diagnosis should be made only after a careful physical examination of the animal(s) and not just after having a first intuition.

Normal Physiological Values for a Healthy Goat

Temperature	103° F
Heart Rate	70-80 per minute; faster for kids
Respiration Rate	12-15 per minute; faster for kids
Rumen Movements	1-1.5 per minute
Onset of heat (age at Puberty)	10-18 months, depending on nutrition and breed
Length of heat	12-48 hours
Heat cycle	17-23 days
Length of Gestation	148-158 days

Clinical Examination Sheet

Identification of the Goats : ....

History ....

Weight .... Temperature .... Pulse............... Respiration ....

Condition (good) (fair) poor)..

	Normal	Abnormal	Not examined	Remarks
Behaviour				
Posture				
Respiratory system				
Skin				
Digestive system				
Uro-genital system				
Eyes and Ears				
Lymphnodes				
Nerves				

Diagnosis..

Differential diagnosis...

Treatment ....

Further examination ....

Postmortem .... Parasitology .... Microbiology .... Others ....

Final diagnosis ....

Cured .... not cured .... Died ....

3.2 History Taking

Very few diseases or health-related problems are randomly distributed in a flock of goats; rather, they are concentrated in specific groups usually by sex, function, production status, or age. Always establish early on what age, sex, breed or group of goats are dying, showing signs of illness, aborting or showing decreased productivity. Detailed history should include a determination of the total flock population and estimation of its breakdown by sex, age, breed, and pregnancy status. Having determined the total animal population, the population at risk, and the number of animals affected and dying, it is possible to determine rates of disease occurrence and case fatality rates. By counting cases, the investigator is also in a better position to determine the actual significance of a problem as compared to the farmer's perception of it. In some cases, the loss of a few animals may be insignificant compared to a more serious unrecognized problem such as endo- or ectoparasitism.Such an epidemiologically based history should aim at identifying not only specific problems but also specific risk factors that appear to be associated with mortality, morbidity, or suboptimal performance. For example, when a primary complaint of kids developing diarrhea after weaning suggests coccidiosis, additional questions are necessary concerning the segregation of kids from adults, the manner that kids are fed, the design of feeders,the frequency and manner of shed cleaning, and details on the use of coccidiostats. In such cases, modification of management practices may stop the spread of additional disease.

Temporal relationships are important to note. Some diseases may occur seasonally, in association with abrupt weather changes, or in relation to specific events such as breeding, pregnancy, parturition and lactation. For example, an unexpected cold snap or heavy rain can increase pneumonia, abortion, and death rates particularly if adequate shelter and supplemental feed have not been provided. Localization of death or disease to specific areas on the premises is helpful information. For example, if losses are seen only in certain areas of the farm, specific pastures or particular shed, then suspicion of poisoning is increased. Other important aspects of the history include questions pertaining to the actual ration being fed and its consumption, methods of feeding, changes in feeding, access to grazing, and water supply type and water availability.

If husbandry procedures or preventive health procedures have been undertaken recently, they should be identified. Drenching, dehorning, spraying or dipping, castration, or vaccination can be associated with increases in morbidity and mortality. When range animals are mobbed for such procedures, sudden close confinement, temporary feed deprivation, and abrupt weather changes can predispose to outbreaks of conditions such as coccidiosis, salmonellosis, hypocalcemia, or starvation as a result of mismothering. When drugs or vaccines are used, the products and dosages, number of treatments, and method of administration should be determined; particularly as many goat farmers traditionally obtain their drugs and biologicals from non-veterinary sources.

If animals have been transported recently, dates, origins, means of transportation, and quarantine times should be determined. Information should also be collected on visits to shows or fairs, and on the origin of purchased animals, be it other farms, stockyards, or specialized goat sales. If animals have come interstate, the relevant health certificates should be examined, and the disease situation in the state of origin reviewed.Finally, the reliability of information obtained should be checked with herdsmen if the owner is not involved with day-to-day management decisions. If the veterinarian has prior knowledge of the local disease patterns in goats, such knowledge should not be used to make hasty and possibly incorrect judgments.

Range and Pastured Goats

Extensively managed animals may not be closely observed, and histories can be sketchy. With large flocks, it should be determined if the animals are managed as a single flock or in smaller, self-contained units. The seasonal pattern of grazing and the length of grazing periods should be noted. Pasture composition, seasonal stocking rates per acre, the degree of pasture subdivision, and length of resting periods between grazing should be established. Note, if supplemental feeding is practiced and the types of feed used. This may be important in terms of meeting specific nutritional needs and in the case of silage, may be associated with diseases such as listeriosis or rumen acidosis. Inquiries should also be made whether crops are fed or grazed, the type and stage of growth, and whether there is a recent history of fertilizer or herbicide application. Knowledge of local trace element deficiencies or excesses may be helpful.

Intensively Managed Goats

A complete history can usually be obtained from the owner or herdsman of intensively managed goats. The patterns of disease are also likely to be different, with pneumonia and enteric diseases of young goats assuming much greater significance than foot rot, helminthiasis, predation, or toxic plant problems more often seen under grazing systems. Feed composition and intake are more regulated but the veterinarian must inquire about episodes of sudden changes, excesses, or deprivations in the feed and water supplies.

Under close quarters, the movement, mixing, or introduction of new animals is more likely to cause an outbreak of disease. Kidding is often assisted in intensively managed operations, and artificial kid rearing methods are commonly used.These procedures should be carefully reviewed when morbidity and mortality are concentrated in young kids. Weather should not adversely affect intensively managed animals. However, extremes in temperature may affect the ventilatory capacity of confinement buildings and extremely cold weather may predispose the animals for pneumonia. Questions about changes in milking procedures or personnel may help to explain mastitis problems.

Inspection from a Distance

It is often useful diagnostically to observe a group of goats from a distance prior to disturbing them for "hands-on" examination. This is especially true at the time of the initial visit to identify the existence of common problems in the flock. The animals should be observed at rest, while eating or drinking, and during spontaneous and forced movement. General impressions of body condition, mental attitude, and social hierarchy may be acquired and abnormal behaviors characteristic of certain diseases may be noted. Estimated prevalence of common disease problems such as kid pneumonia and diarrhea can be roughly assessed by counting coughers and stained hindquarters.

Individual goats that appear listless, separate themselves from the herd, or are not actively feeding when others should be noted and later caught for careful examination as should animals in very poor bodily condition. Reluctance to feed may be due to a wide range of systemic diseases, localized conditions such as dental or pharyngeal problems, or the result of inadequate space or bullying by dominant does. Latent signs of respiratory disease or anemia associated with parasitism can be brought out by forced movement of a flock. Anemia will be manifested by rapid fatigue, increased heart and respiratory rates, and sometimes collapse. Increased respiratory rate, dyspnea, and coughing indicate respiratory problems. Signs of skin irritation or pruritus manifested as hair loss, fleece biting, or rubbing against fences or other solid objects usually suggest ectoparasitism. Goats scratching at their ears with their hind limbs or shaking their heads vigorously probably have ear mites.Any animals with abnormalities of gait or lameness after forced exercise should be carefully examined for evidence of arthritis, fractures, laminitis, and lesions of foot rot or foot scald.

A variety of clinical signs may be observed in goats with neurologic disease. Among the more common are ataxia, posterior paresis, circling, depression, head pressing, unilateral facial paralysis, and blindness. Straining while attempting to urinate particularly in bucks is the possibility of urolithiasis. Cutaneous swellings or discharges may be observed. Draining abscesses associated with lymph nodes are highly suggestive of caseous lymphadenitis in the flock. A high rate of subcutaneous swellings is often associated with injection site reactions in goats particularly after vaccination. When kids are left with does, careful observation of suckling behavior can indicate kids that are not successfully nursing. Their does should be examined particularly for signs of mastitis or other udder problems.

3.3 Physical Examination

The physical examination should be carried out as quietly as gently as possible to avoid disturbing the animal and thus increasing the resting heart and respiratory rates. Because of the small size of the goat, rectal palpation is limited to insertion of a finger into the rectum to assess the pelvic structures and to determine the presence and character of feces. When economics permit, physical examination findings should be supplemented through use of appropriate diagnostic procedures and laboratory tests. When numerous individuals in a herd or flock are showing signs of disease, field necropsy examination may be the most effective way to establish or confirm a diagnosis.

General Inspection

Body condition, mental attitude, and the status of the superficial lymph nodes should be noted. The temperature, pulse, and respiration should be recorded. Digital palpation of the ribs, spinous and transverse processes of the vertebrae, and the loin muscle may be necessary to evaluate condition. A word of caution about condition scoring in goats: scoring systems derived for sheep are not directly applicable to goats because goats as a species tend to deposit stored fat intra-abdominally rather than subcutaneously. Scoring systems for goats combine palpation of the sternal and lumbar regions. In general, any palpable back fat in a goat would classify her as obese.

Normally goats have an alert, attentive, and inquisitive mental attitude. A depressed attitude is characterized by dullness, separation from the flock, and indifference to handling. Depression will be present in a wide variety of septic and toxemic conditions but is a particularly prominent sign in pregnancy toxemia and listeriosis. An anxious or apprehensive state is often associated with urethral obstruction in males, sudden blindness as may occur in polioencephalomalacia, or with persistent irritations such as flies and nasal myiasis. An attitude of extreme excitation is most often associated with neurological diseases such as tetanus and meningitis that may be accompanied by muscular rigidity, or encephalitic conditions such as rabies.

Digital palpation of all superficial lymph nodes should always be a part of the physical examination because of the clinical importance of caseous lymphadenitis in goats. The intermandibular, parotid, retropharyngeal, superficial cervical (prescapular), subiliac (prefemoral), and superficial inguinal (supramammary) lymph nodes should be inspected. Normal size nodes may not be palpable in some of these locations but affected nodes should be readily evident. Any other swellings on the body surface should be noted. Temperature, pulse, and respiration should be measured when the animal is calm because the activity of catching the animal for examination may elevate all three parameters. When taking the goat rectal temperature, an accumulation of brown, waxy material may be noted near the anus. This is the normal secretion of the sebaceous gland located beneath the tail head. The normal body temperature of goats is usually reported in the range of 38.6-40.0°C (101.5-104.0°F). To accurately assess the febrile state of the animal, it is useful to record body temperatures in apparently normal flocks.

The pulse can be measured by stethoscope over the heart or by digital palpation of the femoral artery. Normal pulse rate ranges from 70-90 beats per minute in resting adults, but can be double those in young active kids. Fetal heart rates up to 180 bpm have been recorded by ultrasound. It may be useful to assess respiratory rate both at rest and after exercise. Any abnormalities of respiration should also be noted including flaring of nostrils, extension of head and neck, grunting, abdominal press and so forth. Normal resting respiratory rate is 10-30 per minute in adults and 20-40 in kids.Neonates should be inspected particularly for congenital defects. More commonly observed problems would include brachygnathia, cleft palate, hydrocephalus, atresia ani or rectovaginal fistula and abnormalities of the genitalia associated with the intersex condition.

Inspection of Body Regions

In goats the character of the skin and hair coat is a good indicator of general health. A rough, dry unglossy coat, excessive dander or flakiness and failure to shed out in the spring are all suggestive of poor nutritional status, parasitism or other chronic diseases. The hair or fleece should be parted and the skin examined for lice, ticks, fleas, nodules, swellings, crusts, eczema, necrosis, neoplasia, photosensitization and sunburn and focal or regional alopecia.

Examination of Head

Many conditions can cause general asymmetry or focal swellings around the head and these abnormalities should be noted.

Mucous Membranes

Inspection of the conjunctivae and mucous membranes of the mouth may reveal paleness due to anemia, icterus resulting from hemolysis or hepatic dysfunction or hyperemia and congestion associated with acute febrile or toxemic states.

Oral Cavity

Evidence of brachygnathism, cleft palate, mucosal lesions, dental abnormalities or dysphagia such as drooling salivation, dropping food from the mouth or accumulating food in the buccal space should be noted. Necrotic odors of the breath may occur. They may reflect necrotic stomatitis, alveolar periostitis, pharyngitis or even pneumonia. Thorough examination of the oral cavity requires good restraint, a speculum, a towel, and a penlight. All oral structures should be examined and the molar arcades digitally palpated for missing teeth from outside the mouth. If more direct examination of the teeth is required, extreme caution should be taken as the molars may have sharp, jagged edges and exert powerful grinding motion. Tranquilization is indicated, because fingers can be badly damaged. Wearing gloves during oral examination is a wise precautionary measure particularly if the goat is showing neurologic signs.

Eyes

Blindness is initially assessed by testing the menace response, but facial nerve paralysis can render a sighted animal unable to blink. Intact pupillary light responses in a blind goat suggest a cerebrocortical lesion. Lacrimation and hyperemia of the conjunctiva, cloudiness of the cornea and hypopyon in the anterior chamber should be noted if present.

Nares

Both nostrils should be evaluated for symmetry of air flow. If nasal discharge is observed, determine if it is unilateral or bilateral and note its character. Collapse of a nostril may result from facial nerve paralysis. Crusting of the nares occurs when the sick animal does not clean the nostrils or it may be a sign of specific disease problems.

Ears

If ear mites are suspected, they can be identified by collecting debris from the ear canal on a cotton swab and smearing it on a slide for examination. Goats are often identified by tattoos inside the ear. It may be necessary to deal the inside of the ear with soap and water and use powerful light source to backlight the tattoo to make it readable. Metal and plastic ear tags commonly tear out of goats ears and their use should be avoided.

Horns

Goats may be horned or polled. Horn buds may be present at birth or become palpable within several days of birth. Generally, horned kids have two irregular whorls of hair over the location of the horn buds whereas hornless kids have a smooth poll with a single central symmetrical whorl of hair. It is important to establish and record which offspring are naturally polled because homozygous polled goats have a high incidence of infertility. Deformed horns, or scurs are often seen on older goats as a result of incomplete removal of germinal horn tissue at the time of debudding. The glands partially responsible for the characteristic odor of buck goats in the breeding season and they are located in skin folds just caudomedial to the horn buds.

Examination of the Neck

Traumatic injuries to the pharynx from balling guns and drenching equipment occur in goats. The throat should be palpated for swelling, heat and pain associated with cellulitis from traumatic injury. A number of normal and abnormal structures and swellings in the neck must be differentiated on physical examination including goiters, thymus, branchial cleft cysts, wattles and abscesses. Thorough examination of the neck should include palpation of the jugular furrows for evidence of phlebitis, palpation of the esophagus for evidence of obstruction and auscultation of the trachea. Prominent distension of the jugular vein though possibly suggestive of congestive heart failure.

s of the joint
r structures,
ent of bursae
ot necessarily
umn and the
f fractures in
of physical
sculoskeletal,
y.

anted. Visual
paratus, slack
skin. Digital
nflammation,
n gangrenous
d in time the
e teats should
ay be present.
k plate or strip
lots or flakes.
the California
e normal goat

gns of heat or
vulvovaginitis
harges should
harge, initially
. This is often
nced observer.
mination of a
inal canal and
the source and
does may have
y swell during

se in advanced
plicated births,

respiratory disease is often difficult to
he chances of accurate diagnosis and
hould be given to auscultation. When
e moved to quiet surroundings. The
be auscultated to identify the presence
cough by compression of the pharynx
l the presence of bronchial exudates.
vith a stethoscope placed well forward
nterior ventral pneumonia which are
he intensity of identifiable sounds can
y of the animal before auscultation or
n glove over the nares to increase the
ing of carbon dioxide. When questions
e persist, radiographic examination is
sound can be heard about evenly on
e chest at the fourth or fifth rib space.
lace the heart, resulting in a shift in
ffling of the heart sounds because of

uld be inspected to assess conditions
cy in females, and ruptured bladder
detect abdominal fluid accumulations,
Auscultation of the rumen in the left
ormal rumen contractions occur at a
tion of cud chewing suggests normal

ommon in goats. Lameness and
sult from neurological disease,
ar dysfunction, skeletal trauma,
rthritis and diseases of the foot.
areful physical examination is the
diagnosis. Overgrown hooves must
uately assess the health of the foot.
bove the coronary band should be
ocal infections or systemic disease.

All joints should be carefully palpated. Distension capsule, heat, pain, swelling or fibrosis of periarticul limitations on the range of joint motion and enlargem should all be noted. The degree of joint enlargement may r correlate with the severity of lameness.The vertebral co long bones of the legs should be palpated for evidence acutely lame or recumbent animals. A major goal examination in recumbent goats is to differentiate mu metabolic, toxemic and neurologic causes of recumben

Examination of Mammary gland

Careful examination of the udder is always warr inspection may reveal weakness of the suspensory ap halves, abnormal swellings and discolorations of the palpation of the gland will identify udder edema, active fibrosis, scarring or abscesses in the parenchyma or teats. mastitis, the udder skin may be blue-black and cold a gland may slough if the animal survives. Patency of th be assessed in lactating animals. Supernumerary teats n The milk of lactating does should be observed on a blac cup to assess color, consistency and the presence of Bovine screening tests for sub-clinical mastitis such as Mastitis Test must be used cautiously in goats, becaus milk tends to have higher cell counts.

Vulva

Swelling and hyperemia of the vulva may be si impending parturition, but may also be seen in herpes in conjunction with vesicles or scabs. Any vulvar dis be noted. As females come out of heat, the vulvar disc serous and mucoid, may become white and tenaciou misinterpreted as a purulent discharge by the inexperi Often, there is a sanguinous discharge after ter pseudopregnancy. Speculum examination of the vag cervix should be carried out when there is doubt about nature of the discharge. Occasionally, otherwise norma ectopic mammary tissue present at the vulva which m lactation.

Does may show vaginal eversion or frank prolap pregnancy or immediately post partum. After uncon normal lochia may be discharged for a period of 1 to 3 weeks. It is reddish brown in color and odorless. Placentas are usually passed within 4 hours of parturition and are frequently eaten. It is important to carefully examine the external genitalia of young does, particularly when there is a complaint of infertility, because of the high incidence of intersexes among polled goats. Malformations of the genitalia range from the clinically inapparent, such as a slightly enlarged clitoris, to the overt, such as male phenotypes in genetically female individuals. Accurate record keeping may help to identify homozygous-polled individuals.

Scrotum

The scrotum and its contents should be palpated. Normally there is bilateral symmetry of all structures. A bipartite or split scrotum is a common congenital condition that some breeders consider a fault. Semen samples can be collected for evaluation. Gynecomastia occurs in male goats and it is not extraordinary to detect distended teats anterior to the scrotum during physical examination.

Penis and Prepuce

Examining the penis of male goats, especially breeding bucks, can be difficult and is not ordinarily attempted unless there is a history of urinary or breeding problems. The preputial opening should be routinely examined, particularly in males, for the presence of ulcerative posthitis. The preputial orifice may become occluded in this condition. Crystals or drops of blood may be noted at the orifice in cases of obstructive urolithiasis.

3.4 Examination of the Environment

An examination of the environment where goats are raised should include a detailed review of all feeds used, the feeding facilities, water sources and delivery system, the sheds, pastures, range, or buildings where the animals are kept, and any mechanical or manual equipment used as part of the routine farm procedures. Because of their prolificacy, goat herds tend to expand faster than anticipated. Inadequate feeder space for does and overcrowded kid sheds are two examples of common faults found on environmental inspection. Besides visual inspection of equipment, farmers could be asked to demonstrate how common procedures are carried out to detect if they are using inappropriate techniques. Equipment used for treatment should be examined for its potential to cause injury. Poorly maintained syringes with either contaminated or overly long, large-gauge needles can result in injection abscesses or systemic infections. Drug stocks and facilities for storage should also be examined to determine if they are suitable and also secure. If the farmer blends his own feed or components, the techniques and constituents used should be examined particularly when toxicities are suspected. Accidental inclusion of agricultural or other farm chemicals is not uncommon. Finally, the facilities available and routinely used to dispose of any dead animals should be noted particularly to see that they are consistent with any legal requirements and do not serve as a source of infection for the remainder of the flock.

3.5 Range and Pasture Operations Examination

The quality and quantity of pasture and the availability of supplementary or conserved feed should be assessed. This may require a specialized knowledge of pasture species, weeds, or other potentially toxic plants. If possible, animals should be observed while feeding. The source of water available both years round and seasonally, its quality, and its quantity, should also be inspected. Feed and water samples should, if necessary, be collected for laboratory examinations. On range or pasture, goats may congregate in certain shaded or sheltered areas particularly during periods of high or low temperatures. Provision of adequate shelter should be created. Heat stress can be reduced by ensuring the animals have adequate shade and water when the livestock weather safety index is in the danger or emergency zone. Cold and heat stress are very common during winter and hot humid seasons, temporary shelter and extra feed are necessary for animals. If shelter areas are limited or overused, buildup of infectious agents can occur leading to increases in enteric and other diseases, especially in kids. This is aggravated by poor drainage under feeders and water troughs. Fencing available for these types of operations should also be inspected particularly regarding its effectiveness in keeping goats in and keeping predators out.

3.6 Confinement Operations

In semi confinement operations adult animals and sometimes young stock are grazed on pasture during the comfortable months but confined during the inclement weather. As flocks expand, there is a tendency for small pastures to be overused, with poor nutrition and increased parasitism as the result. Stocking rates should be noted and worming history verified. The type of buildings used for housing should be evaluated in terms of square footage per adult breeding animal, type of flooring, degree of insulation and also the adequacy of natural or mechanical ventilation systems, the source and availability of water and the amount of trough space per goat. The distribution of animals within buildings should be noted. Ideally, a separate kidding facility should exist and be hygienically maintained. Artificially reared kids should be in a separate facility from adult stock and bucks should be housed separately from milking does. Methods and times of feeding for all groups within the herd should be noted.

3.7 Necropsies and Slaughterhouse Checks

The economic value of most individual goats in commercial flocks is such that most owners can be persuaded to have a necropsy undertaken of an ill or obviously moribund animal. This is an extremely useful adjunct to clinical examination, because necropsy can confirm the suspected diagnosis or suggest new avenues of investigation. All necropsies should be carried out following a systematic procedure with special regard to the likelihood of any zoonotic infections being transmitted and the safe and legal disposal of carcasses. Necropsy of all or a random sample of young animals born dead or dying in the first 2 to 3 weeks of life is an extremely valuable service that veterinarians can provide to their clients. Categorization of deaths into pre-parturient, parturient, and post-parturient causes can be done with a minimum of laboratory diagnostic techniques and immediately direct the owner to possible ways of reducing these losses. Whenever possible, goats slaughtered for meat or culled from the flock should be examined post mortem by a veterinarian as an inexpensive and worthwhile means of disease surveillance in a flock. Such inspections can provide information on the efficacy of parasite control measures, the presence of sub-clinical pneumonia and the occurrence of injection site abscesses, visceral caseous lymphadenitis, and hydatid disease.

□□□

4

Collection, Preservation and Dispatch of Materials

To aid in the proper and accurate animal disease diagnosis, it is paramount to collect various clinical materials and dispatch them to the disease diagnostic laboratory for isolation of the causative agents like bacteria, virus, fungi etc., and/or chemical analysis. Many similar diseases are prevalent in the animals and it is not easy to diagnose a disease clinically. Therefore, clinical materials like blood, serum, pus, skin scrapings, biopsy materials from necropsy etc., are required to properly diagnose the actual cause of disease. In intensive system of rearing goats, the morbidity and mortality pattern of a disease are different and baffling than the free-range system, mainly due to the human intervention and mismanagement. In these cases to identify the specific cause of the break in the bio-security, it is paramount that clinical samples have to be sent for laboratory examination. So it is essential for the veterinarians, livestock owners, field assistants, farm personnel to get to the proper know-how of collecting, preserving and dispatching the clinical materials to the nearest laboratory.

The samples should be carefully packaged, labeled and transported to the laboratory by the fastest practicable method with the appropriate temperature control. All samples should be accompanied by a letter or submission form includes name and address of the submitter, origin of the material, relevant history, animal identification and corresponding specimens. In this chapter, collection of the materials, preservation and dispatch to the laboratory for aiding in proper diagnosis has been described. Collection of sample depends on the type of the suspected disease condition. But in general the common type of materials like blood, serum, pus organ and tissues, etc., will be dealt in brief. There are a few facts of great importance which must be observed no matter what type of sample is being submitted to the laboratory and no matter what type of estimation is required.

Generally speaking, samples to be sent to a laboratory should be placed in a water-tight glass or plastic container. The container should be well packed to prevent breakage during transit, e.g. with sawdust, paper or cotton wool. In those cases where refrigeration is advisable, e.g. bacterial or virological specimens, this can be achieved in two ways. Natural ice from a refrigerator can be placed in a water proof container and the specimen container placed into the ice. This will keep the specimen cool for between 8-24 hours depending on the external temperature. If the veterinary surgeon can take the samples to the laboratory then a thermos flask containing ice is an ideal method of refrigeration but such flasks are prone to breakage if sent by post or rail unless packed extremely well. A longer lasting method of refrigeration is to use dry ice (solid CO2). Put the specimen container in a plastic bag and surround this with dry ice wrapped in paper. Never enclose the dry ice in an air-proof box otherwise CO2 volatilization during transit may result in an explosion. A cardboard box is the best type of container.

4.1 Collection of Samples

Before taking samples, careful consideration should be given to the purpose for which they are required. This will determine the type and number of samples needed to provide valid results. When samples are taken from live animals, care should be taken to avoid injury or distress to the animal or danger to the operator and attendants. Whenever handling biological material, from either live or dead animals, the risk of zoonotic diseases should be kept in mind and precautions taken to avoid human infection.

Organs and Tissues

Tissues can be collected for microbiological culture, parasitology, biochemistry, histopathology and/or immuno-histochemistry. Cut the affected organ such that the cut part should have the affected part along with some normal organ with sterilized scissors. Put the organ in a clean, sterilized vial neatly labeled having normal saline or transport medium for bacterial and viruses. If the organ is to be sent to for histopathological examination, then vial should have 10% formal saline.

Serum

After collection of blood it is allowed to clot in sterilized tubes placed in slanting position. After sometime the serum is squeezed out and collected by sterilized Pasteur pipette and transferred in to another sterilized glass vial or eppendorf's tube. Merthiolate @ 1:1000 strength may be used as a preservative. The serum samples can be better harvested from slants which are chilled in refrigerator, which aids in retracting the clot, whereby allowing the serum to come out easily. The serum samples should also be sent in an ice pack to the laboratory.

Pus and Nasal Discharge

The pus and nasal discharge are collected directly on sterilized cotton swab placed in a sterile glass vial. Later the glass vial can be preserved in ice pack and dispatched for further laboratory examination.

Body Fluids and Exudates

Mostly these samples are collected from dead animals from body cavities. These are collected by a sterile syringe or sterile Pasteur pipette and are dispatched in a sterile glass vial under refrigeration conditions.

Blood and Tissue Impression Slides

Thin smears of blood should be made. Pressing the clean non-greasy slide against the cut part of the organ/tissue after blotting with filter paper can make the tissue impression slides. The slides are dried and fixed with methyl alcohol. These slides should be dispatched with proper care in the slide box packed properly.

Skin Scrapings and Hair Samples

These are required for the diagnosis of mite infestation, and are obtained by scraping the skin (not cutting it) with the edge of a sharp scalpel blade. The scraping should be continued until bleeding occurs and all debris collected is then transferred to a clean dry container (waxed ointment box) and dispatched. Scrapings should be made from active lesions and several sites should be used to enhance the chances of locating mites. Hair samples are used primarily for the identification of fungal infections and samples are best obtained from the outer edges of active lesions. The hair must be pulled out, not cut, as the hair roots are essential for diagnostic purposes. Seal the samples in a clean, dry container and dispatch. In the case of *Dermatophilus* infections fresh scabs or smears prepared from the moist underside of a scab should be submitted for examination. In disease producing vesicular lesions, collect, if possible 2 g of affected epithelial tissue as aseptically as possible and place it in 5ml phosphate buffered glycerine or tris buffered tryptose broth virus transport medium. The vesicular fluid can be aspirated using a sterile syringe and collected in a sterile glass vial.

Genital Tract and Semen

Sample may be taken by vaginal or preputial washing or by the use suitable sterile swabs. The cervix or the urethra may be sampled by swabbing. Samples of semen are best collected using artificial vagina or by the extrusion of penis and artificial stimulation. The samples collected in a sterile tube should be sent in ice to the nearest laboratory.

Eye

A sample from the conjunctiva can be taken by holding the palpebra apart and gently swabbing its surface. The swab is then put in to a transport medium. Mucopurulent lacrimal discharges are also useful in some cases.

Pieces of Intestine and Their Contents

The intestine is cut from both the sides of the affected area and these ends are tied with a clean thread. These pieces are then put in an open-mouthed glass bottle and dispatch them in ice. It is necessary to dispatch the samples as early as possible to the laboratory after collection.

4.2 Bacteriological Samples

When the area to be sampled is normally free from bacteria, e.g. the eye, the greatest care should be exercised to ensure that a true sample is obtained which is uncontaminated by organisms from the surrounding tissues. This is particularly true when samples are taken through a contaminated area, e.g. uterine swabs can easily be contaminated by the bacteria which normally inhabit the vagina. If contamination does occur then the bacterial findings will be almost impossible to interpret. When sterile swabs are used to collect material it is important not to allow the swabs to dry out. This can be prevented to some extent by using glass or plastic swab containers rather than cardboard ones and avoidance of delays before the swab is examined will also help. Transport media will help preserve organisms susceptible to drying out. In all cases submit as many properly taken samples from as many sites as possible to ensure that there will be a good chance of the causal organism being isolated. In the case of organisms such as *Salmonella,* take sizeable samples, e.g. 20 g faeces in a sterile container, in order to enhance the chances of isolation.

Bacteriological samples can also be obtained by aspiration techniques using a sterile needle and syringe, e.g. peritoneal fluid, synovial fluid or cerebrospinal fluid. The fluid is withdrawn, transferred immediately to a sterile screw-top container and sent to the laboratory without delay. Milk samples or pieces of fresh tissue should be treated with the same aseptic precautions and it is advisable to refrigerate tissues to prevent undue bacterial multiplication occurring. Small pieces of tissue are prone to invasion following bacterial contamination of the cut surface and so when possible submit whole organs for examination. Cauterization of the surface with a red-hot spatula before taking samples from deeper areas also helps reduce contamination. Always place each tissue in a separate marked container but never use a histological fixative such as formal saline if bacterial isolation is required.

4.3 Virological Samples

Tissue samples for virus isolation must be fresh, collected using aseptic techniques, and refrigerated but not frozen if there is to be any delay before the sample reaches the laboratory. It is important to collect several specimens from different sites or different animals to enhance the chances of virus recovery. Blood samples sent without chilling should be collected in EDTA and 50 per cent sterile glycerol saline. A less reliable method is to collect in OCG (I part to 2 parts blood). OCG consists of:

Distilled water	8,240 ml
Glycerine	5,000 ml
Carbolic acid	50 ml
Potassium oxalate	50 g

Swabs for viral isolation should be placed into tissue culture fluid or isolation media containing antibiotics if they cannot be sent to the laboratory quickly. This is best done by breaking off the head of the swab and putting it in a bottle containing the culture fluid, or transport media. Never allow swabs to dry out as this will destroy any virus present.

4.4 Biopsy Samples

Bone Marrow Biopsy

This technique can only be performed satisfactorily on live animals. Material taken from dead animals is usually valueless owing to the rapid rate at which marrow cells degenerate after death. The bone selected for biopsy must contain active red marrow and strict asepsis must be observed throughout the procedure. All needles must be sterile. The needles used are proper type and their size depends on the species to be sampled and the sampling site. Needles of 20 mm length are suitable for small animals. The site of skin penetration is shaved, thoroughly cleaned and infiltrated with local anaesthetic. It is also advisable to desensitize the periosteum. The biopsy needle has sterile 3.8 per cent sodium citrate drawn through it to prevent coagulation and is then inserted until the bone is reached. It is then pushed into the marrow cavity with a rotating motion. The needle guard prevents over-penetration. The stilette is removed, a syringe attached to the needle and negative pressure applied until 0.5 ml of marrow has been obtained. The syringe is then detached and the needle withdrawn. It is of importance to obtain only a small sample if gross contamination with blood is to be avoided.

Biopsy Sites

In sheep and goat, the third sternebra is punctured with the animal sitting upright supported by an assistant, and penetration of the bone should not exceed 0.5 cm.

Sample Preservation

The sample once obtained can be treated in three ways. A drop of marrow can be placed on a clean grease-free slide, and smears prepared in the same manner as for blood smears using a coverslip technique.These smears are then fixed for two minutes with methyl alcohol or stained. A second method is to place the sample in a watch-glass containing 3.8 per cent sodium citrate in order to wash blood from any marrow flecks present. Now place marrow flecks on a slide, add a drop of serum, crush the flecks with a coverslip and prepare smears immediately. Once prepared, the smears are fixed or stained. A third technique is to transfer the sample into a bottle containing 5 mg of dipotassium sequestrene anticoagulant (sufficient for 0.5 ml of sample), mix gently but well and then prepare smears from this material as soon as convenient (preferably within one hour of collection).

Lymph Node Biopsy

This technique is undertaken in suspected cases of, for example, lymphosarcoma when histological examinations are required. The whole node can be removed surgically or a portion can be removed using a liver biopsy needle.

Liver Biopsy

Successful liver biopsies, which are used to obtain material for histological or chemical estimations (vitamin A, copper), depend on experience and a good knowledge of anatomy. They can be obtained from most species of domestic animal, using a trocar and cannula which is introduced into the liver from high on the right side. The size of the trocar and cannula required depends on the species involved. The point of the trocar must be sharp and the penetrant end of the cannula beveled to a keen cutting edge. The other end of the instrument has a screw thread to which can be attached a large syringe. This screw attachment is essential if sufficient negative pressure is to be obtained. The puncture site is shaved, swabbed clean and the sterilized trocar and cannula then inserted in the direction of the left elbow after the skin has been desensitized and a small incision made over the point of entry. The instrument is inserted until the resistance of the liver surface is felt. The trocar is withdrawn and the syringe screwed into position after which the cannula is pushed about two inches into the liver substance. The cannula is rotated rapidly and negative pressure applied to the syringe. This pressure is maintained while the instrument is slowly withdrawn. The syringe is used to discharge the core of liver tissue from the cannula into the appropriate container. If histological investigations are required the sample is placed into ten times its own volume of 10 per cent formal saline.

For copper estimations the sample should be dried completely before dispatch, and for vitamin A estimation the sample is placed in five or ten times its volume of absolute ethyl alcohol and put in a light-proof container immediately to prevent oxidation. Alternatively the sample can be stored in open tubes in solid CO2 or put into 3.3 per cent hydroquinone. Vitamin A estimations must be done within twelve hours of collection of the biopsy sample. The technique of liver biopsy can be dangerous if performed by unskilled hands and it is essential that asepsis be maintained otherwise there is a risk of peritonitis ensuing.

Skin Biopsy

These biopsies are required for histological examination and should be obtained from one or two representative skin lesions, preferably ones showing active pathological changes. The sample can be obtained either by surgical excision of the diseased area or by means of punch biopsy. In either case the area round the lesion is clipped, cleaned and infiltrated with local anesthetic taking care not to infiltrate the lesion as this will ruin it for histological purposes. The lesion is then excised with a scalpel or it can be excised with a cylindrical, hollow, metal punch, and the wound then closed with a suture. Once obtained, the sample should be placed in a clean, water-tight container. Ten per cent formal saline can also be used, although it tends to harden keratinized tissues. It is important that the samples should be small, otherwise the preserving fluid will not penetrate adequately, and both normal and abnormal tissue should be present in the sample.

4.5 Blood Samples

The method of collection varies with the species but the same method of collection can be employed in anyone species irrespective, of the laboratory investigation required. In all cases the puncture site should be clipped and swabbed with alcohol to remove all excessive contamination. Needles and syringes should preferably be of the sterile disposable variety or self-filling tubes (Vacutainers) can be employed. The use of dirty needles is to be abhorred as is the medieval practice of using one needle to sample several animals. All syringes must be perfectly dry. Blood is obtained from the jugular veins using fine bore 21 gauge and a syringe or using Vacutainers. The vein is raised by pressure with the left hand in the lower half of the jugular furrow and the needle inserted in the centre of the vein and in an upward direction. Gentle traction on the syringe is sufficient to facilitate collection.

Blood sample may be taken for hematology or for culture and/or direct examination for bacteria, viruses or protozoa, in such situation it is usual to anticoagulants like Ethylene Diamine Tetra-acetic acid (EDTA). The blood can be collected from external jugular vein in goats with a sterile needle and syringe. The blood should be transferred carefully in to a sterilized glass vial having some anticoagulant like sodium citrate or Sodium EDTA or Heparin. The vial should be rocked from the base for proper mixing of the anticoagulant with the blood, which otherwise may lead to clotting of blood. The vial should be dispatched in an icepack to the laboratory.

4.5.1 Haematological samples

Blood for haematology should be placed in bottles containing di-potassium sequestrene (EDTA). 5 mg is required for every 2.5 ml of blood. Remember to mix the anticoagulant and the blood to prevent clotting and always put in the correct amount of blood as an excessive anticoagulant/blood ratio will cause PCV errors and cellular distortion. Freshly prepared blood smears should be fixed in absolute methyl alcohol for two minutes to prevent cellular degeneration and then air-dried prior to transit.For prothrombin time estimations blood should be placed in bottles containing 3.8 per cent tri-sodium citrate (1 part to 9 parts of blood) or potassium oxalate.

4.5.2 Serological samples

Submit 2 ml of serum to the laboratory in a chemically clean or sterile, screw top container. Whenever possible it is advisable to send two samples (paired samples), one taken at the onset of the disease and another taken 2-3 weeks later. By this means it is possible to show a rising antibody titre which indicates recent infection. A single sample is not satisfactory unless it reveals a very high titre, e.g. in leptospirosis. In general, serological tests are used as confirmatory tests rather than a means of making an early diagnosis, for detectable antibody only appears two or three weeks after the onset of clinical signs.

4.5.3 Biochemical samples

For blood urea estimations 2 ml of blood is placed in bottles containing potassium oxalate as the anticoagulant. For blood sugar estimations 2 ml of blood is placed in bottles containing potassium oxalate and sodium fluoride, the fluoride being essential to prevent the breakdown of glucose. The majority of other biochemical estimations are done on serum although plasma can also be used and heparin (1 mg/5 ml of blood) is employed as the anticoagulant. Serum is obtained by placing the blood in a wide-necked chemically clean glass or plastic container allowing it to clot at room temperature and then withdrawing the serum after clot retraction and centrifugation. No further preservation of the serum is required but estimations should be performed as rapidly as possible unless the sample is deep frozen.

Bicarbonate, Chloride or pH Estimations on Blood

Take the samples carefully without venous stasis into a disposable syringe containing heparin, mix and expel all the air. Seal the syringe and transfer immediately to the laboratory. Alternatively, the blood can be placed into a bottle containing heparin and mineral oil. The blood is introduced below the oil layer to exclude any air. In all cases the estimation should be performed within three hours of collection.

Copper estimations are done on 2 ml of serum collected without venous stasis, or on 1 g of dry liver tissue. Avoid metal contamination by using plastic containers.

Enzyme Determinations

For the estimation of creatine phosphokinase, glutamate dehydrogenase, isocitric dehydrogenase, lactate dehydrogenase, malate dehydrogenase, ornithine carbamyl transferase and sorbitol dehydrogenase, 2 ml of serum should be submitted. The samples are best refrigerated or they should be frozen to - 20°C if any long delay is to occur between sampling and estimation (e.g. more than 24-48 hours). In the case of CPK and SDH deep freezing is essential if the delay exceeds 2- 4 hours, as there is up to 5 per cent activity loss in the first six hours after collection. Haemolysed samples are not suitable for enzyme estimations.

Histamine: 5 ml oxalated blood.

Ketones (quantitative): 2 ml blood in oxalate/fluoride. This allows sugar estimations to be done on the same sample.

Pyruvate: 5 ml blood collected without pressure on the blood vessel and put into citrate anticoagulant. Do the estimation immediately. Alternatively, blood can be put in four times its own volume of 10 per cent trichloracetic acid or in its own volume of 3 per cent perchloric acid, depending on the method of estimation.

Selenium: 3 g hair, 10 ml whole blood (any anticoagulant) or 1 g tissue.

Thiamine: 2 ml heparinised blood, tissue frozen to -15°C.

Transketolase: 1 ml heparinised blood.

Vitamin A : use 3 ml heparinised plasma shaken with 3 ml absolute ethyl alcohol and stored away from light.

Vitamin E: 20 g tissue, 10 ml serum stored at -20°C.

Vitamin B12: 5 ml serum or 1 g of liver.

Cerebro-spinal Fluid

Sampling

Cerebro-spinal fluid can be collected from two sites: from the cisterna magna by means of the sub-occipital approach, or from the spinal subarachnoid space in the lumbar region. In both cases the puncture site must be shaved and strict asepsis observed. Sterile spinal needles are employed, which contain a stilette, and their size will vary according to the species of animal being sampled. The sampling is done under local anaesthesia if the patient is quiet and can be positioned correctly, but if this is not possible general anaesthesia is employed. In sheep and goats, both approaches are satisfactory, but the animal has to be restrained in the 'dog-sitting' position if the lumbar approach is used, otherwise insufficient fluid will be collected. It is very easy to puncture a blood vessel during needle penetration so if the first part of the sample is blood-tinged discard it and then, after making sure the needle is in the sub-arachnoid space, continue with sampling. If the sample is still blood-tinged the red cells will be from the CSF.

Preservation

CSF is generally examined as soon after collection as possible for some constituents, especially glucose and leucocytes, degenerate rapidly. If there is little delay after sampling, no preservation is required and the fluid only has to be placed in a chemically clean glass or plastic bottle. If some delay is unavoidable, an anticoagulant such as di-potassium sequestrene should be used, or if glucose estimations are required, fluoride has to be added. The same concentrations used for blood samples are satisfactory. CSF delayed in the transport has often degenerated so badly that it cannot be examined and therefore ensuring rapid delivery to the laboratory is important.

4.6 Faeces Samples

Faeces samples should be collected direct from the rectum or a quantity of freshly passed faeces is collected from the floor. It is, of course, essential to know exactly which faeces sample was produced by which animal and to avoid the collection of samples contaminated by straw, sawdust, soil or other extraneous matter. The sample is placed in a screw-topped container which has been thoroughly cleaned, and then dispatched. Screw top containers or sterile plastic bags should be used for shipment; avoid tubes with rubber stoppers as gas generated can result in blowing the stopper off the tube, ruining the integrity of the sample and contaminating the samples in the package. An alternative and sometimes preferable method to take swabs from the rectum taking care to swab the mucosal surface. Care should be taken when collecting swabs from small, delicate animals to avoid injury to the animal; small swabs are commercially available that should be used. Swabs should be transported in appropriate transport medium. Faeces are best stored and transported at 4 °c. If parasitic egg counts are required the development of eggs can be delayed by filling the container and packing the contents tightly to exclude air. If long delays between sampling and examination are likely to occur, a little 10 per cent formalin can be mixed with the sample to prevent embryonation. The identification of adult worms is facilitated if they are submitted in 5 per cent formal-saline. If trypsin digest tests are required, no preservative should be added to the samples. Trypsin tests can only be done on fresh specimens passed less than twelve hours previously. No preservative should be added if toxicological estimations are required, and for lead estimations lead free containers should be used, e.g. plastic containers.

4.7 Histopathology Samples

Samples for histopathology will be obtained from post-mortem material. All material collected in this way must be as fresh as possible and when feasible, each piece should contain both normal and abnormal tissue. Small portions of tissue (approximately 1 cm thick) are removed and placed in water-tight containers holding ten times the samples volume of preserving fluid. It is extremely important to place small amounts of tissue in a large volume of fluid otherwise inadequate preservation takes place and the tissues are ruined by autolysis for histological purposes.The preservation fluids used depend on the type of tissue collected and the sort of microscopic examination which is required, but 10 per cent formal-saline is a useful general purpose fixative. It is composed of:

200 ml 40 per cent Formaldehyde
17 g Sodium chloride
1800 ml distilled water

Do not use carbolic acid, antiseptics or spirit as preservatives. Always place samples in a wide-necked bottle as samples are difficult to remove from narrow-necked containers once they have become hardened by the fixative.

4.8 Milk Samples

Milk Samples should be collected direct from the mammary gland into clean, water-tight, glass or plastic containers. If bacteriological examination is required the container must be sterile and external contamination of the sample should be avoided by cleaning the teats with alcohol prior to sampling. Examinations should be performed as soon as possible in warm weather, but if a delay is unavoidable the milk should be refrigerated when cell counts are required. The investigation commonly carried out on milk samples can all be performed if 25 ml samples are submitted. For milk ring tests avoid colostrum and milk from animals drying off or suffering from mastitis.

4.9 Serous Cavity Fluids

Samples taken from the pleural or peritoneal cavities are generally removed for bacteriological or cytological studies, strict asepsis should be observed and the puncture site shaved and then desensitized with local anaesthetic. Chest puncture is performed between the sixth and eighth intercostal space in the lower third of the chest, keeping the needle close to the anterior edge of the rib. Needles used vary in size and should have a stilette in order to remove fat or tissue debris which may block the needle. No preservatives should be added to the samples.

Synovial Fluids

The puncture site and the quantity of fluid which can be obtained will depend on the joint or joints affected and the type of material they contain. In all cases, however, the site must be shaved and strict asepsis observed. Local anaesthetic is infiltrated around the skin and joint capsule before inserting the needle into the joint cavity. Short needles of 18 gauge should be used in conjunction with a stilette, for narrow needles are prone to become occluded with thick exudates. The synovia is collected with the help of a syringe. Synovial fluid will often coagulate unless mixed with anticoagulant. Di-potassium sequestrene is ideal, especially for cell counts (2 mg/ml), although 3.8 per cent sodium citrate (0.1 ml/ml of fluid) is satisfactory, The containers must, of course, be clean, or preferably sterile, one clotted sample (2-3 ml) and one unclotted sample (2 ml) should be submitted if a full range of investigations is required.

4.10 Urine Samples

These are collected either during micturition or after catheterisation. If catheterisation is undertaken, the catheters must be sterile and not lubricated excessively with liquid paraffin otherwise fat will appear in the sample. Sterile water-tight glass or plastic containers should be used to hold the samples. Never use old medicine bottles or drug containers unless they have been scrupulously cleaned. The sample should be frozen if the estimation is to be delayed. Urine creatine estimations also require 24-hour urine samples.

4.11 Toxicological Investigations Samples

The samples required depend on the type of toxin or poison suspected and whether the animal is alive or dead. The veterinarian must be quite clear as to what samples will be needed and state clearly on his request form the estimations required. Whenever possible submit the suspected source of the poison. Any tissue submitted should be at least 50 g in weight and submitted in a chemically clean plastic container.

Arsenic : Submit kidney, liver, vomitus, stomach contents, or, if the poisoning is chronic and the animal alive, submit a hair sample. Urine samples are valuable in both acute and chronic cases.

Copper : Submit liver tissue.

Cyanide : Stomach or rumen contents and portions of liver and muscle should be submitted for examination. Stomach contents alone are not conclusive. Cyanide is lost rapidly, especially from the liver, after death, and therefore all samples should be preserved in a solution of 1 per cent mercuric chloride before being dispatched. It is important that the tissues submitted should be fresh.

Lead : If the animal is alive, 10 ml of whole blood should be sent. Use heparin as the anticoagulant. Never use citrate. The estimation can also be done on serum. If the animal is dead, kidney and liver are the best tissues to take. Intestinal or rumen contents should be examined and anything suspicious sent to the laboratory. At least 20 g of tissue is required and all containers must be lead free, e.g. plastic.

Mercury : Submit fresh liver and kidney tissue. Kidney tissue fixed for histology is also useful. Stomach and intestinal contents are also valuable.

Nitrate-nitrite : Stomach contents and a clotted blood sample.

Organo-phosphorus compounds : Many of these are anticholinesterases, and cholinesterase estimations will help confirm a tentative diagnosis of poisoning although no precise indication of the poison involved can be obtained. Whole blood is required. Send at least 1 ml of heparinised blood and ensure that it reaches the laboratory quickly, or if a delay is likely, pack the sample in ice in a vacuum flask to prevent deterioration. In a freshly dead animal cholinesterase estimations can be done on nervous tissue, muscle or liver.

Phosphorus : Submit liver, stomach contents or 30 ml oxalated blood.

Sodium Chloride : Submit liver, serum, and the whole brain (for histology).

4.12 Collection of Samples in Specific Disease Condition

Peste Des Petits Ruminants (PPR)

It is an acute viral disease of goat, sheep. In live animals, swabs are made of the conjunctival discharges and from the nasal and buccal mucosae. During the very early stage of the disease, whole blood is also collected in anticoagulant for virus isolation, polymerase chain reaction (PCR) and haematology. At necropsy (two to three animals), lymph nodes, especially the mesenteric and bronchial nodes, lungs, spleen and intestinal mucosae should also be collected aseptically, chilled on ice and transported under refrigeration. Fragments of organs collected for histopathology are placed in 10% formalin. At the end of the outbreak, blood can be collected for serological diagnosis.

Contagious Caprine Pleuropneumonia (CCPP)

Contagious caprine pleuropneumonia (CCPP) is a severe disease of goats caused by *Mycoplasma capricolum* subspecies *capripneumoniae (Mccp)*. Typical cases of CCPP are characterized by extreme fever (41-43°C), high morbidity and mortality rates in susceptible herds affecting all ages and both sexes, and abortions in pregnant goats. The necropsy samples of choice are lung lesions, particularly from the interface between consolidated and unconsolidated areas, pleural fluid, and mediastinal lymph nodes. If microbiological examination cannot be performed immediately, samples or whole lungs can be stored at -20°C for considerable periods (months) with little apparent loss of mycoplasma viability. During transport, samples should always be kept as cool as possible, as mycoplasma viability diminishes rapidly with increasing temperature. Lung samples can be dispatched to other laboratories in frozen form. Pleural fluid and tissue suspension swabs can also be collected and suspended in 2-3 ml of culture medium and sent in ice.

Goat Pox

Goat pox is a viral disease of sheep and goats characterized by fever, generalized papules or nodules, vesicles (rarely), internal lesions (particularly in the lungs), and death. Material for virus isolation and antigen detection should be collected by biopsy or at post-mortem from skin papules, lung lesions or lymph nodes. Samples for virus isolation and antigen-detection enzyme-linked immunosorbent assay (ELISA) should be collected within the first week of the occurrence of clinical signs, before the development of neutralising antibodies. Buffy coat from blood collected into heparin or EDTA (ethylene diamine tetra-acetic acid) during the viraemic stage of capri pox (before generalization of lesions or within 4 days of generalization), can also be used for virus isolation. Samples for histology should include tissue from the surrounding area and should be placed immediately following collection into ten times the sample volume of 10% formalin.

Blue Tongue

Bluetongue virus (BTV) infection involves domestic and wild ruminants such as sheep, goats, cattle, buffaloes, deer, most species of African antelope and various other Artiodactyla as vertebrate hosts. This disease is not severe in goats as in sheep. This noncontagious, insect-borne viral infection is unapparent in the vast majority of infected animals but causes fatal disease in a proportion of infected sheep, deer and wild ruminants. During acute pyrexia, viraemic whole blood can be collected without anticoagulants and packed and preserved in ice and transported to the laboratory. In endemic areas periodic collection of serum can be done for screening antibodies in the goats.

Foot and Mouth Disease

If suspected for FMD, the tissue of choice is epithelium or vesicular fluid. Ideally at least 1 g of epithelial tissue should be collected from an unruptured vesicle, usually from the tongue, buccal mucosa or feet. Epithelial samples should be placed in transport medium composed of glycerol phosphate buffer, preferably with added antibiotics (1000 IU penicillin, 100 IU neomycin sulphate, 50 IU polymixin sulphate and 100 IU mycostatin) and should be sent to the laboratory as early as possible.

Brucellosis

Recently aborted animals can be screened for *Brucella* by antibody titer in the serum by standard tube agglutination test. A titer of 1:40 and above is considered positive and the serum samples can be sent in ice packed in sterile glass or plastic vials. While sending aborted contents, animal should not be handled without proper safety precautions, as *Brucella melitensis,* is zoonotic in nature. Use of gloves and masks is recommended with safety precautions for handling the aborted contents. The abomassal fluid of the aborted fetus can be collected using a sterile syringe and packed sent in ice to the laboratory for bacteriological examination.

Blackleg

Excise a cube of affected muscle and send to the laboratory under refrigeration for bacterial identification.

Enterotoxaemia

Enterotoxaemia is an acute to sub-acute bacterial disease in goats and sheep causing high mortality and economic losses. In dead animals, after post mortem, the intestine can be cut and lighted at both the ends and sent in ice to the laboratory for identification of possible toxins produced by *Clostridium perfringens* type D. Alternatively the intestinal contents can be stripped out into a bottle and preserved with a few drops of chloroform. In cases of pulpy kidney the terminal small bowel should be treated in the same manner.

Listeriosis

Submit half the brain under refrigeration for bacterial isolation and the other half fixed in 10 per cent formal saline for histology.

The sample or samples must be as fresh as possible and obtained and preserved in the correct manner. Each sample must be labelled and easily identifiable. The clinician should include the following information with the sample:

1. Name of the owner/sender.
2. Address and telephone number of the owner/sender.
3. Species, breed, sex, age and identity of animals sampled.
4. Date samples were collected and submitted.
5. List of samples submitted.
6. A complete history would be beneficial for the laboratory and should be included if possible. Some of the components of the history are
   a) List and description of the animals examined and the findings of the postmortem examination.
   b) Length of time the sick animals have been in the farm; if they are recent arrivals, from where did they originate.
   c) The date of the first case and of subsequent cases or loss with any appropriate previous submission reference numbers.
   d) A description of the spread of infection in the herd or flock.
   e) The number of animals in the farm, the number of dead, the number showing clinical signs, and their age, sex and breed.
   f) The clinical signs and their duration including the temperature of the sick animals, condition of mouth, eyes and feet and milk production data.

g) The type and standard of husbandry including the type of feed available, possible contact with poison or poisonous plants.

h) Medication given to animals if so when given.

i) Vaccines given to animals and when given.

j) Other observations about the disease, husbandry practices, and other disease conditions present.

□□□

5

Bacterial Diseases

5.1 Enterotoxemia (Overeating Disease)

Enterotoxemia is one of the very important disease and in some areas it is the most prevalent disease of goats. Enterotoxemia is also known as overeating or pulpy kidney disease. Clinically, the disease is characterized by diarrhea, convulsions, paralysis and sudden death.

Etiology

This is an acute toxaemia of ruminants caused by the proliferation of *Cl.perfringens* type C and D in the intestines and the liberation and absorption of a large amount of toxins from the intestines. *Clostridium perfringens* types C and D is bacteria normally found in the soil and as part of the normal microflora in the gastrointestinal tract of a healthy sheep and goats.

Epidemiology

The bacteria are normally present in the soil and the intestinal tract in relatively small numbers. Under certain conditions the organisms proliferate (reproduce billions of their own kind) in the intestine and produce toxin in lethal quantities. These conditions are those which (1) provide an ideal environment and food for bacterial proliferation and (2) slow down the normal movement of material through the intestinal tract; they are often satisfied by ingesting large amounts of starch when the intestinal tract is not accustomed to it. The disease is often associated with lush fast growing pasture or cereal crops, heavy grain feeding or access to a lot of milk. Illnesses which slow down the intestinal tract may predispose to the accumulation of dangerous quantities of the toxin. The predominant predisposing factors in goats have to do with sudden exposure to grain or increases in quantity of milk consumed without gradually increasing the amount over several days. This leads to indigestion with slowing of the intestinal tract motility. This probably plays a large role in the disease in goats by allowing more time for toxin to accumulate within the intestinal tract. Sheep and goats are more likely to produce too much *Clostridium perfringens* types C and D in the gut, and suffer from enterotoxemia, during conditions such as the following:

- When kids and lambs excessively consume milk or feed with high quantities of grain.
- While recovering from an illness or distress; when natural immunity is compromised.
- As a consequence of heavy infestations of gastrointestinal parasites, such as nematodes and coccidian.
- When animals have a diet rich in grains and low in dry matter (hay or green grass).
- When animals have any condition or disease that slows the peristalsis (motility of the gastrointestinal tract).

Pathogenesis

The toxins cause enterocolitis (inflammation of the intestine), increase the permeability of the blood vessel, and become absorbed in the blood. They circulate in the bloodstream, promoting swelling in the lungs and kidneys, giving the condition the name "pulpy kidney disease". Young animals are most susceptible. Sudden and high mortality rates are concentrated in lambs and kids. Although adult animals are also susceptible to enterotoxemia, they develop immunity due to frequent exposure to these toxins. The Type D infection is probably far more common than Type C. The latter type produces a toxin called "Beta Toxin" which causes intestinal necrosis and severe intestinal hemorrhage. It occurs in adult goats. Epsilon toxin is produced by the Type D bacteria. It produces vascular damage and increases the permeability (openness) facilitating its own absorption. In the animals that die with neither signs nor tissue changes, an extremely large amount of toxin was absorbed very rapidly. When fewer toxins are produced, the animal lives longer and there are more time for clinical signs and pathological changes to develop.

Clinical Signs

In the Per-acute disease course, sudden death may be seen in kids. It may occur after consuming excess feed or after sudden access to highly palatable feed or after prolonged hunger and a normal quantity of feed. The acute course of disease lasts 4-20 hours and usually ends in death. Initially the temperature may go to 105°F with severe abdominal pain. There may be a rapid course of hypothermia, severe abdominal pain, depression, ataxia, convulsions and greenish diarrhoea. The diarrhoea may become haemorrhagic and contain sloughed mucosa. Profuse diarrhea (watery consistency with or without blood), loss of appetite, abdominal discomfort as evidenced by kicking at the belly and arching the back are the common signs of the disease (Fig 5&6). Depression, wobbly gait, recumbancy (lying down on side often with head down) generally seen in early stage of the disease. Convulsions often occur intermittently and may be accompanied by continuous or intermittent opisthotonos (head thrown straight over back). The animal may slip into a coma before death or die groaning or even crying. These signs occur in kids but can occur in adult goats in the

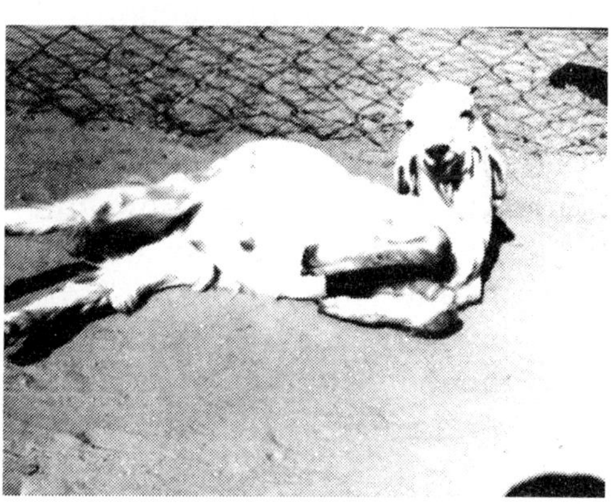

Fig. 5 Abdominal discomfort (Pain)

Fig. 6 Arching of back

disease caused either due to Type C or D bacteria. The Sub-acute disease is more common to occur in older kids and adults. They may be ill for several days or weeks and show anorexia (refusal to eat) and intermittent severe diarrhea occasionally with epithelial shreds in the feces. They will occasionally eat and with time and appropriate treatment, they will usually recover. The chronic form is characterized by intermittent illness lasting several weeks. The goat (usually an adult) will have a dull, starry look, loose feces and an irregular appetite and if a milker, drop in production.

Post Mortem Findings

Type C is associated with acute hemorrhagic inflammation and necrosis of the mucosa of the omasum and small intestine. Type D causes mild to moderate (occasionally severe) inflammation and even hemorrhage of the small intestinal mucosa (Fig. 7). Petechial hemorrhages may be present anywhere in the body but especially on the epicardium and endocardium. The pericardial sac may contain slight excess of yellow fluid. Microscopic examination of the brain may reveal degeneration of the vascular endothelium with perivascular and intercellular edema with foci of necrosis in several subcortical areas.

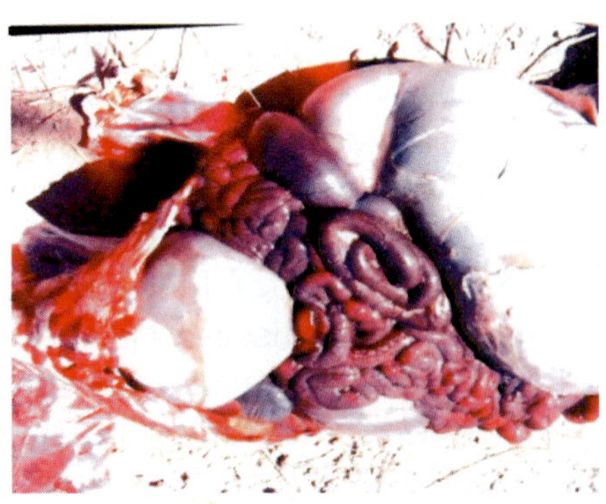

Fig. 7 Severe hemorrhage in intestine

Diagnosis

Diagnosis is based on clinical signs and history of sudden death that can be confirmed by necropsy. Diagnosis can be confirmed by positive identification of enterocolitis (inflammation of the intestine), *Clostridium perfringens* types C and D from the feces and gut content and kidneys cultured and isolated from the affected animals. The presence of glucosuria (high levels of sugar in the urine) can indicate enterotoxemia. Postmortem data are important for the diagnosis of enterotoxemia. Therefore, dead animals should be submitted to necropsy and their tissues sent to a diagnostic laboratory for confirmation of the clinical diagnosis. A postmortem examination of the large and small intestines can identify a high collection of watery blood and fibrinous clots and numerous ulcers on the mucosa (internal layer of the intestines). The brain and kidney tissues may show softening. Specific DNA testing assays can help detect enterotoxemia in postmortem animals. The per-acute and acute signs are helpful but can also occur with acute salmonellosis or intestinal torsion. Individual or first cases of salmonellosis would probably be diagnosed by post mortem bacteriological examination but if a herd problem exists the history, signs and lesions would justify a presumptive diagnosis. Intestinal torsion is an individual and uncommon event and would rarely be diagnosed ante mortem. Sub-acute or chronic cases could resemble coccidiosis, salmonellosis, rumen impaction. Fecal examination, culture and smears would aid in diagnosis of the first two and abdominal palpation, the latter. The petechial hemorrhages, especially on the epicardium should make one think of enterotoxemia. However, one should look for at least two other signs which together give good presumptive evidence of enterotoxemia; these are glucosuria and the presence of many short, plump gram positive rods on an intestinal smear. Ante-mortem diagnosis is made early if one can demonstrate a distinct, though transient, improvement of signs after the intravenous injection of 40-100 ml of Type C and D *Cl. perfringens* antitoxin. Definitive diagnosis, however, can only be made in the laboratory. Intestinal contents should be preserved by adding 1 ml of chloroform to 10 ml of contents which have been collected in glass tube within 12 hours of death. After preserving toxin, neutralization test can be carried out for the identification of the toxin in the ileal contents. Intravenous inoculation of mice or intradermal injection into white shaved guinea pigs using commercial antitoxins to *Cl. Perfringens* type A to E can be used. This confirms the presence of specific toxins in the ileal contents.

Treatment

Treatment is usually ineffective because of the severity of the disease, but if available, specific hyperimmune serum is indicated and oral administration of antibiotics may be helpful. Recommended treatments include the following:

- Administering C & D Antitoxin 5 mL subcutaneously.
- Administering penicillin.
- Orally administering an antacid.
- Administering anti-bloating medication.

- Reducing pain by applying Banamine.
- Administering thiamin (vitamin B1) intramuscularly.
- Replacing fluids intravenously or subcutaneously and using corticosteroids.
- Using probiotics after treatment with antibiotics to encourage repopulation of the microflora in the rumen and guts.

Prevention and Control

The disease is best controlled by vaccination of the pregnant dam during the last third of pregnancy: initially, 2 vaccinations 1 month apart, and annually thereafter. When outbreaks occur in newborn animals from unvaccinated dams, antiserum should be administered immediately after birth. All animals in a flock should be vaccinated against enterotoxemia. Vaccination will reduce the chances that animals will contract enterotoxemia. Vaccinate pregnant animals with multicomponent clostridial vaccine during the fourth month of pregnancy. This procedure will enrich the colostrum (first milky secretion produced between one and three days after birth) with antibodies (specific proteins) that will protect the newborn against enterotoxemia. All young animals should be vaccinated at four weeks of age and then 30 days later. Administer another booster dose at the time of weaning. Vaccinate all adult animals once a year. Do not vaccinate goats within 21 days before slaughter. Avoid causing stress to the animals during vaccination. Handle the herd in a peaceful manner; avoid accidents and noise when working with goats. In case of anaphylaxis (body reaction) caused by a vaccine, epinephrine should be used.

5.2 Infectious Necrotic Hepatitis (Black Disease)

Infectious necrotic hepatitis is an infectious disease of sheep and goats that is sometimes seen in cattle and is rare in pigs and horses. It is caused by bacterium *Clostridium novyi* in association with immature fluke invasion of the liver.

Etiology

The etiologic agent, *Clostridium novyi* type B, is soil borne and frequently present in the intestines of herbivores; it may be present on skin surfaces and is a potential source of wound infections. Fecal contamination of pasture by carrier animals is the most important source of infection. Traumatic damage to liver due to migrating fluke produces tissue damage and anerobic conditions suitable for spore germination and multiplication of organism in areas of liver necrosis and produces a powerful necrotizing toxin. The disease is worldwide in distribution, wherever goat and sheep and liver flukes are found. The alpha toxin is produced by C. *novyi* type A and B. It is lethal and necrotizing. Beta toxin is a phospholipase C and is produced by C. *novyi* type B only.

Clinical Signs

C *novyi* has been suspected but not yet confirmed as a cause of sudden death in goats fed high-level grain diets, and in which preexisting lesions of the liver were not detectable. The lethal and necrotizing toxins (primarily á toxin) damage hepatic parenchyma, thereby permitting the bacteria to multiply and produce a lethal amount of toxin. Usually, death is sudden with no well-defined signs. Affected animals tend to lag behind the flock, assume sternal recumbency, and die within a few hours. Most cases occur in the rainy season when liver fluke infection is at its peak. The disease is most prevalent in 1 to 4 years old animals and is limited to animals infected with liver flukes. Differentiation from acute fascioliasis may be difficult, but peracute deaths of animals that show typical lesions on necropsy should arouse suspicion of infectious necrotic hepatitis.

Post Mortem Findings

The most characteristic lesions are the grayish yellow, necrotic foci in the liver that often follow the migratory tracks of the young flukes. Other common findings are an enlarged pericardial sac filled with straw-colored fluid, and excess fluid in the peritoneal and thoracic cavities.

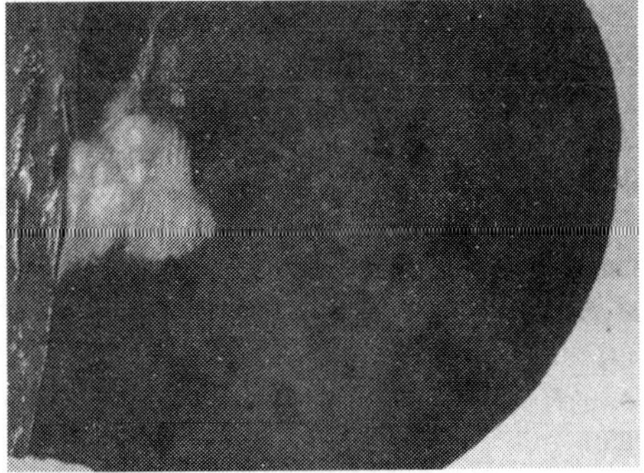

Fig. 8 Dark brown swollen liver showing necrotic areas (1–2 cm)

Usually, there is extensive rupture of the capillaries in the subcutaneous tissue, which causes the adjacent skin to turn black (hence the common name, black disease).

Prevention and Control

The incidence may be lowered by reducing the numbers of snails, usually *Lymnaea* spp , that act as intermediate hosts for the liver flukes or by otherwise reducing the fluke infection of goats. However, these procedures are not always practical, and active immunization with C *novyi* toxoid is more effective. Long term immunity is produced by vaccination. After this, only new introductions to the flock (lambs and sheep brought in from other areas) need to be vaccinated. This is best done before the late summer.

5.3 Tetanus

Tetanus is a common, highly fatal disease in sheep and goats caused by a widespread bacterium.

Etiology

Tetanus is caused by spores of the bacteria *Clostridium tetani.* The causative agent of tetanus, is found widely in both soil and animal feces. When this spore-forming rod is confined to an oxygen-deprived area, such as a deep puncture wound, a potent neurotoxin is released. *Clostridium tetani* produces the exotoxin tetanospasmin (a neurotoxin) is plasmid coded and responsible for the disease.

Epidemiology

Infection normally occurs from a wound contaminated by dirt, dust or faeces. Castration, ear-marking, tail-docking, dehorning and debudding is all routine surgical procedures creating wounds that can be ideal growth sites for tetanus. Infection can also occur through non-surgical wounds eg dog bites; shearing cuts.Goats can contract tetanus through puncture wounds, fights between bucks, castration, tattooing, dehorning, and kidding difficulties (dystocia). Maintaining a clean environment, particularly in barns where horses are or have been kept, is essential. Horse feces are a well-known repository of tetanus. Tetanus spores accumulate in the soil in vast numbers where livestock is crowded and kept under intensive management conditions.

Clinical Signs

The incubation period for tetanus can be from a few days to several months, but is usually 10 to 20 days. Early symptoms include a rigid gait, mild bloat, and anxiety. Tetanus quickly progresses to the animal's being unable to open its mouth (hence the term "lockjaw"), a rigid extension of the legs (front legs extended forward and together, with back legs extended backwards and together in a rocking horse- like stance), excessive salivation, constipation, inability to stand, neck stiffness with the head pulled hard to one side and accompanying tail and ear rigidity, and seizures. Once the goat is down and can't get up, death occurs quickly (usually within 36 hours or less).

Diagnosis

Gram stained smears of material from wound may reveal the characteristic 'drumstick' sporing forms of *C. tetani*. Protection test can be performed in mice. Animal are given antitoxins at least 2 hours before inoculation with the material containing toxin. The control mice show typical signs of tetani spasm in the region of inoculation. Diagnosing tetanus can be complicated by the fact that some symptoms resemble those of other diseases. Polioencephalomalacia (goat polio), strychnine poisoning, nutritional muscular dystrophy (white muscle disease), and even laminitis have similar symptoms at certain stages of these diseases.

Treatment

Because tetanus anti-toxin is not usually effective once the toxin has reached the spinal cord, hence required immediate administration of tetanus anti-toxin, before the wound or infection site is located and cleaned; this is because disturbing the sight while cleaning it can actually result in spreading the toxin. Then flush the wound with hydrogen peroxide as hair, dirt and other debris are removed from it. Penicillin injections for five consecutive days at a rate of 5 ml/50 kg body weight will inhibit the release of more toxin. Tetanus anti-toxin should be continued every 12 hours for at least two injections and longer if the infection site has not been located or is not easily reachable (i.e., internal infection resulting from kidding difficulties). Keep the goat isolated, quiet, and in darkened surroundings. Milk of Magnesia (15 ml/30 kg body weight given orally every 4-6 hours) or an enema may be used to relieve constipation. Electrolytes should be generously given and probably will have to be orally drenched into the goat's mouth, since it is not likely to be able to drink or eat on its own. Intravenous administration of glucose is recommended. The goat must be turned from side to side every hour to prevent skin ulcerations. Complete recovery in severe cases can take up to several weeks.

Prevention and Control

Prevention is easily accomplished by regular vaccination with tetanus toxoid, combined with maintaining clean facilities where the goats live. Vaccinate pregnant does one week expected to kid; this will provide passive immunity to the newborn. If the goat has already had both toxoid injections, then give a booster of the toxoid vaccine when castrating. Annual boosters are necessary for all goats. Do not assume that recently purchased goats, whether they are adults or kids, have been vaccinated. Instead, give them the entire two injections one month apart to maximize protection. It is easy to confuse toxoid and anti-toxin. Toxoid is the vaccine used to prevent the disease; it requires weeks to become effective, must be boostered with a second injection after 28 days, and one vaccination per goat must be given annually thereafter single-injection of anti-toxin is given when immediate protection is needed. If the goat survives, wait at least five days from the last anti-toxin injection and again given two injections of toxoid.

5.4 Anthrax

Anthrax is an acute infectious disease of almost all warm-blooded animals including humans. In animals, the disease is characterized, in most instances, by sudden death.

Etiology

Anthrax is caused by the bacterium *Bacillus anthracis* that belongs to a group of bacteria that have the capability of forming spores. Spores are microscopic, resistant particles, which enable the micro-organism to survive adverse environmental conditions. In most countries, it is required by law that outbreaks of anthrax are reported to the appropriate regulatory officials.

Epidemiology

The disease occurs world-wide. In some countries, specific areas are known to favour the survival of anthrax bacterium spores in the soil and are thus subject to occasional outbreaks. Alkaline soils are favourable for the survival of the spores. In these areas, the spores of the organism multiply when conditions of the soil such as temperature, moisture and nutrition are favourable. Re-occurrence of the disease is experienced in such locations periodically for many years. Cattle, sheep, goats, horses, donkeys, pigs, dogs, etc. are affected. Wild ruminants such as antelopes, gazelles, impalas are known to be equally susceptible.

Elephants and hippopotamus have been reported to have died from the disease in outbreaks in some parts of Africa. Wild carnivores such as lions, hyenas, jackals, etc. are also susceptible. Birds seem to be resistant to anthrax.

Outbreaks of anthrax tend to occur in association with climatic and weather features such as heavy rainfall, flooding and drought. In anthrax prone areas, the close grazing of animals of fresh shoots of grass after rainfall often leads to outbreaks of the disease due to the ingestion of organisms picked from contaminated soils. During severe outbreaks, biting flies may transmit the disease from one animal to another but this is a very minor mode of transmission. The principal mode of transmission is ingestion of infective micro-organisms. Non-biting blowflies may contaminate vegetation by depositing vomit droplets after feeding on a carcass infected with *B. anthracis*. Animals feeding on such vegetation then become infected. Wild carnivores become infected through the consumption of infected animals that have died from anthrax.

Clinical Signs

The incubation period of anthrax is 3-7 days with a range of 1-14 days. A common feature of anthrax is that animals in apparently good condition die suddenly without showing overt signs of ill health. Acute cases in goat, sheep and wild herbivores are characterized by fever, depression, difficult breathing and convulsions. Animals may die within 2-3 days if not treated. It is common to see bloody discharges from natural openings such as the anus and vagina. In few instances, anthrax can manifest itself as a mild disease characterized by general malaise.

Diagnosis

In areas prone to this disease, anthrax should be suspected in any animals that die suddenly. A veterinarian should therefore be called in to rule out anthrax before the dead animal(s) are handled by others. Animals suspected to have died from anthrax are not opened for post mortem examination. The reason for this is that the tissue form of anthrax bacterium on coming into contact with air (oxygen), changes to the environmentally resistant spores which then serve as the means for re-infection of susceptible hosts. The diagnosis of anthrax is made by making blood smears on a microscope slide from superficial blood vessels or natural openings of dead animals and staining the slide with a suitable stain. The slide is examined using a simple light microscope for typical morphology of the anthrax bacilli. The anthrax bacteria can be detected by growing the micro-organism in artificial media. A thermo-precipitation test (Ascoli test) is done if viable *B. anthracis* can not be demonstrated in tissues. About 2-3 gm of homogenized material in a small amount of saline is boiled and passed through filter paper. This filtrate is used as the antigen in a ring precipitation test or gel diffusion test. Polymerase chain reaction test (PCR) is also available for a rapid diagnosis of anthrax. Anthrax must be differentiated from other conditions that cause sudden death in animals such as clostridial infections, plant or chemical poisoning, bloat, lightening stroke, fevers caused by ticks, and other disease conditions.

Prevention and Control

In view of the fact that anthrax is often fatal in domestic animals, a preventive vaccination strategy should be adopted by annual vaccination of susceptible animals in areas prone to the disease. This is usually done 2-4 weeks before the onset of known period of outbreaks. In situations where animals show clinical signs of the disease, antibiotic treatment is recommended for sick and exposed animals. The non-encapsulated Sterne strain vaccine is used. This live attenuated vaccine is incapable of causing clinical disease but produces immunity (protection) against the disease. A single vaccination produces an immunity that lasts for an average of 9 months. Annual vaccination of susceptible animals is sufficient to control outbreaks of anthrax in defined localities. The Sterne vaccine strain is not hazardous to humans. Consequently, no specific protective measures (e.g. vaccination) are prescribed for those who handle this vaccine such as veterinarians, veterinary technicians and farmers. Other measures to be adopted in addition to immunization and treatment are:

- Enforcement of quarantine regulations.
- Prompt disposal of dead animals, bedding and contaminated materials by incineration or deep burial.
- Disinfecting of pens, stables and contaminated equipment and areas.
- Control of scavengers that feed on animals that have died from anthrax because of the possibility of spread of infection.
- Observation of general hygiene by people who have come in contact with diseased or dead animals.

Public Health Hazard

Anthrax is an old disease of humans and animals. It is not contagious; however, one reason it causes concern is that the spore form of the bacterium can persist in the environment for a long period of time if conditions are favourable. Humans may develop localized skin lesions or cutaneous anthrax (malignant carbuncle) from contact of broken skin with infected blood or tissues, or acquire the highly fatal form from inhalation of spores (woolsorters' disease). Humans may also acquire the intestinal form by consuming poorly cooked contaminated meat.

5.5 Brucellosis

Brucellosis is a sub-acute or chronic disease which may affect many species of animals. In sexually mature animals the infection localizes in the reproductive system and typically produces placentitis followed by abortion in the pregnant female and epididymitis and orchitis in the male.

Etiology

Brucella melitensis causes Brucellosis in goats and Malta or Mediterranean fever in man. The brucellae are somewhat host specific but cross-species infections occur, especially with *B. melitensis*.

Epidemiology

Brucellosis is a disease of many animal species but especially of those that produce food: sheep (especially milk-producing), goats, cattle and pigs and on a more localized scale, camels, buffaloes, yaks and reindeer. The recently recognized types associated with marine animals may also have the capacity to cause human disease. Infections in many wildlife species have been reported but those that obviously affect population fecundity and result in human infections are quite rare. *B. melitensis* infections in goats, however, have severe economic and public health implications. Infections in sheep and goats are highly contagious because of the pathogenicity of *B. melitensis* and because of close contact caused by the density of the flocks or herds. Animal-to-animal transmission occurs as a result of the large number of organisms shed in the environment. Humans are often infected due to direct animal contact or ingestion of contaminated dairy products. Human cases may be a useful indicator of the presence of disease in animal populations and may be the only source of information for surveillance. *Brucella* is usually transmitted from animal to animal by contact following an abortion. Pasture or shed may be contaminated and the organisms are probably most frequently acquired by ingestion but inhalation, conjunctival inoculation and skin contamination are other possibilities. Sexual transmission usually plays little role in the epidemiology of brucellosis. However, artificial insemination can transmit the disease and semen must only be collected from animals known to be free of infection. In sheep and goats, *B. melitensis* is nearly always the infecting species. *B. ovis* can also infect sheep but is of little significance in relation to human disease. The mode of transmission of *B. melitensis* in sheep and goats is similar to that in cattle but sexual transmission probably plays a greater role. The transmission of disease is facilitated by commingling of flocks and herds belonging to different owners and by purchasing animals from unscreened sources. For all species, embryo transfer is safe provided that recommended procedures are followed. Transmission to people can occur through the usual routes. However, ingestion of raw or undercooked reindeer bone marrow has also been implicated as a source of human infection.

The bacteria are found in tissues and fluids associated with pregnancy, the udder and the lymph nodes. Most infections result from ingestion of bacteria either from diseased animals or contaminated feedstuffs. However, infection may also be acquired by respiratory exposure and by contamination of abraded skin and mucosal surfaces.

Clinical Sign

Characteristic but not specific signs of brucellosis in most animal hosts are abortion or premature births and retained placenta. In some areas, abortion is relatively uncommon. There is lowered milk production due to premature births. Interference with fertility is usually temporary and most infected animals will abort only once and some are unaffected. The udder is often permanently infected, especially in the case of goats. Shedding of organisms in milk is frequent. Localized infections in sheep

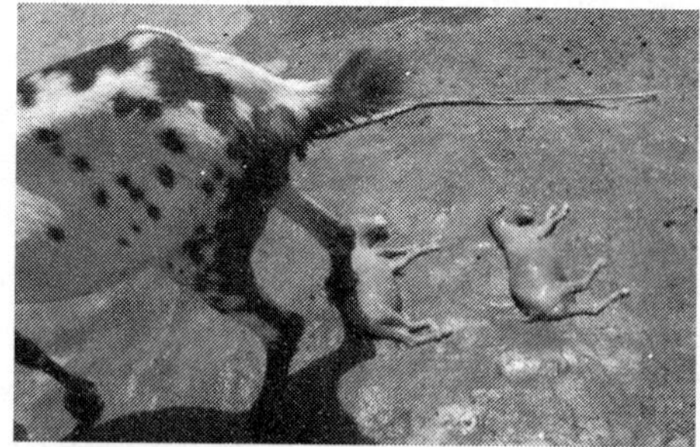

Fig. 9 Aborted fetuses in third trimester of pregnancy

result in orchitis or epididymitis in the case of *B. melitensis.* Arthritis may also be a rare sign in *B. melitensis* infected sheep and goats. Clinical effects of the disease depend upon many factors such as previous vaccination, age, sex and management such as herd or flock size and density. Abortions are more prevalent in unvaccinated animals and numbers of organisms shed are much greater.

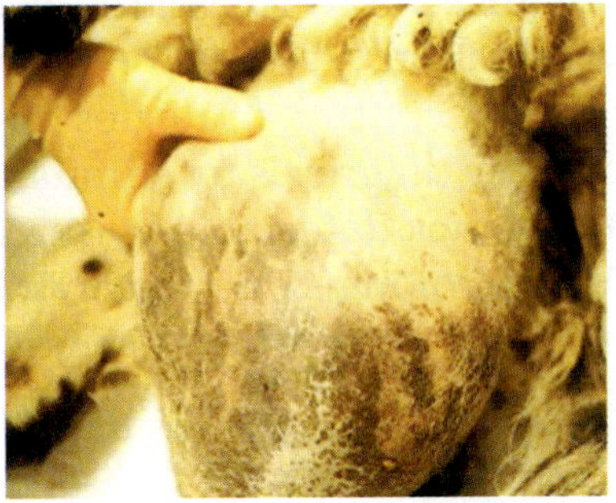

Fig. 10 Oedema and swelling of scrotum

Susceptibility to brucellosis is greatest in sexually mature animals. Young animals are often resistant although it should be noted that latent infections can occur and such animals may present a hazard when mature. Breed may also affect susceptibility, particularly in sheep. The milking breeds seem to be the most susceptible to *B. melitensis.* Latent or inapparent infections can occur in all farm animal species. These usually result from infection *in utero* or in the early post-natal period. Such animals can retain the infection for life and may remain serologically negative until after the first abortion or parturition.

Diagnosis

Diagnosis and control of the disease in animals must be carried out on a herd basis. There may be a very long incubation period in some infected animals and individuals may remain serologically negative for a considerable period following infection. The identification of one or more infected animals is sufficient evidence that infection is present in the herd, and that other serologically negative animals may be incubating the disease and present a risk. Diagnostic tests fall into two categories: those that demonstrate the presence of the organisms and those that detect an immune response to its antigens. The isolation of *Brucella* is definitive proof that the animal is infected but not all infected animals give a positive culture and the methods and facilities that must be employed are not always readily available. The detection of antibody or a hypersensitivity reaction provides only a provisional diagnosis, but in practice is the most feasible and economic means of diagnosis. False positive reactions to serological tests can occur through a number of factors including vaccination, and this must be borne in mind when interpreting results. Similarly, dermal hypersensitivity only indicates previous exposure to the organism, not necessarily active infection, and may also result from vaccination. Vaccination is an extremely important and effective facet of most control strategies but has the disadvantage that its use may confuse diagnosis by stimulating the production of hypersensitivity or antibodies detectable by serological tests. Antibody titres may persist for a prolonged period in a small proportion of vaccinated animals and this proportion is increased with age at vaccination. There is currently no widely available test that is able to distinguish vaccinated from infected animals although some are under evaluation. It is of utmost importance that the use of vaccination is strictly controlled, that it is used at the correct age, that vaccine of sufficient quality is used and that vaccinated animals are correctly identified. If this is not the case, correct serological diagnosis is confused. The vaccination programme can be suspended when the prevalence of the disease reach a very low level, when the disadvantages of vaccination outweigh any benefit that it may bring on the basis of cost: benefit and cost: effectiveness analysis.

Bacteriological Methods

The isolation and identification of *Brucella* offers a definitive diagnosis of brucellosis and may be useful for epidemiological purposes and to monitor the progress of a vaccination programme.

Stained Smears

Smears of placental cotyledon, vaginal discharge or fetal stomach contents may be stained using modified Ziehl-Neelsen (Stamp) or Kosters' methods. The presence of large aggregates of intracellular, weakly acid-fast organisms with *Brucella* morphology is presumptive evidence of brucellosis. Care must be taken as other infectious agents such as *Coxiella burnetii* or *Chlamydia* may superficially resemble *Brucella*.

Culture

Brucella most readily be isolated in the period following an infected abortion but isolation can also be attempted from post mortem materials. *Brucella* is excreted in large numbers at parturition and can be cultured from a range of material including vaginal mucus, placenta, fetal stomach contents and milk using suitable selective culture media. It is of the utmost importance that faecal and environmental contamination of the material is kept to a minimum to give the greatest chance of successfully isolating *Brucella*. If other material is unavailable or grossly contaminated, the contents of the fetal stomach will usually be otherwise sterile and is an excellent source of *Brucella*. In some circumstances it may be appropriate to attempt the isolation of *Brucella* from post mortem materials. Suitable material includes supramammary, internal iliac and retropharyngeal lymph nodes, udder tissue, testes and gravid uterus. Milk samples should be allowed to stand overnight at 4°C before lightly centrifuging. The cream and the deposit are spread on to the surface of at least three plates of solid selective medium. Placental samples should be prepared in the field by selecting the least contaminated portion and cutting off pieces of cotyledon. In the laboratory, the portions should be immersed in alcohol which should be flamed off before cutting with scissors or scalpel and smearing the cut surface on three plates of selective medium. Other solid tissues can be treated in a similar manner, or, ideally, they should be macerated mechanically following flaming before plating out. The tissues may be ground manually or homogenized in a blender with a small proportion of sterile water. Foetal stomach contents are collected, after opening the abdomen, by searing the surface of the stomach with a hot spatula and aspirating the liquid contents with a Pasteur pipette or syringe. Bacterial colonies may be provisionally identified as *Brucella* on the basis of their cultural properties and appearance, Gram staining, and agglutination with positive antiserum. If available, a PCR-based molecular identification method may be used.

Serological Methods

The detection of specific antibody in serum or milk remains the most practical means of diagnosis of brucellosis. The most efficient and cost effective method is usually screening all samples using a cheap and rapid test which is sensitive enough to detect a high proportion of infected animals. Samples positive to screening are then tested using more sophisticated, specific confirmatory tests for the final diagnosis to be made. It is absolutely essential that only internationally recognized tests using antigens standardized against the 2nd International anti-*Brucella abortus* Serum are used. Appropriate quality control (QC) sera should be included with each batch of tests, and tests should be repeated if the QC criteria are not met. Serological results must be interpreted against the background of disease incidence, use of vaccination and the occurrence of false positive reactions due to infection with other organisms.

Rose Bengal Plate Test

The Rose Bengal plate test (RBPT) is one of a group of tests known as the buffered *Brucella* antigen tests which rely on the principle that the ability of IgM antibodies to bind to antigen is markedly reduced at a low pH. The RBPT and other tests such as the buffered plate agglutination tests and the card test play a major role in the serological diagnosis of brucellosis worldwide. The RBPT is a simple spot agglutination test where drops of stained antigen and serum are mixed on a plate and any resulting agglutination signifies a positive reaction. The test is an excellent screening test, but may be oversensitive for diagnosis in individual animals, particularly when vaccinated. The procedure can be automated but this requires custom-made equipment.

ELISA

The ELISA tests offer excellent sensitivity and specificity whilst being robust, fairly simple to perform with a minimum of equipment and readily available from a number of commercial sources in kit form. They are more suitable than the complement fixation test (CFT) for use in smaller laboratories and ELISA technology is now used for diagnosis of a wide range of animal and human diseases. Although in principle, ELISA can be used for the tests of serum from all species of animal and man, results may vary between laboratories depending on the exact methodology used. For screening, the test is generally carried out at a single dilution. It should be noted, however, that although the ELISAs are more sensitive than the RBPT, sometimes they do not detect infected animals which are RBPT positive. It is also important to note that ELISAs are only marginally more specific than RBPT or CFT.

Serum Agglutination Test

The serum agglutination test (SAT) has been used extensively for brucellosis diagnosis and although it is simple and cheap to perform, its lack of sensitivity and specificity mean that it should only be used in the absence of alternative techniques.

Complement Fixation Test

The sensitivity and specificity of the CFT is good, but it is a complex method to perform requiring good laboratory facilities and trained staff. If such are available and it is carried out regularly with good attention to quality assurance, then it can be very satisfactory. It is essential to titrate each serum sample because of the occurrence of the prozone phenomenon whereby low dilutions of some sera from infected animals do not fix complement. This is due to the presence of high levels of non-complement fixing antibody isotypes competing for binding to the antigen. At higher dilutions these are diluted out and complement is fixed. Such positive samples will be missed if they are only screened at a single dilution. In other cases, contaminating bacteria or other factors in serum samples fix or destroy complement causing a positive reaction in the test, even in the absence of antigen. Such "anti-complementary" reactions make the test void and a CFT result cannot be obtained.

Supplementary Tests

Many other serological tests have been employed. Some, such as the Rivanol or 2-mercaptoethanol test are variations of the SAT and although more specific, share many of its disadvantages. At present, the use of such procedures in the place of the standard test is not advised.

Milk Testing

Milk is an ideal medium to test as it is readily and cheaply obtained, tests can be repeated regularly and give a good reflection of serum antibody. Bulk milk can be screened to detect the presence of infected animals within the herd which can then be identified by blood testing. This method of screening is extremely effective and is usually the method of choice in dairy goats.

Milk ELISA

The ELISA may be used to test bulk milk and is extremely sensitive and specific, enabling the detection of single infected animals in large herds in most circumstances.

Fluorescence Polarization Assay

This technique, which requires special reagents and reading equipment, is claimed to have advantages in sensitivity and specificity over other methods. Evaluation has been limited however, and the procedure is not widely available. Further information is required before its overall value can be assessed.

Intra-Dermal Test

This procedure, using a standardized antigen preparation such as Brucellin INRA or Brucellergene OCB, can be used for monitoring the status of herds in brucellosis-free areas. It is sensitive and specific but false positive reactions can occur in vaccinated animals.

Comparative Importance of Diagnostic Tests in Sheep and Goats

The RBT is useful for screening sheep and goat sera for antibodies to *B. melitensis.* An antigen suspension adjusted for highest sensitivity against a panel of control sera is recommended. The test is less sensitive than in cattle, however, and may not detect some infected animals. It is best used in combination with the complement fixation test. The SAT using 5% sodium chloride as diluents has been widely used but has low sensitivity and specificity. The micro-agglutination variant has higher sensitivity and specificity. It is only recommended in situations when more sophisticated tests are not available. The antigen is standardized to give 50% agglutination with a 1:650 dilution of the second International Standard for *Brucella abortus* antiserum. It should be noted that agglutination methods are particularly sensitive to non-specific agglutinins and cross-reacting antibodies. They tend to be more sensitive in the acute phases of infection and are severely affected by vaccination. They should only be used if no alternative is available. The CF test is superior to agglutination methods but its sensitivity and specificity are limited and it should be regarded as a complementary test rather than confirmatory test. It may be used for screening if automated methods are available. Sera should be inactivated at 62°C for 30 min. Vaccination produces sero-conversion in the CF test but antibody titres decline much more rapidly than those resulting from infection. ELISA is reported to give superior results to other tests in sensitivity and specificity but experience is limited. The MRT is not suitable for use on sheep or goat milk but ELISA can be used. The CF test is also usable on whey samples but is technically demanding and no longer recommended. It should be understood that no currently available serological test can be considered reliable for the detection of ovine or caprine brucellosis at the level of individual animals. Diagnosis and control should be applied at the herd/flock level. The intradermal test for delayed hypersensitivity to *Brucella* antigens is useful as a flock test. It is affected by vaccination however a purified antigen preparation such as "Brucellergene OCB R", which contains a mixture of *Brucella* proteins free of smooth LPS, should be used to avoid compromising serological tests. It is useful for monitoring the status of brucellosis-free flocks, especially if vaccination is not practiced.

Prevention and Control

The aim of a control programme is to reduce the impact of a disease on humans health and the economic consequences. The elimination of the disease from the population is not the objective of a control programme, and it is implicit that some "acceptable level" of infection will remain in the population. Control programmes have an indefinite duration and will need to be maintained even after the "acceptable level" of infection has been reached, so that the disease does not re-emerge. In many countries, methods for the control of brucellosis are empowered through governmental regulation/legislation. In others, no authorities exist. Therefore, the procedures for management of infected flocks may vary widely. Nevertheless, certain principles apply, namely: 1) the reduction of exposure to *Brucella* spp. and 2) the increase of the resistance to infection of animals in the populations. These procedures may be further classified under the general categories of test and isolation/slaughter, hygiene, control of animal movement, vaccination.

Test and Isolation/Slaughter

There are no pathognomonic signs of brucellosis in animals at individual level; however, the occurrence of abortion storms in native herds/flocks is usually a strong indicator of infection. Therefore, serological (and sometimes allergic) tests are the usual method of identifying possible infected animals. Bacteriological procedures are very useful for confirming test results and for epidemiological studies. The decision about slaughter of test positive animals is made after regulatory, economic and prevalence factors are considered. In most cases, test and slaughter of positive animals is only successful in reducing the incidence if the flock prevalence is very low (2%). Retention of positive animals is less hazardous if the remaining animals have been vaccinated. The isolation of test positive animals is recommended, especially during and after parturition. The immediate slaughter of test positive animals is expensive and requires animal owner cooperation. Compensation is usually necessary. Furthermore, the application of test and slaughter policies is unlikely to be successful with brucellosis of sheep and goats where the diagnostic tests are less reliable than in cattle. Repeated flock tests are necessary to further reduce the incidence of brucellosis and to confirm elimination.

Hygiene

The goal in the application of hygiene methods to the control of brucellosis is reduction of exposure of susceptible animals to those that are infected, or to their discharges and tissues. This is a classical procedure in disease control. Factors such as the methods of animal husbandry, prevalence of clinical signs, type of facilities, and degree of dedication of the owners of animals, will also determine success. Owners are often poorly informed about disease transmission and recommendations, such as separation of parturient animals, can be difficult or impossible to implement. Antibiotic treatment of known infected animals, or those which are potentially exposed to them, has not been commonly used in the control of brucellosis. A limited number of studies have shown rapid reductions in the incidence of brucellosis when the flock was treated but this procedure is considered to be restricted in practicality. Treatment is sometimes used in animals of special breeding value, but because of the uncertain outcome it is not generally recommended.

Control of Animal Movement

This may be regarded as an aspect of hygiene. However, it is essential in any programme to limit the spread of brucellosis. Animals should be individually identified by tattoo or ear tag. Unauthorized sale or movement of animals from an infected area to other areas should be stopped. Similarly, importations into clean areas must be restricted to animals that originate from brucellosis-free areas, that have a flock history of freedom from the disease and that have given negative reactions to recently performed diagnostic tests. In practice, it is much more difficult to control the movement of small ruminants kept under nomadic or semi-nomadic conditions. The owners of flocks may be accustomed to seasonal migrations which may cross national boundaries.

Vaccination

Vaccination of sheep and goats with *B. melitensis* Rev 1 can reduce susceptibility by a thousand fold or more to the homologous species. For the present, Rev 1 vaccine is the most effective vaccine available against *B. melitensis* and in many countries has given very good results. Its use is recommended when uncontrolled *B. melitensis* infection exists in ruminant populations.There is general agreement that the most successful method for prevention and control of brucellosis in animals is through vaccination. While the ideal vaccine does not exist, the attenuated strains of *B. melitensis* strain Rev.1 for sheep and goats have proven to be superior to all others. The doses and methods of administration, especially with Rev.1, vary and these can affect the results. It is often recommended that vaccination with strain Rev.1 should be limited to sexually immature female animals. This is to minimize stimulation of postvaccinal antibodies which may confuse the interpretation of diagnostic tests and also to prevent possible abortions induced by the vaccines. However, field and laboratory studies have demonstrated that intra-conjunctival administration of these vaccines makes the vaccination of the flock a practical and effective procedure. Rapid herd immunity is developed and application costs are minimized. The lowered dose results in lower antibody titres and these recede rapidly.

Several diagnostic tests have been developed which are useful in differentiating antibody classes. Of these, the complement fixation test and ELISA are currently the most widely used. Vaccination of animals usually results in elimination of clinical disease and the reduction in numbers of organisms excreted by animals which become infected. Furthermore, animal owners are more likely to accept vaccination as a method of control since they are accustomed to this form of disease control. In many countries, vaccination is the only practical and economical means of control of animal brucellosis.

Eradication

Eradication means the elimination of a pathogenic agent from a country or a zone (i.e. part of the territory of a country with a distinct animal health status). A highly organized effort is needed to reach eradication in either a territory and in a population. Eradication is conceptually very different from control: it is neither a casual nor an automatic consequence of a control programme, no matter how well planned and implemented the control programme. It is based on sanitary measures and an organisation of activities completely different from those implemented for a control programme. Crucial factors for the success of an eradication programme are the implementation of an effective surveillance system and the understanding and sharing of objectives for eradication by the decision-makers, farmers, and all other stakeholders. To keep an unaffected population free from an infection, prevention measures must be implemented to segregate an infectious organism from a geographical area and its human and animal populations. On the long-term basis, eradication programmes in general are more economically advantageous compared to control programmes. This advantage, however, can not always be translated into practice. In fact, an eradication programme involves the mobilization of an amount of resources (in terms of financial and human resources) that may not be available or whose returns for the investment may require a time span longer than it is affordable by any decision-making authority. Cost-benefit and cost effectiveness analysis can be used to support decisions on control strategies. However, no in depth analysis is possible in absence of epidemiological surveillance. There is also little doubt that very often failures of control and eradication efforts are due to the absence of an adequate epidemiological surveillance system sustaining both technical and political decision making.

Brucellosis is a true zoonosis and the link between animals and humans is strong. While public health measures such as pasteurization and education have varying degrees of success, it remains primarily a veterinary responsibility to control brucellosis, including application of principles of epidemiology. It is nearly always more economical and practical to prevent diseases than to attempt to control or eliminate them. For brucellosis, the measures of prevention include:

1. Replacement animals should be carefully selected. These whether purchased or produced from existing stock, should originate from brucella-free herds or flocks.

2. Pre-purchas e tests are necessary unless the replacements are from populations in geographically circumscribed areas that are known to be free of the disease.

3. Purchased replacements should be isolated for at least 30 days. In addition a serological test prior to commingling is necessary. Prevent of contacts and commingling with herds of flocks of unknown status or those with brucellosis.

4. If possible, laboratory assistance should be utilized to diagnose causation of abortions, premature births, or other clinical signs. Suspect animals should be isolated until a diagnosis can be made.

5. F locks should be included in surveillance measures such as periodic testing of with simple screening serological procedures such as the Rose Bengal test.

6.Proper disposal (burial or burning) of placentas and non-viable fetuses.

7.D isinfection of contaminated areas should be performed thoroughly.

8.C ooperation with public health authorities is necessary to investigate human cases. Animal brucellosis, especially when caused by *B. melitensis*, can often be identified through investigations of cases in humans.

5.6 Campylobacteriosis

Campylobacteriosis is characterized by late abortion,stillbirths and weak lambs and kids in sheep and goats. Infections in sheep are sometimes followed by metritis and occasionally deaths.

Etiology

C. jejuni, C. fetus subsp. *venerealis,* and *C. fetus* subsp. *fetus* (also known as *C fetus* subsp. *intestinalis* and *Vibrio fetus* var *intestinalis*) cause infertility and abortions in sheep, goats and cattle. *Campylobacter* are Gram negative, microaerophilic, curved or spiral rods in the family Campylobacteriaceae. *Campylobacter jejuni* (formerly known as *C. fetus* subsp. *jejuni*) and *C. coli* are associated with enteritis in domestic animals and humans. The isolation rate in diarrheic sheep was 33.3% and the identified species were *C. jejuni* (14.%), *C.coli* (4.0%) and *C. lari* (2.0%).

Epidemiology

Campylobacter jejuni and *C. coli* are transmitted by the fecal-oral route; they can be spread by direct contact and on fomites including food or water. *C. jejuni* may also be present in the vaginal discharges, aborted fetuses and fetal membranes of aborting goats. House flies are mechanical vectors. Humans can be infected after ingesting undercooked meats, raw milk, contaminated foodstuffs or unchlorinated water, and after contact with infected pets or livestock. Asymptomatic carriers are seen in many species of domestic animals; humans do not usually become carriers.*Campylobacter fetus* subsp. *fetus* is transmitted by ingestion in sheep and goats. *Campylobacter* species are susceptible to many disinfectants, including 1% sodium hypochlorite, 70% ethanol, 2% glutaraldehyde, iodine-based disinfectants, phenolic disinfectants and formaldehyde. *Campylobacter* is also sensitive to gamma irradiation and UV radiation. *Campylobacter* species do not tolerate drying or heating but can often survive for a time in moist environments. *C. fetus* can survive in liquid manure for 24 hours and soil for up to 20 days.

Clinical Signs

Campylobacter species cause enteritis, abortions and infertility in various species. *C. fetus* subsp. *fetus* and *C. jejuni* can cause late term abortions, stillbirths and weak lambs in sheep and goats. The incubation period for *Campylobacter* infections is generally short. Infections in goats are sometimes followed by metritis and occasionally deaths. Recovery, with immunity to re-infection is typical. Goats can become persistently infected and continue to shed bacteria in the feces. *Campylobacter* species are readily transmitted between animals or from animals to humans. Organisms are present in feces, vaginal discharges and aborted materials and can be spread by direct contact, on fomites and by arthropods acting as mechanical vectors. Contaminated food and water is often the source of infections. Gastrointestinal *Campylobacteriosis* is usually self-limiting in mammals. Mortality is also low in adult goats affected by abortions and infertility. Morbidity may be up to 90% in outbreaks in sheep but is usually around 5 to 50%.

Post Mortem Findings

In goats, the fetus is usually autolyzed after *C. fetus* abortions. One to two cm orange/ yellow necrotic foci can sometimes be found in the liver. Placentitis may be evident, with hemorrhagic necrotic cotyledons and edematous or leathery areas between the cotyledons.

Diagnosis

Enteritis can be diagnosed by isolating the causative organism in fresh fecal samples; however, *Campylobacter* is fragile and cannot always be found. Forty-eight to 72 hour colonies are raised, round, translucent and sometimes mucoid. Biochemical testing, antigen testing and restriction endonuclease DNA analyses are used for species and strain identification. A presumptive diagnosis can also be made by observing the characteristic motility in darkfield or phase contrast preparations. Gram negative, curved or spiral rods are seen in Gram stained preparations. Serology on paired titers may be helpful in some cases. Darkfield and phase contrast preparations of samples from the placenta, fetal abomasum and uterine discharge are used to diagnose *Campylobacter* abortions in goats. *Campylobacter* antigens can also be detected by immunofluorescence. Vaginal cultures can also be collected immediately after abortion or infection, but this method may be unreliable because *Campylobacter fetus* is fragile and usually present in low numbers. Systemic antibody responses are not useful in genital *Campylobacteriosis,* as they can be directed against nonpathogenic species.

Treatment and Prevention

Antibiotics may be useful for some cases of enteritis; however, information on efficacy is limited. Antibiotics may also prevent exposed sheep and goats from abortion during an outbreak. Vaccines are not available for enteritis, but can prevent abortions in sheep and goats.

5.7 Pasteurella Pneumonia

Pneumonic pasteurellosis is a cranioventral fibrinous bronchopneumonia. The disease occurs in goats throughout the world.

Etiology

Both *Pasteurella multocida* and *P. haemolytica* cause pneumonia in goats. Both species are Gram-negative tiny ovoid rods that do not form spores. They grow well on blood agar, where only *P. haemolytica* causes hemolysis. Colonies are 1 to 2 mm in diameter. *Pasteurella multocida* but not *P. haemolytica* produces indole. Although recent work suggests that *P. haemolytica* should be subdivided into two species corresponding to the current biovars A and T, how this might clarify the epidemiology of pasteurellosis in goats remains to be determined. A biotype T isolate of *P. haemolytica* obtained from an acute fatal case of caprine pneumonia has been used to experimentally produce a proliferative and exudative pneumonia. When typing has been performed, *P. haemolytica* type A2 has been reported most frequently.

Epidemiology

Both of these organisms commonly reside in the upper respiratory tract of normal goats. While a previous viral infection would increase the probability of lung invasion by *Pasteurella,* field conditions often provide enough stress for the organism to be a primary pathogen. Poor ventilation is a major factor permitting invasion of the lungs, but crowding, parasitism, and malnutrition all, contribute to development of disease. Some affected goats have been recently stressed by transport, and steroid administration appears to increase the proliferation of *P. haemolytica* in the nasal passages of transport-stressed goats. The virulence *of* the *Pasteurella* organism is either very high or increases during an outbreak, so that the disease may spread to unstressed herd members. In other instances, pneumonia is recognized only in animals exposed to newly introduced goats.

Pathogenesis

A ruminant-specific leukotoxin produced by *P. haemolytica* is believed to be very important in the pathogenesis of pasteurellosis. This toxin impairs and lyses alveolar macrophages and neutrophils that arrive in the lung to fight the infection. Enzymes released by the dying neutrophils cause additional injury to lung tissue.

Clinical Signs

In the acute case, there is typically a fever of 104 to 106° F and also mucopurulent nasal and ocular discharge. Lethargy, anorexia, dyspnea, and a moist painful cough are noted. Auscultation may reveal crackles, areas of consolidation (increased bronchial tones), or pleuritis (friction rub early, muffled sounds later). Mortality rates may be 10% or more. Commonly, one goat is found suddenly dead before any are noticed to be ill.

Diagnosis

Diagnostic cultures may be obtained from tracheal washes or from necropsy specimens. Cultures taken from the nasal passages are not acceptable substitutes. Kids with acute pasteurellosis often have a septicemia, and bipolar organisms may be visible in stained blood smears. A radiographic examination would document the cranioventral distribution and rule out the presence of a penetrating metallic foreign body from the reticulum, which rarely acts to introduce infection into the thoracic cavity.

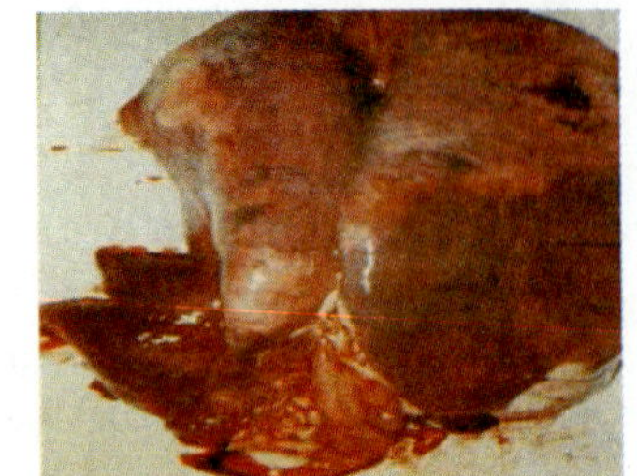
Fig. 11 Lobar Pneumonia

Post Mortem Findings

The cranioventral lung lobes are usually affected bilaterally. There is a red to purple consolidation of these lobes, sometimes accompanied by a fibrinous pleuritis.

Histologic changes are typical of pasteurellosis in other species and include hemorrhage, necrosis, and exudation of fibrin, edema fluid and neutrophils or macrophages into airways. A *Pasteurella* infection may be secondary to a chronic progressive viral pneumonia. Thus if poor body condition suggests that a dead goat has been ill for weeks or months, the dorsal lung lobes should be palpated very carefully. If they seem firmer than normal, histological examination to detect concurrent interstitial pneumonia should be requested.

Treatment

Antibiotics commonly used for parenteral treatment include penicillin (20,000 to 40,000 IU/kg once daily), ampicillin (5 to 10 mg/kg twice daily), tetracycline (5 mg/kg once or twice daily), and tylosin (10 to 20 mg/kg once or twice daily). A culture test obtained by tracheal wash or at necropsy should be submitted for sensitivity testing when dealing with herd outbreaks. Ventilation should be corrected to decrease humidity in the shed so that there is no condensation on windows and walls. Newly introduced goats should be kept isolated from the herd for at least two weeks.

Prevention and Control

In some outbreaks, young kids appear to be especially susceptible to a septicemic or pneumonic form of pasteurellosis. Colostral transfer of antibodies may be very important in preventing these infections of the neonate. No bacterin has been proven effective against pasteurellosis in goats. This is partly because of a multitude of serotypes possible and partly because the antibodies produced in response to vaccination may contribute to damage of pulmonary tissue when infection occurs. A toxoid vaccine directed against the leukotoxin produced by *P. haemolytica,* as developed for cattle, may prove more useful but needs to be evaluated in goats. Recently, *P. haemolytica* organisms embedded in microagar beads and deposited in lung tissue by the transthoracic route have induced significant immunity to challenge exposure in goats.

5.8 Leptospirosis

Leptospirosis occurs in all animals and is an important zoonosis. It causes septicemia, interstitial nephritis, hemolytic anemia, and abortion.

Etiology

Leptospirosis is caused by various species of *Leptospira,* a spirochete in the family Leptospiraceae, order Spirochaetales. The classification of this organism is complex. Before 1989, all of the pathogenic isolates belonged to the species *Leptospira interrogans,* which contained more than 200 serovars in 23 serogroups. More recently, the genus *Leptospira* has been reclassified into 16 or more species. Pathogenic serovars are now found in the species *Leptospira interrogans, L. noguchii, L. santarosai, L. meyeri, L. borgpetersenii, L. kirschneri, L. weilii, L. inadai, L. fainei* and *L. alexanderi.* Serovars associated with disease in sheep and goats include *hardjo, pomona, grippotyphosa* and *ballum.* The new classification system can be confusing because both pathogenic and non-pathogenic serovars and serogroups occur in the same species and a single serovar or serogroup can occur within multiple species.

Epidemiology

Leptospira species are found worldwide; however, the predominant serovars vary by geographic region. Leptospirosis can be transmitted either directly between hosts or indirectly in the environment. *Leptospira* spp. can be ingested in contaminated food or water, spread in aerosolized urine or water, or transmitted by direct contact with the skin. The organisms usually enter the body through mucous membranes or abraded skin. They may also be able to penetrate intact skin that has been immersed for a long time in water. *Leptospira* spp. is excreted in the urine, and can be found in aborted or stillborn fetuses, as well as in normal fetuses or vaginal discharges. They can be isolated from the male reproductive organs. *Leptospira* spp. does not multiply outside the host. In the environment, they require high humidity for survival and are killed by dehydration or temperatures greater than 50°C. They can remain viable in water for several months under laboratory conditions, but do not survive as well in river water under natural conditions.

All mammals appear to be susceptible to at least one species of *Leptospira.* Disease is less common in sheep and goats than cattle. The primary reservoir hosts for most *Leptospira* serovars are wild mammals, particularly rodents. Reservoir hosts among domestic animals include cattle, pigs, sheep and dogs. The specific reservoir host(s) varies with the serovar and the geographic region. Disease in reservoir hosts is more likely to be asymptomatic, mild or chronic. *Leptospira* species can be inactivated by 1% sodium hypochlorite, 70%

ethanol, glutaraldehyde, formaldehyde, detergents and acid. This organism is sensitive to moist heat (121° C for a minimum of 15 min) and is also killed by pasteurization. Leptospirosis is particularly prevalent in warm and humid climates, marshy or wet areas, and in regions with an alkaline soil pH. This disease is often seasonal.

Clinical Signs

Leptospira infections may be asymptomatic, mild or severe, and acute or chronic. The clinical signs are often related to kidney disease, liver disease or reproductive dysfunction. Chronically infected animals are often asymptomatic. Leptospirosis in sheep and goats is similar to the disease in cattle. It is characterized by fever and anorexia and, in some animals, jaundice, hemoglobinuria or anemia. Abortions, stillbirths, weak lambs or kids and infertility can also be seen, either with or without other clinical signs. Clinical disease is relatively uncommon in sheep.

Diagnosis

Leptospirosis can be diagnosed by culture, detection of antigens or nucleic acids, or serology. The location of the organisms varies with the form of the disease. In acute infections, *Leptospira* may be found in the blood, milk, and cerebrospinal, thoracic or peritoneal fluids. During chronic infections, they are sometimes found in the urine. The liver, lung, brain and kidney are collected at necropsy from acute cases, and the kidney and genital tract are tested in chronic cases. Organisms can also be found in the body fluids or tissues of aborted fetuses. *Leptospira* species can be cultured on a variety of media but are fastidious and grow slowly on primary isolation. Special transport media may be required for shipment to the laboratory. Depending on the serovar, culture may take up to 13 to 26 weeks. Identification to the species, serogroup and serovar level is done by reference laboratories, using genetic and immunologic techniques. *Leptospira* can also be identified in clinical samples by immunofluorescence and immunohistochemical staining, as well as DNA probes and polymerase chain reaction (PCR) techniques. Darkfield microscopy can be used but is non-specific. Silver staining is sometimes useful as an adjunct technique. These organisms stain poorly with the Gram stain. Serology is also used for diagnosis. Paired acute and convalescent samples are preferred from most animals, but a single positive sample from an aborted fetus is diagnostic. Herd tests are often used in ruminants. The most commonly used serological tests are the microscopic agglutination test (MAT) and enzyme-linked immunosorbent assays (ELISAs). Serovar-specific ELISAs are available in veterinary medicine, and cross-reactions are less common in animals than in humans. Other serologic tests include radioimmunoassay, the microcapsule agglutination test, immunofluorescence, counterimmunoelectrophoresis and thin-layer immunoassay.

Treatment

Antibiotics used to treat leptospirosis include the tetracyclines, penicillin/ampicillin, dihydrostreptomycin, streptomycin and the fluoroquinolones. The efficacy of treatment may depend on the serovar. Fluid therapy, blood transfusions and other supportive care may also be necessary. In infected flocks, further abortions may be prevented by vaccination and treatment of all animals with antibiotics, if leptospirosis is diagnosed early during an outbreak.

Prevention and Control

Leptospirosis vaccines are available only for pigs, cattle and dogs. Although the vaccines prevent disease, they do not completely prevent infection or the shedding of the organisms. Immunity is largely serovar specific: vaccines are protective only against the included serovars or closely related serovars. Prophylactic treatment of exposed animals with antibiotics can also prevent disease. Sanitation and the prevention of contact with contaminated environments or infected wildlife, particularly rodents, can decrease the risk of infection. Animals should not be allowed to drink from or enter contaminated bodies of water. Good sanitation can reduce the risk of infection in areas where livestock are bred or give birth. Infections are usually introduced in an infected animal, through the environment, or by contact with infected animals. Replacement stock should be selected from herds negative for leptospirosis. Animals not known to be *Leptospira* free should be quarantined for 4 weeks and tested before being added to the herd. Other control measures include vaccination, isolation and treatment of infected animals, rodent control, prevention of contact with wildlife, sheep and pigs, and prevention of access to contaminated bodies of water.

5.9 Listeriosis

Listeriosis is an infectious disease caused by *Listeria monocytogenes* and characterized by abortion and encephalitis.

Etiology

Listeriosis is caused by the bacteria *Listeria monocytogenes.* Five serotypes and number of subserotypes have been identified. Serotype 5 has been identified as a cause of abortion in sheep but is apparently of low pathogenicity.

Epidemiology

Listeria monocytogenes is found in soil, water, silage, and even in the goat digestive tract. The bacteria are known to multiply well in cold temperatures. The organism can be shed in the milk of both carrier and sick goats. Its zoonotic potential (ability to be transmitted to humans) is of growing concern. Listeriosis is most often seen in intensive management situations. Listeriosis is more common in adult animals than in kids. It is entirely possible to buy infected animals and introduce this disease into a previously uninfected herd, because some goats are carriers who never display any symptoms. Listeriosis is generally occur when feeding silage, sudden changes in kind of feed, parasitism, weather changes and advanced stages of pregnancy. The encephalitic form is most common, causing inflammation of the nerves in the goat brain stem.

Clinical Signs

There are two forms of Listeriosis: one form results in abortions, while the other causes encephalitis. Both types are seldom seen simultaneously in the same herd. Symptoms include depression, decreased appetite, fever, leaning or stumbling or moving in one direction only, head pulled to flank with rigid neck (similar to symptoms of tetanus), facial paralysis on one side, slack jaw and drooling. Diarrhea is presently only in the strain of Listeriosis which causes abortions and pregnancy toxemia. Listeriosis can be mistaken for rabies.

Post Mortem Findings

Visceral lesions occur as multiple foci of necrosis in the liver, spleen, endocardium and myocardium especially in the septicaemic form and in aborted fetuses. Gross lesions suggested as being almost pathognomonic for listeriosis in aborted lambs are small yellow foci of necrosis in the liver,small abomasal erosion and yellow orange meconium.

Diagnosis

Listerial meningoencephalitis may be confused with the pregnancy toxaemia in sheep and goats. However Pregnancy toxaemia is observed only during late pregnancy and usually in association with multiple pregnancy and declining nutritional status. Listerial abortion must be differentiated from the other causes of abortion. Acute septicemia of new born animals are due to many causes but association with abortion and stillbirths may suggest the possible presence of listeriosis.

Treatment

Treatment involves administration of high doses of Procaine Penicillin every six hours for three to five days, then daily for an additional seven days. Forty thousand IU per kg of body weight of procaine penicillin is needed to cross the blood brain barrier and put sufficient amounts of the antibiotic into the tissue of the goat central nervous system.

Prevention and Control

Most attempts to produce a satisfactory killed vaccine have been unsuccessful although field trials with a killed bacterin are reported to reduce the incidence in sheep flocks. Living attenuated bacteria are also under investigation as immunizing agents and appear to exert significant protection. Silage feeding should be reduced or stopped. No moldy feed or hay should be allowed to feed to the animals. Lots of free-choice quality roughage, particularly in the latter stages of pregnancy should be given.

5.10 Neonatal Diarrhea Complex

Diarrheal disease is very important cause of kid losses throughout the world including India. It is caused due to inflammation of the intestinal mucosa. There is varying degrees of dehydration and acid-base imbalance.

Etiology

Colibacillosis, caused by *E. coli* is generally considered to be the most common cause of diarrhea in young kids. Currently, enteropathogenic *E. coli* are identified as a cause of neonatal kid diarrhea. The adherence pilus, *K99*, most frequently associated with enteropathogenic *E. coli* in calves and lambs, was some time identified in kid. The enteropathogenic *E. coli* that infect goats have antigenically distinct adherence pili. A large Indian survey result has identified 22 different serogroups of *E. coli* associated with diarrhea in kids. However, some of these were isolated from heart blood and were responsible for colisepticemia rather than localized enteric infections.

The few epidemiologic studies actually carried out in kids, however, suggest that the nature and frequency of the various etiologic agents of diarrhea may differ considerably in the goat. The generally held presumption that *E. coli* is a major cause of diarrhea in kids is challenged by several studies. Many workers have identified cryptosporidiosis and reported the most common cause of diarrhea in kids younger than 15 days of age. The organism is rarely isolated from non diarrheic kids. Cryptosporidiosis can occur alone or in conjunction with other presumed pathogenic bacteria, viruses and protozoa.

Rotavirus and coronavirus are commonly identified from diarrheic kids but their significance is unclear, as the viruses are also identified readily from non-diarrheic kids with almost equal frequency. In a report, rotavirus particles were identified from the diarrheic feces of 4 month old kids. If the virus was indeed responsible for the diarrhea, this would be in marked contrast to rotavirus infection in calves and lambs, where clinical diarrhea is unusual in individuals older than 3 weeks of age.

Adenoviruses were identified in diarrheic goats and *Yersinia* spp., *Giardia,* and *Campylobacter jejuni* were confirmed as causes of diarrhea. Adenoviruses also have been isolated from the intestine of goats that died of peste des petits ruminants. Because the latter virus is a major, known cause of diarrhea, it is difficult to ascribe significance to the adenovirus. *Campylobacter jejuni* has also been isolated from diarrheic feces of kids. Again, the role of this organism remains equivocal. Caprine herpesvirus infection has been associated with diarrhea in kids. Herpesvirus infection was recognized in association with a severe generalized infection of 1 to 2 week old kids causing increased morbidity and mortality rates.

Coccidiosis is likely the most common helminth and is the second most common prevailing identifiable causes of diarrhea in weaned, growing kids worldwide. Other well-known causes of diarrhea in older kids also include enterotoxemia caused by *Clostridium perfringens* type D, salmonellosis, yersiniosis, peste des petits ruminants and indigestion associated with the transition to solid feeds. These are either more sporadic or more geographically restricted than coccidiosis and helminthiasis.

Infectious agents may not always be involved in diarrhea of young kids. Many cases may be nutritionally based and difficult to diagnose at post mortem without adequate feeding and management history. Nutritional causes of diarrhea in young kids are often associated with husbandry and feeding practices. Problems identified include overfeeding of milk, use of poor quality milk replacers, sudden changes in feeding. These situations may produce noninfectious diarrhea, or predispose to the proliferation of pathogenic bacteria in the abomasum, rumen or intestine. The common causes of death in kids with age have been summarized in table 3.

Table 3: Common causes of death in newborn kids

Cause of death	Age
Preparturient still- birth	-
Parturient still- birth	-
Hypothermia exposure	1-5 hr
Hypotheri starvation	10 hr +
Colibacillosis	1-20 days
Joint- ill	3 days +
Navel- ill	3 days +
Contagious ecthyma (orf)	1 days +
Pneumonia	1 day +
Enterotoxaemia	7 days +
Coccidiosis	20 days +
Congenital defects	birth
Watery mouth	12- 72 hr
Liver abscess	4 days
Trauma	-
Predation	-

Epidemiology

A high incidence of diarrheal disease is associated with intensive rearing of kids under conditions of overcrowding and poor sanitation. Under extensive management conditions, the incidence is increased when kidding seasons coincide with periods of extreme weather conditions, especially excessive heat, excessive cold or heavy rains.

Mortality and morbidity among the neonates have always been a bottleneck in successful goat husbandry in India. Kid mortality varies from 1.9 to 45.5% in various breeds of goats at different situation in India. Kid mortality during different stages indicated that 50.0% of the deaths are inflicted during the first month and about 25% during the first week of birth. Most deaths occurred during the first month of life and death within first week of life in kids was 53.95%. An overall mortality in kids in different India breeds presented in Table 4. Kid mortality is generally affected by various factors such as season, age, sex, litter size, birth weight and housing. Various conditions responsible for mortality could be infectious and non-infectious. The important infectious disease that affect the kids constitute the enteric (coli septicemia and colibacillosis, enteropathogenic viruses, enterotoxaemia, coccidial infection) and disease of respiratory system.

Table 4: Percentage of kid mortality in different Indian goat breeds due to various causes

Breed	Age (0-3 M)	Remark
Sirohi	1.90	Semi-arid free range grazing with no supplementation
Marwari	23.10	-
Jhakrana	3.10	Farmers flock (based on inquiry)
Beetal	37.60	Organized farm conditions (stall- fed)
Beetal	25.40	Semi-arid free range grazing
Barbari	41.50	Organized farm conditions (Mostly stall- fed)
Jamunapari	45.50	Organized farm conditions (Mostly stall- fed)
Jamunapari	5.75	Farmers flock (based on inquiry)
Gohilwadi	6.10	Farmers flock (based on inquiry)
Mehsana	13.00	Farmers flock (based on inquiry)
Zalawadi	22.00	Farmers flock (based on inquiry)
Kutchi	15.00	Farmers flock (based on inquiry)
Sangamneri	28.90	Organized farm conditions; mostly stall-fed with limited grazing
Osmanabadi	18.00	Organized farm conditions
Malabari	32.80	Organized farm conditions, mostly stall-fed with limited grazing
Gaddi	2.19	Organized farm conditions
Changthangi	5.70	Organized farm conditions
Chegu	44.20	Organized farm conditions
Ganjam	27.50	Farmers flock (based on inquiry)
Black Bengal	16.80	Organized farm conditions

Pathogenesis

In colisepticemia, virulent strains of *E. coli* gain entrance to the blood stream and produce clinical disease through the effects of bacteremia and endotoxemia. Diarrhea may be part of the clinical syndrome, but is not the primary problem. In enteropathogenic *E. coli* infection, the bacteria remain in and colonize the gut, aided by the action of adherence pili. They then produce a severe hypersecretory diarrhea through the action of enterotoxins. Affected kids are usually 3 to 10 days of age and show severe generalized weakness, distension of the forestomachs with undigested milk and oftentimes yellowish and greenish diarrhea. The condition is thought to be a form of colisepticemia similar to the condition known in lambs.

Diarrhea, as a pathophysiologic process results in dehydration, acidosis, and electrolyte depletion and in neonates, is also associated with hypoglycemia. These physiologic derangements when unchecked are often fatal. Certain agents, such as *Salmonella* spp. and *Clostridium perfringens,* though primarily enteric pathogens, can cause marked, concurrent systemic effects in addition to diarrhea as a result of bacteremia and toxemia respectively. Certain agents may produce clinical effects in multiple organ systems as exemplified by the pneumoenteritis syndrome seen in peste des petits ruminants.

In herpesvirus infection outbreak, the disease is characterized by fever and ulcerative enteritis with high mortality in kids was recorded in goats. Affected kids show signs of weakness, dyspnea, cyanosis, abdominal pain, diarrhea, nasal discharge, and oral erosions. Not all signs are present in all individual kids. The disease has a fatal course of 1 to 4 days. Necrosis and ulceration of the digestive tract, especially the caecum and colon are prominent necropsy findings.

Clinical Signs

In the case of diseases limited to the gastrointestinal tract such as cryptosporidiosis, enteropathogenic colibacillosis and the enteric viruses, diarrhea is the predominant clinical sign. The character of diarrheic feces in kids can vary considerably from pasty white yellowish to watery brown and it is inadvisable

Fig. 12 Pale, yellowish watery diarrhea and staggering gait in colibacillosis in kids

to ascribe too much diagnostic significance to the color or consistency of feces in a given case, especially because mixed etiologies are possible. Streaking of feces with blood may also be seen with a variety of causative agents. Regardless of cause, there may be some degree of abdominal discomfort, possible abdominal distension, and probable loss of appetite in association with diarrhea.

If diarrhea is prolonged or severe, then dehydration ensues. Dehydrated animals become depressed, the mouth is dry and they gradually lose the sucking reflex. They grow progressively weaker until they are unable to stand. Loss of body fluid produces hypovolemia and poor peripheral circulation. Extremities feel cold, the eyeballs become sunken and the body temperature may become subnormal. Electrolyte and acid-base imbalances resulting from diarrhea may lead to neurologic disturbances including hyperaesthesia and opisthotonos or terminal convulsions.

Diagnosis

Definitive diagnosis of the causes of diarrhea in young kids may require exhaustive and costly bacteriologic, virologic and parasitologic testing in the diagnostic laboratory. Measurement of packed cell volume, total plasma protein, serum sodium, potassium, bicarbonate, blood glucose and blood gases are useful in determining the severity of electrolyte deficits, acidosis and hypoglycemia in neonatal diarrhea cases. In the field, however, useful estimates of these deficits can also be gauged based on the severity of clinical dehydration.

Submission of diarrheic feces to a diagnostic laboratory for etiologic evaluation is important. Samples from kids younger than 3 weeks of age should be evaluated for cryptosporidia, enteropathogenic *E. coli,* salmonella, rotavirus, and coronavirus. In kids older than 3 weeks of age, additional evaluations should include coccidiosis, enterotoxemia, and yersiniosis. Samples should be collected within 12 to 24 hours of the onset of diarrhea, as the viruses may be rapidly cleared from the gut. A volume of 5 to 10 ml of fresh, diarrheic feces should be submitted under refrigeration. If such volumes are unavailable and rectal swabs are used, they should be shipped to the lab in a suitable bacterial transport medium. Also, fresh fecal smears prepared on glass slides can be of value in the diagnosis of cryptosporidiosis and enteric viruses by acid fast staining and fluorescent antibody testing, respectively. Survival of cryptosporidial oocysts in fresh feces is poor. While viruses can be identified in fresh feces by electron microscopy, such equipment is not always available.

Post Mortem Findings

At necropsy, a loop of bowel can be ligated and removed and submitted to the laboratory unopened under refrigeration. Formalinized tissues from major organs and various portions of the gastrointestinal tract should also be submitted for histopathologic evaluation. Gross lesions may be found at necropsy in the case of coccidiosis, salmonellosis, yersiniosis, enterotoxemia and some forms of gastrointestinal nematodiasis. The remainder of the causes of diarrhea in young kids may produce little or no gross abnormalities beyond some gaseous distension of the bowel and slight hyperemia.

Treatment

Many of the common causes of diarrhea are self-limiting and the major goals of therapy are to keep the kid physiologically intact while the diarrhea runs its course. Implementation of this strategy includes moving the kid to a warm, dry place to maintain body temperature and to provide appropriate fluid therapy to counter dehydration, electrolyte depletion, hypoglycemia and acidosis. Oral or parenteral therapy can be used depending on the severity of disease. Many commercial preparations are manufactured for oral rehydration therapy in diarrheic animals and are increasingly available worldwide. Effective preparations should contain sodium chloride, potassium chloride, sodium bicarbonate, glucose and glycine.

The feeding of milk or milk replacer products should be approached cautiously during the acute phase of diarrhea. Because there likely will be some derangement of digestive function, lactose in milk-based feeds may remain undigested in the gut and promote a hyperosmotic diarrhea on top of the existing enteritis. Alternating oral electrolyte solutions with small feedings of milk is a reasonable approach. Kids should be brought back on milk gradually as diarrhea improves. In diarrhea that lasts for more than several days, it will be necessary to feed milk products to avoid starvation. The strategy then becomes to decrease the volume fed per feeding and to increase the number of feedings. In this manner, the residual digestive capacity of the gut may not be overwhelmed. A variety of oral antidiarrheal medications and intestinal protectants such as bismuth subsalicylate or kaolin / pectin have been used in kids. They may be helpful. Similarly, oral lyophilized *Lactobacillus* cultures are given empirically to restore a more normal gut flora. When bacterial causes of diarrhea are strongly suspected or confirmed, then broad spectrum oral or parenteral antibiotic therapy such as trimethoprim/sulfa may be indicated. Antibiotics are also indicated prophylactically in scouring kids to protect against secondary bacterial pneumonias that are common in debilitated kids.

Prevention and Control

Regardless of the etiologic cause, one of the most effective means of controlling the spread of diarrhea in kid populations in the face of an outbreak is to immediately and strictly isolate affected kids at the first sign of diarrhea. When possible, other exposed kids should be moved to a clean shed or at least temporarily removed from diarrhea contaminated shed to permit disinfection. In the case of viral enteritis and cryptosporidiosis, where no effective therapeutic agents exist, the use of isolation and sanitation as control measures becomes critical.

Fig. 13 Hygienic rearing of kids

Diarrhea is likely to be more common in intensive goat rearing. In these situations, kids should be born in clean, well ventilated sheds and not allowed to be exposed to cold. They should receive adequate colostrum as early in life as possible, at least before 1 hours of age. Colostrum banks may be helpful in insuring an adequate supply of colostrum. Kids should be housed separately or in small groups of approximately equal age. Bedding should be clean and dry.

Fig. 14 Colostrums feeding immediately after birth

Regardless of feeding system, all utensils and equipment should be cleaned and disinfected between feedings. In *ad libitum* systems, a small amount of formalin has been added to milk to inhibit bacterial overgrowth in the milk and the kids gastrointestinal tract. Kids should be offered access to hay as soon as possible and a grain ration offered by week 2 to initiate good rumen development, a normal flora, and to avoid subsequent weaning shock. When possible, kid pens should be organized so feed can be fed in trays or buckets outside the pen to avoid fecal contamination of feedstuffs.

Fig. 15 Spraying of lime on the floor of pens

Vaccination of the doe in the 4 to 6 weeks before kidding may improve the concentration and specificity of protective antibodies available in the colostrum. Vaccination of the dam with *Clostridium perfringens* types C and D is already justified. Caprine *K99+ E. coli* vaccines have been demonstrated to produce increased *K99* specific antibody in doe serum and colostrum. The use of such vaccines in goats is likely to increase where the prevalence of *K99+ E. coli* is established. However, confirmation of *K99+ E. coli* in outbreaks of kid diarrhea is currently proved. It is also unknown if bovine rota virus and coronavirus vaccines are effective in the control of diarrhea in young goats. In endemic areas, vaccination of goats for peste des petits ruminants is a must.

5.11 Foot Rot

It is an infectious disease of goat and sheep that causes severe lameness and economic loss from decreased flock production.

Etiology

Foot rot is caused by an interaction of two anaerobic gram negative bacteria, *Bacteroides nodosus* (formerly *Fusiformis nodosus*) and *Fusobacterium necrophorum* (formerly *Sphaerophorus necrophorus*). *Fusobacterium necrophorum* is a normal inhabitant of the ruminant digestive tract and in wet weather may interact with other bacteria, *Corynebacterium pyogenes*, to produce foot scald, an infection of the skin between the toes. This infection sets up the foot for invasion by *Bacteroides nodosus* in conjunction with the *Fusobacterium.*

Epidemiology

Bacteroides can only live in the hoof of an infected animal or in the soil for not more than 10-14 days, it is possible, through careful management procedures to keep from introducing foot rot into a flock and to successfully control and/or eliminate the disease, if the flock is infected. The bacteria that causes foot rot, *Bacteriodes nodosus*, is spread from infected animal to the ground, manure, bedding, etc., where it is then picked up by non-infected goat. Foot rot is introduced by purchase of an infected animal or by simply using facilities or trucks that have been contaminated by infected goat. Spread occurs mostly when temperatures are from 40-70° F and the environment is wet. Since the organism doesn't survive long in the environment (< 2 wks), carriers in the flock will continue to re-infect the flock unless the animal is either culled or the organism is eliminated by proper treatment.

Diagnosis

Lameness is usually the major sign of an infected animal, although goat with an early infection may not exhibit lameness. The area between the toes first becomes moist and reddened. Then the infection invades the sole of the hoof, undermining and causing separation of the horny tissues. The infection causes a characteristic foul odor and may infect one or more feet at the same time. Not all lame goats have foot rot. Other diseases that may be confused with foot rot are foot abscesses, foot scald, laminitis, corns, traumatic injuries and foreign bodies lodged between the toes.

Treatment and Prevention

Prevention of caprine foot rot is based on several management practices that decrease pre disposing factors, and on the treatment and immunization of infected and susceptible goat. The best results are obtained when several of the following methods are used for the prevention of the disease in the flock.

Foot Trimming

1. This reduces the number of cracks and crevices where bacteria can hide, removes infected hoof, and exposes the organism to air and various medications. All affected tissue should be trimmed away. Many times, this involves removing a large portion of the hoof wall as well as the overgrown portion. This is necessary if the medication and oxygen are to reach the bacteria and kill them. Foot trimming should be done at least one to two times per year as a part of normal management practices and more often in conjunction with footbaths in the control of foot rot. When trimming feet, it is important to disinfect the trimming

instruments between animals to prevent spreading of the infection. During a severe outbreak, trimming without any other treatment may actually increase the severity of the disease.

2.F ootbaths

There are two different types of solutions commonly used in foot baths: zinc sulfate and copper sulfate. For treatment, they should be used 1-2 times per week for several weeks. They may also be used routinely after foot trimming and as a preventative.

(a) Zinc sulfate (10%) is perhaps the most effective and least toxic of the two baths. A surfactant or wetting agent (detergent) can also be added to the baths to increase their penetration into the cracks and crevices of the hoof. Use of zinc sulfate or copper sulfate solutions as a foot soaks (30-60 minutes of contact) increases their efficacy in a treatment program. When designing the foot bath area, it is important that length of contact with the solution be kept in mind. Sufficient bath size is necessary to completely dip the foot and allow sufficient contact time with the solution. Zinc sulfate (dry) can also be used and placed in an area to allow the goats must walk through it. This will not treat infected animals, but will help decrease the spread of the disease. Lime disinfectant may be used around feed or water troughs to reduce moisture and decrease the spread of the disease.

(b) Copper sulfate (10%) is also useful but is toxic if consumed by the goat. Hot water helps to dissolve it. Addition of some vinegar also helps to dissolve the copper sulphate. It is very corrosive so it should not be used in metal foot baths.

1.Peni cillin and streptomycin combinations are administered daily up to ten days for the treatment of foot rot. Procaine Penicillin G or long-acting penicillin may also be effective. Single injection of long-acting tetracycline have also been successful.

2.There are several different medications that can be applied to the hoof immediately after paring that is helpful in controlling foot rot.

(a) Zinc sulfate (10%)

(b) Copper sulfate (10%) in vinegar

(c) Copper sulfate in pine tar - 2 parts CuS04 in one part pine tar.

(d) Oxytetracycline solution.

(e) Potassium penicillin G 5 million units.

It is always easier and less expensive to prevent foot rot than to treat it after it has become established. To remain disease free, there are five management principles that will help keep foot rot from being introduced into a clean flock.

1. Never buy animals with foot rot or from a flock infected with foot rot, even if the animal(s) appear unaffected.
2. Avoid buying goat from the livestock markets where clean and infected animal may have been commingled or run through the same area.
3. Avoid using facilities (trails, corrals, and dipping areas) where infected goat may have been in the last two weeks.
4. Never transport goat in a vehicle that has not been properly cleaned and disinfected.
5. Trim and treat the feet of all new arrivals, then re-examine them periodically during the isolation period.

Vaccination

Vaccines for *Bacteroides nodosus* is available in some countries and can be used for the prevention of disease. The vaccine works not only as a preventative but has been shown to be fairly effective as a treatment. A regimen of two vaccinations given subcutaneously on the neck just behind the ear 4-6 weeks apart is used. Vaccination before the start of the wet season is recommended, followed by a booster each year prior to the wet season if eradication efforts have not been successful. Abscesses are common at the injection site but should not be treated. These will usually break and drain on their own with no ill effects to the goat. In the eradication programme, vaccination can be done six weeks prior to the start of the program and the booster can be given when process is started. This can increase the immunity and some healing may be taking place by the start of trimming. Some labor savings can be made by doing the first vaccination at the start of the eradication program. Also, there will be savings on vaccine because the clean group will not have to be vaccinated a second time.

5.12 Caseous Lymphadenitis

This is a chronic contagious disease of sheep and goats manifested by abscesses in the lymph nodes. Caseous lymphadenitis has a worldwide distribution and causes great economic losses to the sheep and goat industry.

Etiology

Corynebacterium pseudotuberculosis (previously known as C. *ovis)* is the causative agent of caseous lymphadenitis.

Epidemiology

Disease is transmitted through the discharge from the lymph nodes, via wounds caused by shearing, castration and docking, contaminated sheep dips, skin abrasions or traumatized oral mucosa. Animals with open abscesses should be segregated in order to prevent the spread of the disease. Soil is known as a natural reservoir of C. *pseudotuberculosis,* capable of transmitting CLA in the event of skin injury. The incidence of the disease is low in India varying from 0.92% in sheep and 0.15% in goats.

Clinical Signs

The goat without internal abscesses typically shows no clinical signs other than enlargement and abscessation of one or more peripheral lymph nodes. Many of the external lymph nodes are involved in caseous lymphadenitis. Exactly which nodes are infected depends on the location of the wound that allowed entry of the organism into the body. Thus, most goats (75%) with external abscesses have lesions on the head and neck (parotid, mandibular, and prescapular nodes), presumably because injuries from thorns and scratching due to lice infections. A popliteal lymph node would only become enlarged if infection occurred in the distal hind limb. Animal is lagging behind the flock. There is dyspnea, purulent ocular and nasal discharge. Generalized disease is associated with weight loss, depression and loss of appetite.

Pathogenesis

The organism enters the animal body through small breaks in skin or mucous membranes and eventually becomes localized in a regional lymph node. Cell wall lipid permits the organism to resist digestion by cellular enzymes and C. *pseudotuberculosis* can survive as a facultative intracellular parasite, even in activated macrophages. Sphingomyelin-specific phospholipase D exotoxin produced by the organism is largely responsible for its spread.

The incubation period until abscesses are noted in superficial lymph nodes is typically 2 to 6 months or longer. These abscesses may rupture and drain spontaneously. The environment becomes contaminated but the initially infected goats abscess heals. It is common for one or more additional lymph nodes, following the lymphatic drainage pattern, to develop abscesses several months later. Experimental intradermic inoculation with small numbers of bacteria has demonstrated that spontaneous cures, without abscess rupture but accompanied by development of antitoxin titers can occur. Internal (visceral) abscesses, especially in the lungs, may develop if the organism reaches the thoracic lymph duct or if it is inhaled. Other adverse economic effects of the infection in a herd include decreased market value of stock for sale and of hides from slaughtered goats. Milk production may be decreased.

Diagnosis

Diagnosis of the condition is based on the presence of a firm to slightly fluctuant subcutaneous swelling in the anatomic location of a lymph node. In a herd with a history of caseous lymphadenitis, the clinical findings alone are considered presumptive evidence. When a herd history is lacking then laboratory assistance may be indicated for confirming the diagnosis. A sample for culture is obtained by inserting a sterile needle through shaved, disinfected skin into the mass. If aspiration yields nothing, the needle is withdrawn and saline is flushed through it to obtain material for making smear or culture.

The pus in an abscessed node in a goat may be creamy white, yellowish, or greenish; it is typically odorless and is more pasty (less inspissated) than in sheep. A lamellar "onion ring" configuration as seen in sheep with this disease is rarely present in goats. The organism grows readily but slowly on blood agar. The colonies are very tiny or invisible after 24 hours and a longer incubation period is recommended before declaring that no growth has occurred. The colonies are still tiny and button shaped after 48 hours, but are surrounded by a narrow zone of hemolysis. The colonies can be easily moved around on the surface of the agar plate and splatter when placed in a flame, because of high content of lipid. The organism is catalase-positive. Gram-stained smears show Gram-positive or Gram-variable small coccoid rods. A longer rod form is sometimes seen in smears of pus.

Serologic tests such as agglutination test and synergistic hemolysis-inhibition test are of value in identifying goats with early (no abscess yet developed) or internal forms of caseous lymphadenitis. Gel precipitation and ELISA were used for seroprevalence infection in ovine and caprine population. The modified gel-precipitation test using broth saline buffer made the sero-survey more economical, quicker and more satisfactory. ELISA detects specific antibodies and has a greater sensitivity than most of the other sero-tests. The increases in total serum proteins and gamma globulin that are reported for goats with caseous lymphadenitis are also nonspecific. An intradermal allergic test using a water-soluble protein 'lymphadenin' has been used to differentiate goats with abscesses from uninfected goats. Injections were made in the shoulder region and the maximum increase in skin-fold thickness occurred after 48 hours.

Post Mortem findings

Caseous abscesses in the superficial lymph nodes and carcass musculature, firm and dry abscess in the kidney (Fig. 16) and other organs were the lesions generally observed. Soft pasty abscess in the early stages changes to firm and dry with a characteristic laminated appearance in the later stages of disease. Abscess content is creamy and pasty in goats.

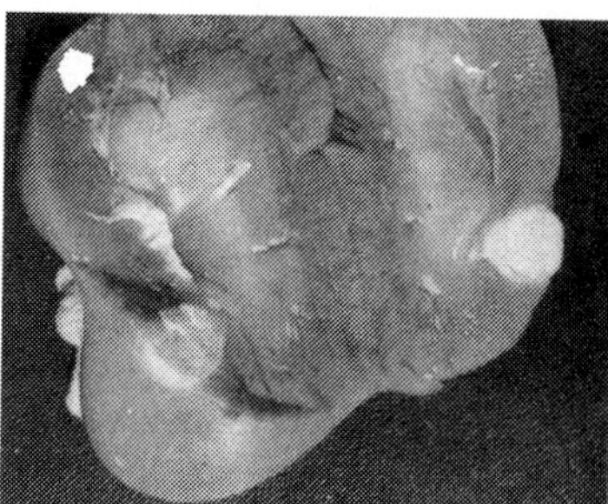

Fig. 16 Firm and dry abscess in the kidney

Treatment

Treatment of individual animals involves either draining or surgically removing the abscessed nodes. Ripened abscesses can be incised generously at a ventral point and the cavity is then flushed with dilute disinfectant. Because the infection is potentially zoonotic, people performing this procedure should wear gloves. The pus should be collected and burned and the goat should be kept strictly isolated until the lesion is completely covered by healthy skin. Surgical removal of the encapsulated abscess offers the advantages that the treated animal does not need to be quarantined afterwards, and spread to other nodes is less likely to occur, but veterinary expertise is required and there are dangers associated with anesthesia and dissection near large blood vessels and cranial nerves. Extirpation of abscessed parotid nodes is particularly dangerous.

Administration of penicillin or tetracycline for a few days after spontaneous rupture or lancing of an abscess is suggested to prevent dissemination of the organism to other lymph nodes, but the value of this treatment has not yet been determined in controlled clinical trials. Antibiotic sensitivity testing is of little benefit, as most antibiotics do not penetrate into the abscessed lymph node and the organism itself may be intracellular. An oral dose of 10 to 20 mg/kg rifampin might be useful. Erythromycin at the dose rate of 4 mg/kg administered intramuscularly or subcutaneously is beneficial. Because erythromycin is highly irritating, a rifampin/penicillin combination might be preferable. Treatment should be continued for 4 to 6 weeks.

Prevention and Control

Valuable infected animals may be kept isolated; their kids should be removed at birth to avoid transmission of the disease as the newborn searches for the udder. Division of an enzootically infected herd into infected and apparently uninfected groups has also been proposed as a means of limiting spread. The housing facilities should be free of nails, wire, and other objects that might induce breaks in the skin. Control of external parasites is very important because pruritic goats will rub themselves on nails and posts. All wounds should be treated promptly with a disinfectant and kids should have the umbilical cord dipped in iodine at birth. Animals with chronic respiratory disease or wasting should be culled, or at least isolated from the herd. Natural transmission from lung lesions via discharge of pus into airways has been demonstrated in sheep. An environment contaminated with pus may be the source of new infections for weeks or months. Recovery of the organism from wood surfaces, straw, hay, and soil has been demonstrated by various workers. Thus it is desirable that the isolation facilities used to contain a goat after rupture or lancing of an abscess have a concrete floor. The bedding should be burned and the pen thoroughly cleaned and disinfected.

Numerous studies using mice have evaluated immune responses to the organism. Cell-mediated immunity has been demonstrated to restrict bacterial proliferation. Neutralization of exotoxin (phospholipase D) produced by the organism is believed to limit spread from the primary site of infection. The value of vaccination as an aid in controlling the disease in ruminants has been frequently questioned. Where autogenous bacterins have been used, a decrease in the prevalence of abscesses in the herd has often been noted. An autogenous vaccine, if not properly prepared and tested against this organism may contain enough free toxin to kill the vaccinated goat.

A commercial vaccine developed in Australia and available in Canada has been evaluated in goats. This product is a formalized exotoxin with incomplete Freund's adjuvant. In one study, challenge by swabbing a live culture onto abraded skin revealed that either one or two doses of vaccine gave good protection. Colostrally derived immunity has protected young kids against challenge. Vaccination should probably be performed before 4 months of age. Field evaluation of crude filtrated *C. pseudotuberculosis* toxoid combined with whole killed cells has been disappointing in goats, whereas another experimental whole cell vaccine gave a no statistically significant trend in vaccinated goats in a field trial. A formalin-inactivated whole cell vaccine with aluminum gel phosphate adjuvant gave partial protection in a field trial. Whole cell preparations of the organism adjuvanted with mycobacterial components are currently being evaluated. A modified live intradermal vaccine has been developed.

5.13 Johne's Diseases

Johne's disease (Paratuberculosis) is a chronic granulomatous enteritis primarily infecting ruminants, caused by *Mycobacterium avium* subspecies *paratuberculosis* (Map) and is endemic in most of the countries. In all ruminants, clinical disease is characterized by a progressive, afebrile weight loss that leads to emaciation and poor coat quality (roughness, loss of pigmentation, alopecia, 'wool-slip'), despite maintaining a good appetite. A major feature of illness is chronic, intractable diarrhea, although in some animals this may be intermittent. In contrast to cattle, diarrhea is not a feature in small ruminants.

Etiology

Mycobacterium avium subspecies *paratuberculosis (Map)*

Economic Losses in India

Important genetic material is wasted due to JD and finds their way to slaughter houses. In heavily infected herds, this leads to a high rate of wastage in animals in the prime productive life at 2 to 4 years of age. It has been recognized as serious health problem of goats and sheep, particularly in organized herds. It causes high mortality and morbidity in young and adult goats. Economic losses are significant through lost of productivity, infertility and susceptibility to other disease, lost export markets and direct expenditure on diagnosis and control. JD affected goats showed marked reproductive inefficiency.

Epidemiology

Johne's disease has been reported on every continent. Sweden and some states in Australia are the only regions of the world that can claim freedom from Johne's disease based upon a reliable disease reporting system and extensive surveys using laboratory tests. Johne's disease is fairly common in small ruminants like sheep, goats in many countries but precise survey data have not been published. Only a small percentage of sheep flocks in Australia have been found to have Johne's disease prompting the sheep industry there to move aggressively to contain the infection. The cause of Johne's disease, *Mycobacterium paratuberculosis* sub species *avium (MAP)*, has a broad host range. The type of animals most commonly infected are ruminants: hoofed herbivores with four-chambered stomachs that chew their cud. Examples of ruminants include: cattle, sheep, goats, deer, elk, antelope, bison, etc. (Camelids such as llamas, guanacos and alpacas, while considered "pseudo-ruminants", are also susceptible to the infection). Johne's disease has been reported in a number of wild ruminant species, both captive and free-ranging. In addition, there have been a few reports of *M. paratuberculosis* infecting pigs, horses, and nonhuman primates. One report found that the same strain of the bacterium was found to infect dairy cattle plus numerous rabbits in the region and a number of different carnivore species (fox, stoat, badger, raven) that may have been preying on the rabbits.

MAP has been recorded in all sheep and goats rearing states in India. Organized farms continue to suffer heavy losses due to high morbidity, mortality and premature culling caused by paratuberculosis. The incidence of JD was ranging from 4.1- 44.4% on the basis of fecal examination in different goat breed in different climatic and rearing practices. It was reported 18.9% and 2.5% JD incidence in goat's herds located in semi-arid and arid regions. Post mortem based study on the status of paratuberculosis in goats, showed overall incidence of 3.5% in goats which was lower than the earlier reports based on the retrospective study on the necropsy data (8.6%) and fecal cultures (18.9% and 2.5%) at two separate organized goat farms.

It occurs by fecal oral route, but milk is an important source of transmission of disease in suckling animals. MAPs were isolated from milk of clinically infected goat with Johne's disease. Significant effect of parturition and lactation was reported on the precipitation of JD and showed that 80.0% of parturient animals precipitated in disease and became shedders within 6 months (optimum 1-2 month). Incidence of JD was directly proportional to the herd size. Small herd sizes with extensive system of management were apparently useful to reduce the incidence of JD in goats. Timely culling of weak and unthrifty kids reduced the burden of JD in herds. Low incidence of JD has been reported in goat herds located in arid regions as compared to semi-arid regions in India.

M. paratuberculosis bacteria are obligate pathogenic parasites of animals. This means that the only place they can multiply in nature is inside an animal. Most accurately, it is inside cells that are part of the animal's immune system called macrophages. When *M. paratuberculosis* leaves an animal, for example in the feces, it can survive for a long time in environments such as soil and water, but it can not multiply once it is outside the animal. Consequently, the primary source of infection is infected animals. Herds or flocks of animals acquire the infection by being exposed to the bacteria from an infected animal. When an infected animal is introduced to a herd, the opportunity for transmission of *M. paratuberculosis* bacteria to other animals increases the longer that animal remains in the herd. As the infection progresses in the animal, the frequency and number of *M. paratuberculosis* bacteria being excreted increases. *M. paratuberculosis* infects the intestine, thus feces (manure) is the most common mode of exit of the bacterium from the animal. Ingestion of *M. paratuberculosis* containing feces, or fecal contaminated feed or water is the most common way animals become infected. In manure, *M. paratuberculosis* can remain alive for over a year, depending on environmental conditions. Milk from infected female animals is a second source of *M. paratuberculosis* infection. The likelihood of *M. paratuberculosis* being excreted into milk increases with time as the infection progresses. The probability of young animals becoming infected by drinking milk from infected cows, does, or ewes is a direct function of the time spent with the mother and/or how often they are fed milk from infected females. In animals where husbandry practices allow young animals to remain with the dams and nurse naturally, the chance for transmission of the infection from mother to off-spring is greatest. *M. paratuberculosis* may be excreted directly into the mother's milk and/or the surface of the teats might be contaminated with infected manure. Pond water contaminated with manure carrying *M. paratuberculosis* is another potential source of infection. A less likely, but possible, infection source is pasture land contaminated with the bacteria. Johne's disease has been reported in free ranging wildlife, but their role in the ecology of *M. paratuberculosis* is not known.Animal age is perhaps the well recognized factor affecting *M. paratuberculosis* transmission. In cattle, there is an age-dependent increase in resistance to *M. paratuberculosis* infection. This means it takes a larger dose of the bacterium to infect an adult (over 2 years-old) than it does to infect a young animal (0 to 6 months old). This may also be true for small ruminants, but it is not as well studied and clinical reports suggest there is a greater susceptibility of sheep, goats and deer to this infection, even as adults. The extent and duration of exposure to contaminated manure and milk from infected adult animals directly affects the likelihood of *M. paratuberculosis* transmission.

Pathogenesis

Pathology caused by *M. paratuberculosis* infection varies among animal species and, to a lesser extent, between individual animals. The primary site of *M. paratuberculosis* infections is most often the ileum, the most distal portion of the small intestine. Consequently, although secondary sites of infection may have pathologic lesions, the ileum is generally the tissue most severely affected and has the highest burden of *M. paratuberculosis* bacteria. Lymph nodes draining the ileum are also sites of secondary infection where pathologic changes are commonly seen. Tertiary sites of infection are the liver, spleen, and lymph nodes distant from the gastrointestinal tract.

Diagnosis

It is a common misconception that there are no accurate diagnostic tests for Johne's disease. There are, in fact, 8 available tests: 3 methods for detecting the bacterium that causes Johne's disease (*Mycobacterium paratuberculosis* sub species *avium*), 4 blood tests for detecting serum antibody, and 2 assays for cellular immunity. Five of these tests are commercially available from multiple companies for use by veterinary diagnostic laboratories worldwide. Some tests are simple enough to be able to be done in a veterinary clinic. In addition, factors such as the amplitude of a test result (numerical value) and herd infection rate (prevalence) can influence interpretation of Johne's disease test results. Consequently, it is recommended test interpretation be made by veterinarians who are experienced in laboratory diagnostics.

In general, tests for Johne's disease have a high specificity : low rate of false-positive results. To avoid making management decisions based on false-positive blood tests, fecal culture can be completed as a "confirmatory" test for blood test positive animals. Test sensitivity (percentage of infected animals that test positive) varies among tests, but due to the biology of this slowly progressing chronic disease is often less than 50%. In the very early stages of *M. paratuberculosis* infection, before animals start shedding the bacterium in feces or begin an immune response to the infection, all animals will be test-negative. As the infection progresses, most tests will eventually become positive as the animal begins to produce the signals of infection detected by the tests (e.g. antibody in the blood, the organism in manure). Tests have maximum sensitivity when used for animals with clinical signs of the infection: diarrhea and/or weight loss. Use of two or more different types of tests at the same time is recommended for maximum infection detection sensitivity. An excellent way of diagnosing Johne's disease is to detect the bacterium that causes the infection. Since this bacterium is not normally found in the manure of uninfected animals, if detected, it is assumed to be causing an infection and pathology in the animal. There are three ways of detecting *M. paratuberculosis*: two for isolating the entire organism with culture (standard or BACTEC) and one for detecting the genetic material (DNA probe/ PCR).

Bacterial Culture

Standard bacterial culture has been used for almost 100 years. Many laboratories around the world can perform this test. The standard culture method for *M. paratuberculosis* sub species *avium* requires little equipment and is technically simple to perform. This test takes at least 16 weeks to complete because of the extremely slow growth rate of *M. paratuberculosis* sub species *avium*. A problem with this test is that strains of *M. paratuberculosis* sub species *avium* from sheep and perhaps other non-bovine species may fail to grow on standard culture media.

Tests for Antibodies to *M. paratuberculosis* sub Species *avium* in Serum (blood tests)

Detection of serum antibody to *M. paratuberculosis* is good evidence the animal is infected. There are three techniques for detection of serum antibodies in common use today: Complement fixation (CF), Agar-gel immunodiffusion (AGID), and Enzyme-linked immunosorbent assay (ELISA). All three tests are fundamentally measuring the same thing, antibodies to *M. paratuberculosis,* but use different types of technology and reagents. Differences among these three serological methods for diagnosis of Johne's disease are too complex. There is a large body of scientific literature, evaluating each of these tests and comparing their accuracy. These three assays have primarily been assessed for cattle, plus sheep and goats to a lesser extent.

Complement Fixation (CF)

This test is one of the oldest tests used for blood testing cattle for Johne's disease. It is falling out of favor and being replaced by newer methods that are more accurate and more easily standardized. Some countries, however, still require this test before allowing importation of cattle to their country.

Agar-Gel Immunodiffusion (AGID)

The AGID was developed in several laboratories and sold as a commercial kit. It is a simple test to run and requires no special equipment. The AGID also can be used on a wider variety of animal species although it has been more thoroughly evaluated on cattle and goats. It rarely makes a mistake when it is positive (high specificity). However, it has a lower sensitivity and its use is primarily restricted to confirmation of a clinical diagnosis: i.e. to verify a diagnosis on animals with clinical signs of disease that look like Johne's disease (diarrhea and weight loss).

Enzyme-Linked Immunosorbent Assay (ELISA)

ELISA is a technology for detection of antibodies. It is more sensitive than either the CF or AGID in goats. It has a high specificity and is almost as sensitive as fecal culture at detecting infected animals. It is designed for testing large numbers of samples quickly and this makes it a low-cost test. Another advantage is that the test produces numerical results that are a measure of the quantity of antibodies to *M. paratuberculosis* in the serum sample tested. Research has shown that the level of antibody is useful in predicting the stage of infection. A disadvantage of the ELISA is that it requires skilled laboratory technicians and specialized laboratory equipment. The ELISA has only been validated for goats in India but work is underway to adapt this test to other animal species. When choosing among available tests there are many considerations such as (1) the animal species being tested (2) the age of the animal being tested (3) the clinical history of the herd (4) the goals of testing (control? eradication? diagnosis of an individual animal?) and the economics of the herd involved (5) the accuracy of each test when used on individual animals or herds/flocks (6) which would be more costly to you: false-positive or false-negative test results ? (7) the epidemiology of Johne's disease.

DNA Probes

DNA probes allow detection of *M. paratuberculosis* without having to grow the bacterium. Therefore, they are much faster. A DNA probe test can be done in less than 3 days. The earlier versions of this test have not been as sensitive as culture methods for detection of *M. paratuberculosis* in fecal samples (even though it uses a technique known as PCR amplification).

Clinical Signs

Overt clinical disease rarely occurs before 1 year of age and is most common in goats 2 to 3 years of age. Clinical disease is often triggered by some episode of stress, such as parturition or recent introduction into a new herd. Affected animals begin a course of progressive weight loss, which may extend from weeks to months and can lead to dramatic emaciation (Figure 17). Appetite may remain intact initially, but decrease later on, and the animal becomes increasingly lethargic and depressed. A rough hair coat and flaky skin are common. Unlike cattle, goats rarely show persistent watery diarrhea, except possibly in the terminal stages of disease. Diarrhoea may occur

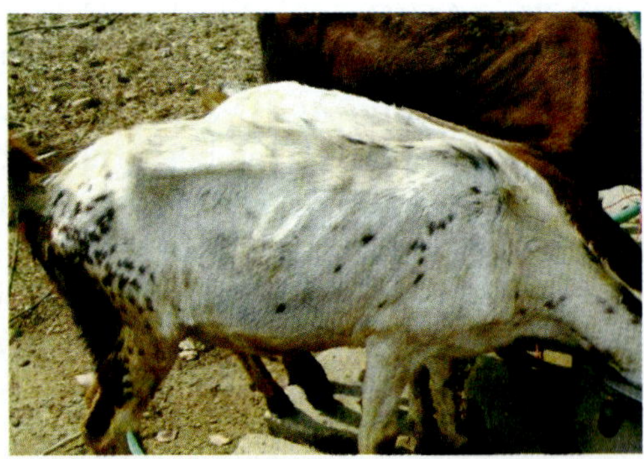

Fig. 17 Severe emaciation and intermittent diarrhea in goat

intermittently through in the course of disease. As the disease progresses a moderate anemia of chronic infection may develop and clinical evidence of hypoproteinemia such as intermandibular edema may be seen.

Post Mortem Findings

Because clinical caprine paratuberculosis is invariably a terminal disease, necropsy provides the opportunity for definitive diagnosis. Gross lesions are more variable in occurrence in the goat as compared with cattle. Therefore, histologic examination of tissues and bacterial culture of selected organs also should be performed. The prominent accordion like corrugated thickening of the intestinal wall and mucosa commonly seen in cattle is not often observed in goats. When gross lesions are present in small ruminants, they usually present as focal or diffuse thickening or edema of the ileum, cecum or spiral colon. Adjacent lymph nodes may be enlarged and edematous with focal calcification occurs more often in goats than cattle. Calcification of the aorta often occurs in goats with prolonged paratuberculosis, though the pathogenesis is unknown. Histologically, paratuberculosis produces characteristic granulomatous lesions in the intestinal tract, lymph nodes and possibly liver. At a minimum, tissues microscopic examination submitted should include section of the ileocecal junction and adjacent ileac lymph node, ileum, spiral colon and another mesenteric node. Acid-fast stains should be done on these samples. Submission of fresh or frozen tissue in addition formalized sections will allow for culture confirmation of *M. paratuberculosis* when microscopic lesions a scarce.

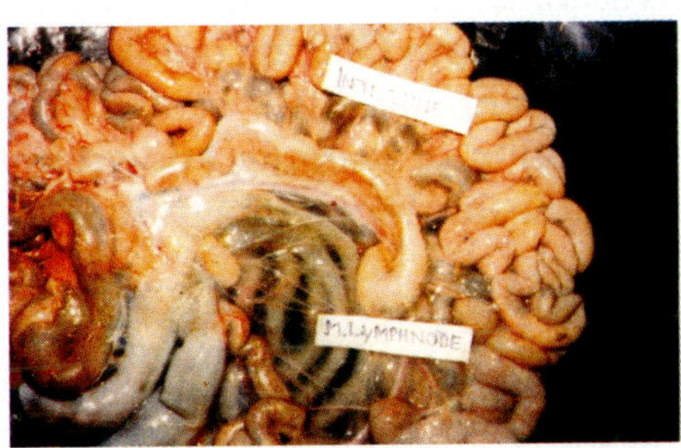

Fig. 18 Enlarged calcified mesenteric lymph nodes

Treatment

There are no known effective treatments for the elimination of *M. paratuberculosis* infection. Various antimycobacterial drugs have been used with limited effect in goats. Attempted unsuccessful therapies have included isoniazid, isoniazid plus rifampin, isoniazid plus ethambutol, or all three drugs together. However, some treatment regimens have alleviated clinical signs in individual cases as long as therapy is continued. Nevertheless, the organism can still be isolated from cultures prepared from tissues at necropsy. One such regimen is daily streptomycin sulfate (0.5 g intramuscularly) plus isoniazid (25 mg orally) plus sodium aminosalicylate (850 mg orally) for 6 months. A second regimen is daily dihydrostreptomycin (0.5 g intramuscularly) plus rifampin (300 mg orally) and isoniazid (300 mg orally) twice daily. Treatment has little economic application in goats except for the possibility of keeping known infected; high quality breeding does in good health long enough to achieve sufficient production.

Prevention and Control

Paratuberculosis is most commonly a herd problem rather than an individual animal problem. When a clinical case is diagnosed from an existing herd, the herd almost always also includes sub-clinical shedders and infected, non shedders. There are three basic elements to consider in developing a strategy for elimination of paratuberculosis from infected herds: one, identification and removal of infected individuals from the herd; two, reduction in the rate of new infections in susceptible kids through improved sanitation and modified kid-rearing techniques; and three, vaccination to increase host resistance to new infections. It is unlikely that anyone of these components of the control program taken alone will effectively eliminate paratuberculosis, but the emphasis on each component in the overall control program will vary with local conditions. On the national level, paratuberculosis has been controlled in Norway largely through vaccination. At the local level, elimination of paratuberculosis in a large herd was achieved without vaccination by testing and culling in conjunction with altering management procedures.

The best techniques currently available for identification of infected goats are the bacterial culture of feces and the AGID test. Because neither test will identify infected non-shedders, testing must be carried out at frequent intervals to assure that infected; non-shedders are rapidly identified as they convert to sub-clinical, in apparent shedder status. Testing at intervals of at most 6 months is recommended. Improved management and sanitation will help reduce the incidence of new infections in young stock. In intensive operations, kidding pens should be cleaned properly between kiddings. Kids should be taken immediately from the does and raised in separate facilities on heat treated colostrum and pasteurized milk or milk replacer. Kids should not commingle again with adult animals until they themselves have kidded. Adult animals should not be overcrowded, manure should be removed regularly from the shed and feed and water should be provided in a manner that does not allow for fecal contamination. If manure is spread on pastures, goats should not be allowed to graze those pastures for at least 1 year. Flock-wise testing by culture of feces or AGID tests to identify infected dams should be performed before the kidding season begins. Infected dams should either be culled or segregated from the non-infected animals before kidding.

Vaccination of goats for paratuberculosis in Norway has been extremely effective in reducing the national prevalence of caprine paratuberculosis. The vaccine employed is a live, attenuated, adjuvanted containing two strains of *M. paratuberculosis* and is given to goat kids between 2 and 4 weeks of age. Post mortem evaluation of vaccinated and unvaccinated goats over a 15-year period indicated that the infection prevalence was reduced from 53% to 1 % primarily through the use of vaccination. Caprine paratuberculosis in India is involving a unique caprine-specific strain of *M. paratuberculosis* (S-5). The preparation of vaccine using S-5 stain isolated from organized flock of goats is underway to establish effectiveness of this vaccine. Currently, data on extensive field trials of this vaccine in under generation.Two potential drawbacks to the use of all vaccines containing *M. paratuberculosis* are the appearance of a granulomatous nodule at the site of vaccination and the fact that animals vaccinated against *M. paratuberculosis* may develop cross-reactivity to standard tuberculosis tests, thus complicating the testing procedures for regulatory programs designed to control tuberculosis.

5.14 Bacterial Pneumonia

Pneumonia is one of the most common respiratory problems in sheep and goats throughout the world. Goats are very susceptible to damp, draughty, dusty and ill-ventilated conditions. Although pneumonia often occurs in kids, illness and deaths also occur in adult animals.

Etiology

Pneumonia occurs when infectious and non-infectious agents cause the lungs of goats to become inflammed. *Mycoplasma* spp. and *Pasteurella* spp. are frequent causative organisms. *P. haemolytica* may cause sudden death and an acute fibrinous bronchopneumonia with pleurisy. The most frequent causes of respiratory infection and death are *Pasteurella multocida* or *Mannheimia haemolytica* (previously called *Pasteurella haemolytica*). *P. multocida* and *M. haemolytica* are commonly found in the upper respiratory tract of healthy goats. *M. haemolytica* are subdivided in two groups, A and T. Type A is most prevalent and is associated with a severe form of pneumonia. Goats that survive an acute stage may recover or become chronically infected with reduced lung capacity. Pneumonia caused by *P. multocida* and *M. haemolytica* can lead to significantly decreased growth performance. These two pathogens (agents that cause disease) cause outbreaks of acute pneumonia in goats of all ages. Respiratory infections from these pathogens are associated with poor management practices, occur as a secondary infection, or occur as a consequence of severe stress. Transportation stress, viral infections (parainfluenza-3 virus), lung parasites, overcrowding, poor housing conditions, sudden environmental changes, and other stressful conditions increase goats susceptibility to *P. multocida* and *M. haemolytica* pneumonia.

Pathogenesis

Bacterial, viral, and parasitic infections break down goat tissue defense barriers. Loss of this natural protection increases the goat susceptibility to secondary infection by *P. multocida* and *M. haemolytica.* Animals whose lungs are already weakened from previous diseases will suffer from leukotoxins and lipopolysaccharides, both potent toxins that, in high levels, promote inflammation and severe lung damage. In kids, acute outbreaks can occur with low morbidity but high mortality rates. Typically, several kids that appear healthy die suddenly. Signs of the disease may not be noticed until several animals have died. Optimal conditions for pneumonia caused by *P. multocida* and *M. haemolytica* includes viral diseases, sudden environmental changes, high air humidity, poor hygiene and shed ventilation, severe internal parasites, poor nutritional management, undernourishment or sudden change in diet and transportation stress can cause outbreaks.

Clinical Signs

Rapid, shallow respiration is the cardinal sign of early pneumonia, dyspnea occurring in the later stages when much of the lung tissue is non-functional. The important symptom includes high temperature ranging from 104 to 106°F, moist painful cough and dyspnea. Examination of the lungs may reveal crackle like sounds along with nasal and ocular mucopurulent discharge. Bronchopneumonia is usually accompanied by a moist painful cough, interstitial pneumonia by frequent dry hacking coughs. A nasal discharge may or may not be present. In acute bacterial bronchopneumonia, toxaemia, anorexia, depression, tachycardia and a reluctance to lie down are common symptoms.

Diagnosis

Diagnosis is based on clinical signs and herd history. Dead animals can be used for a definitive diagnosis. Isolation and culture of the bacteria harvested from a tracheal wash and from pulmonary secretions can be used to isolate and identify the causal agent. At necropsy, lungs are hemorrhagic. Recent diagnostic techniques can be used such as polynucleotide chain reaction (PCR) and immunohistochemical for the diagnosis of cause of pneumonia.

Treatment

Medicines effective in treating pneumonia in goats include penicillin, ampicillin, tetracycline, oxytetracycline, tylosin, florfenicol, and ceftiofur. ceftiofur is used at the dose rate of 0.5 to 1.0 mg/lb body weight injected intramuscularly for three days. Probiotics are recommended after antibiotics to promote regrowth of thê normal rumen microflora (bacteria and protozoa populations).

Prevention and Control

A systematic vaccination of the entire herd is advised. *P. multocida, M. hemolyticum* vaccine is used in goats. This vaccine may cause temporary limping in a few goats. Improve management practices by providing optimal sanitation and air quality in housing. Minimize transportation stress. Quarantine of new animals before introducing them into the existing herd is important. Trace minerals, such as Cu, Se, and Zn are needed to enhance immune function. Adding vitamin E in feed at the rate of > 1,000 IU/animal daily also seems to be beneficial. Provide good quality hay and water and supplement as appropriate. Keep sick goats in a dry, well-ventilated location away from the rest of the herd.

5.15 Staphylococcal Dermatitis

Staphylococcal dermatitis is characterized a severe ulcerative skin infection affecting particularly the face of sheep and goats.

Staphylococcal folliculitis is an infection and inflammation of hair follicles. This may be primary or secondary.

Etiology

Coagulase-positive species of *Staphylococcus intermedius* and *Staphylococcus aureus* are the etiological agent involving in dermatitis in goats.

Clinical Signs

The primary lesion is a follicular papule that develops into a pustule. Lesions may enlarge or coalesce, discharge purulent or serosanguinous exudate, and become encrusted. Alopecia and scaling are prominent in the chronic or healing stage. Multiple small pustules frequently appear on the teats and udder or perineum and underside of the tail. Because these pustules may be preceded by vesicles and followed by scabs, they may be confused with lesions of contagious ecthyma. When lesions remain localized they are relatively benign and self-limiting. In some goats, the infection becomes a general one, involving the skin of the abdomen, inner thighs and even the neck and back. Distributions of lesions in other area are possible, especially if the staphylococci are secondary to another condition, such as chorioptic mange. Periocular alopecia and crusts comprise yet another possible manifestation.

Diagnosis

A presumptive diagnosis is often based on critical examination of the clinical case. Gram stains and cultures will document the presence of staphylococci, while evaluation of skin biopsy specimens should help to rule out other diseases that might have been primary and still require therapy. Confusion with mycotic infections or nutritional deficiencies is possible.

Treatment and Prevention

Localized lesions on the udder may be washed with an iodophor or chlorhexidine solution, dried and then coated with an antiseptic or antibiotic ointment. Affected does should be milked in the last. Use of fresh towels and hand washing by the milker will decrease the risk of spread of infection to other does. Rubber gloves would protect against transmission of infection to humans. Similar treatment of lesions around the tail is possible, but these seem to be of lesser importance and frequently heal spontaneously. If a generalized infection is suspected, culture of the organism and determination of antibiotic sensitivity are recommended. Systemic antibiotic treatment with penicillin for 1 to 2 weeks will give encouraging results. Autogenous bacterins may be tried for control of chronic or epizootic infections but bacterins have not received scientific evaluation in caprine dermatology.

5.16 Bacterial Polyarthritis

Bacterial polyarthritis are also known as joint ill, occurs predominantly in neonates as a sequel to omphalophlebitis or bacteremia.

Etiology

A variety of bacteria have been isolated from joints of goats with arthritis. The most commonly reported isolates from kids includes *Actinomyces (Corynebacterium) pyogenes, Escherichia coli, Streptococcus* spp and *Staphylococcus* spp., all common environmental contaminants that can gain entry into kids via the umbilicus. *Erysipelothrix rhusiopathiae,* a soil-borne pathogen has been responsible for a high incidence of polyarthritis in young lambs after castration and tail docking or as a result of navel infection.

Epidemiology

Bacterial polyarthritis is most often seen in newborns and young kids after bacteremia. Factors that predispose these goats to bacteremia include kidding in contaminated environments, poor sanitation in housing and the development of omphalophlebitis after failure to disinfect the navel at birth and failure of passive transfer of immunity via colostrum from the dam. Improved hygiene and disinfection in lamb management have reduced the incidence of this disease.

Pathogenesis

Polyarthritis arises secondary to bacteremia, when circulating bacteria localize in joints and produce a destructive inflammatory response. Underlying immunodeficiencies can predispose to the development of bacterial polyarthritis.

Clinical Signs

Affected kids will have hot, swollen, painful joints and may be reluctant to rise or walk. They usually have a fever in the range of 103 to 105°F. Carpi, stifles and hock joints are most often affected but other joints may also be affected. Any number of joints can be involved. Concurrent or previous episodes of diarrhea, pneumonia or navel abscess are common. The umbilical stump should be palpated for evidence of omphalophlebitis.

Post Mortem Findings

At necropsy, multiple joints may contain purulent synovial fluid and there is evidence of inflammation in the synovium. Evidence of infection may be also present in other organ systems, especially in the lungs, gastrointestinal tract and central nervous system. The umbilical remnant is often thickened and may contain focal abscesses.

Diagnosis

Bacterial polyarthritis and mycoplasma arthritis may occur in young goats under similar circumstances and with similar clinical presentations. Confirmation of the diagnosis and differentiation from mycoplasmosis depend on culture and identification of the causative agent. A neutrophilic leukocytosis and leukopenia may be generally seen in case of Gram-negative sepsis. Arthrocentesis will reveal an abnormal, turbid and flocculent synovial fluid with increased total protein and elevated cell counts consisting primarily of neutrophils . Serum immunoglobulin levels are often low, reflecting the predisposing failure of passive transfer of colostral antibody.

Treatment

Successful therapeutic management of bacterial polyarthritis is difficult in goats, especially when economic constraints are present. Whenever possible, bacterial culture of affected joints and antibiotic sensitivity testing should be performed. In extremely valuable goats, joint lavage and intra-articular antibiotic therapy may be attempted. In most cases, practical therapy is limited to the parenteral administration of broad spectrum antibiotics and anti-inflammatory drugs. Little information is available about the concentration of various antibiotics in the joints of goats and antibiotic selection is largely empirical. The prognosis for complete recovery in these cases is not satisfactory.

Prevention and Control

Bacterial polyarthritis is best managed by prevention. Extensively managed goats should have access to fresh, well-drained areas for kidding. Bedding should be changed frequently in kidding sheds for intensively managed goats and pens should be disinfected regularly. Adequate and early ingestion of colostrum should be insured. Umbilical cords should be left at least 4 cm long and dipped in disinfectant as soon after birth as possible. Tincture of iodine or strong (Lugol's) iodine solution is commonly used and effective. In problem herds, re-dipping navels on the second day may be helpful in controlling infection.

5.17 Kerato-Conjunctivitis

Infectious keratoconjunctivitis is known as "pinkeye." A number of etiologic agents have been associated. When only one or two animals are affected, it becomes more difficult to distinguish between an infectious and an irritant cause. In fact, many sources of irritation, such as bright sunlight, dusty particles and transport in an open vehicle can predispose the goat for the development of an eye infection. Flies can work as the mechanical carrier and get contaminated by ocular secretions which can spread the infection to other goats. A herd outbreak may follow introduction of a carrier animal.

Etiology

Mycoplasma and Chlamydia are currently believed to be the most common causes of keratoconjunctivitis in goats however there are other etiological agents involved in causing keratoconjunctivitis in goats but the incidence are very rare.

Mycoplasma

Mycoplasma conjunctivae has been isolated from naturally occurring cases of pinkeye in goats and in sheep and has induced the disease in experimental studies. It also can exist in a carrier state in the conjunctival sac of clinically healthy eyes. Mild forms of the disease are self-limiting and last approximately 10 days, although clinical signs have been reported as persisting as long as 12 weeks. Other mycoplasma species isolated from goats with keratoconjunctivitis include *Mycoplasma agalactia, M. mycoides* subsp. *mycoides,* large colony type, *M. capricolum,* and *Acholeplasma oculi.*

Chlamydia

Chlamydial conjunctivitis has been recorded as a contagious disease of young goats and adults. In sheep, recurrence of chlamydial keratoconjunctivitis within a few weeks after clinical outbreak have been recorded. Lymphoid follicles are reported to develop early in the course of the disease.

Rickettsia

Colesiota (Rickettsia) conjunctivae has been reported as an etiologic agent for pinkeye in goats and sheep. Most reports were based on cytology and predated the knowledge that chlamydia and mycoplasma are commonly involved and thus rickettsia have fallen out of favor as being involved in the syndrome. The reported clinical signs, including occasional prolonged outbreaks are indistinguishable from the disease produced by the other agents. However, experimental subconjunctival injection of the Q fever agent *(Coxiella burnetii)* has produced severe keratoconjunctivitis in goats with rickettsia demonstrable in scrapings.

Other Possible Etiological Agents

The isolation of a bacterium from the conjunctival sac of a goat with keratoconjunctivitis is not adequate proof of infection. *Branhamella (Neisseria) ovis* is sometimes involved in the disease pathogenesis. Conjunctivitis has been reproduced experimentally in goats using a strain of *Branhamella ovis* isolated from an outbreak of keratoconjunctivitis, but this same organism appears to be common in normal eyes of goats. *Staphylococcus aureus* has also been isolated from ocular swabs of healthy goats. Infectious bovine rhinotracheitis virus has been isolated from a goat that developed severe keratoconjunctivitis. Paramyxoviruses cause high fever and erosive lesions in the alimentary tract of ruminants. Ocular signs may include increased lacrimation and a serous conjunctivitis in Peste des petits ruminants is the common sign in goats. Goats with Capri pox infections mostly develop oculonasal discharge, conjunctivitis, and keratitis, in addition to fever and anorexia. *Thelazia* spp, the eyeworms, can cause conjunctivitis or keratitis in sheep and goats. A single *Setaria cervi* worm longer than 1 inch was found in the aqueous humor of a goat in India. The goat exhibited continuous unilateral lacrimation. Migrating larvae of *Gedoelstia hassleri* (nasal botfly) cause keratoconjunctivitis and panophthalmitis in goats and other species. *Trypanosoma brucei* and *T. rhodesiense* cause blepharoconjunctivitis and keratitis in goats. In many tropical and subtropical countries, screw-worm larvae may invade fresh wounds on the head or ocular tissues infected with keratoconjunctivitis.

Clinical Signs

Early or mild keratoconjunctivitis results in lacrimation; the side of the face is wet below the eye. The conjunctiva is also red and swollen. Over several days, hyperemia of the conjunctiva increases, follicle formation occurs and revascularization of the cornea may develop. The cornea may be slightly hazy at the limbus or entirely opaque. A few animals develop a corneal ulcer and that ulcer may perforate. The eye is painful and held partially closed; blinking is frequent. If both eyes are opaque or ulceration occurs, the goat will lose body condition and totally blind animals in flock may die.

Diagnosis

The various etiologic diagnoses for infectious keratoconjunctivitis cannot be distinguished on the basis of clinical signs; laboratory assistance is required. A dry sterile cotton swab is robbed briskly across the conjunctiva and placed in transport media. Then a suitable media should be used for the isolation of chlamydia, mycoplasma, and rickettsia because it is difficult to isolate on normal media. It is also important to take samples from early lesions; secondary bacteria, leukocytes and various products of immune mechanisms will interfere with isolation efforts. Scrapings from the palpebral conjunctiva (especially lymphoid follicles) can be made. The tissue thus obtained is spread on several microscope slides. Fluorescent antibody testing for chlamydia is easier than culturing this agent in embryonated eggs or tissue culture cells. Interpretation of cytological preparations from conjunctival scrapings is difficult. Superficial cells can be harvested with minimal distortion by rolling a dry swab across the conjunctiva and then across the slide. New methylene blue, Wright, Giemsa, or Gram stain is typically used. Large basophilic, Gram-negative cytoplasmic inclusion bodies in epithelial cells occur with chlamydial conjunctivitis but are difficult to find after the first week. Smaller basophilic coccobacillary and signet-shaped bodies are found attached to or within epithelial cells in mycoplasmal infections. Rickettsia are described as small, Gram-negative, highly pleomorphic bodies that may be both intracytoplasmic and free. Pigment granules or stain precipitates can be confusing and visible evidence of an etiologic agent.

Treatment

The eye should be irrigated with physiologic saline or clean (preferably previously boiled) water to remove exudates and dust or other foreign matter. Although antibiotic drops are theoretically better than ointments, it is inconvenient to apply drops every 2 hours. Ointments are generally effective if given at least twice a day. Several antibiotics have been clinically effective, but it must be kept in mind that many animals heal uneventfully without treatment. Given the spectrum of agents associated with keratoconjunctivitis, a tetracycline eye ointment is a reasonable and give good response. Chloramphenicol ointment should also be effective.

Powders and aerosols are irritating to the eyes and ideally should not be used but they have been reported to give good results. When economics become an important consideration as when many goats are affected or the whole herd is being treated simultaneously then it may not be possible to use ophthalmic ointments. Experimentation may identify a non-irritant ointment containing a suitable antibiotic that is better than no treatment at all. Under similar circumstances, intramuscular injections of long acting tetracycline have prevented relapses in sheep affected with *Mycoplasma conjunctivae* and other agents. Intramuscular tylosin (200 mg/ goat/ day) has proved good results in goats treated early in the course of chlamydial keratoconjunctivitis. Corticosteroids can also be used in many different topical and injectable preparations.

5.18 Mastitis

Mastitis is an inflammation of the mammary gland (udder).It can be caused by physical injury or infectious agents such as bacteria which invade the mammary gland. It is one of the more common health problems affecting sheep and goats. Severe cases can result in death, but more often it takes its toll in the form of treatment costs and premature culling. Sub-clinical mastitis may be defined as mammary gland infection as revealed by laboratory examination of milk samples. Clinical mastitis is characterized by signs of inflammation such as swelling, pain, fever and abnormal milk secretion. Clinical cases may be acute, where animals clearly show all the characteristic signs of inflammation and chronic, where the infection remains in a more or less quiescent state with recurrent mild to severe attacks.

Etiology

A wide variety of organisms have been associated with clinical and sub-clinical mastitis in goats. Haemolytic *Staphylococcus aureus,* non-haemolytic *Staphylococci* and a variety of *Streptococci* are the most important pathogens. Haemolytic *Staphylococcus aureus is* the most common cause of gangrenous mastitis in the goat but the effects of infection may vary from an in apparent infection through sub-clinical mastitis to acute clinical mastitis. Non haemolytic *Staphylococci* cause an often unobserved, insidious, progressive mastitis with eventual induration and atrophy of the gland. The milk may be normal but with a high somatic cell count. Other organisms isolated from mastitic glands include *Streptococcus agalactiae, Strep. dysgalactiae, Strep. uberis, Pasteurella haemolytica, Klebsiella, Escherichia coli* and *Mycoplasma* spp.

Clinical Sign

Mastitis is usually observed shortly after kidding until the post-weaning period. It can take on several forms. Clinical mastitis (chronic or acute) involves physical changes in the udder. The udder becomes swollen and warm, sometimes painful to the touch. In severe cases, blood supply to the udder is affected and a blue discoloration may result of the udder and called "blue bag" mastitis (Fig. 19). Goats affected with mastitis become feverish, off feed and become depressed. Goats with sub-clinical mastitis usually appear quite healthy, but there is a reduction in their milk supply and development of lumps in their udders called "hard bag" mastitis. This is probably the most serious form of mastitis, since it often goes undetected. Keen observation is necessary to pick out these cases and prevent the potential damage. Clinical acute cases result when infected udders are injured and they are characterized by severe inflammation which may rapidly become gangrenous, with fever, intoxication and gross changes in milk secretion. The milk secretion of a clinical mastitis flare-up in a gland or the whole udder may become yellow, thick and greatly reduced in quantity. In per-acute cases, gangrene quickly develops, often within a few hours and the affected animal may die unless the entire gangrenous gland is surgically removed. *Corynebacterium pyogenes* mastitis in goats is characterized by the presence of firm round abscesses

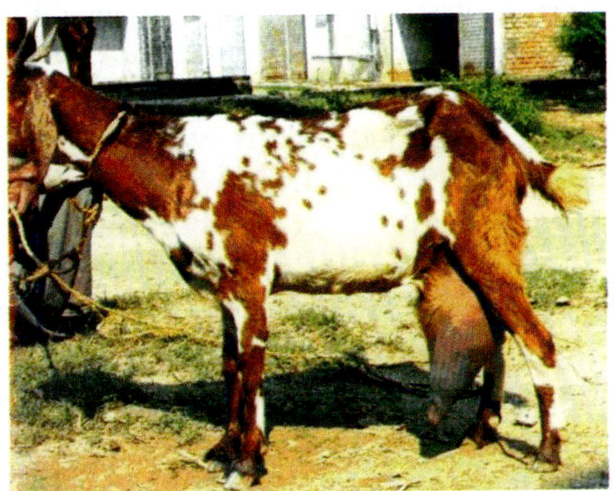

Fig. 19 Mastitis in goat (Blue bag)

in the milk producing tissue. The disease is usually progressive; advanced cases of the disease reveal multiple abscess formation with nearly complete destruction of milk secreting tissue. Mycoplasma mastitis is rare in goats. It may be found in animals suffering from systemic mycoplasma infection. This form of mastitis is not rare in countries where contagious caprine pleuropneumonia occurs in goats, but other mycoplasma infections have been suspect in severe cases of arthritis and pleuropneumonia.

Diagnosis

The diagnosis of mastitis in the goat is difficult since apparently normal goats may have high somatic cell counts in their milk. These cell counts may be higher than 750 000/ml (21% lymphoid cells, 69% neutrophils, 0.6% epithelial cells). 368 milk samples from 185 goats were examined and it was revealed 23% and 35% of these had cell counts of 500000 to 1×10^6/ml and $> 1 \times 10^6$/ml, respectively, even though bacteria could be cultured from only 4.3% of the samples examined. Similarly, it has been demonstrated that 78% of 422 milk samples which had a California Mastitis Test (CMT) reaction of 3 or more produced no bacterial growth while 95% of milk samples with a CMT reaction of less than 3 produced no bacterial growth. It has been suggested that counts of $< 1 \times 10^6$/ml represent normal glands or slight mechanical irritation; counts of 500000 to 2×10^6/ml represent infection with non-pathogenic organisms or weakly pathogenic organisms; and counts of $> 1.5 \times 10^6$/ml indicate mastitis caused by a pathogenic organism. These counts mean that apparently normal goats may give a 1+ or even 2+ CMT reactions. Modified Whiteside Test was more effective in the diagnosis of caprine mastitis than the CMT. The California Mastitis Test (CMT) and Somatic Cell Counts (SCC) of milk are useful monitoring tools to detect the presence of mastitis in the mammary glands of goats. The California Mastitis Test is a simple rapid means for detecting mammary gland infection. It has had wide acceptance and use for routine mastitis detection. There is widespread belief that a higher CMT is normal for goats than for cows.

Direct Somatic cell count and Teepol test were used to estimate the Somatic cell count in milk. There was relationship between Somatic cell count and bacterial load in Goat milk. There was 58% correlation observed between the direct microscopically SCC and Teepol test. These tests don't take the cytoplasmic particles in consideration. There was a steep increase in SCC's, bacterial load (Total aerobic and facultative anaerobes) during the third week of lactation in goats. Most common bacteria isolated from milk were *Bacillus* spp. and *Staphylococcus* spp. *Staphylococcus* was differentiated into coagulase negative (CNS) and *Staphylococcus aureus* on the basis of coagulase tests using Rabbit Plasma. Out of the total Staphylococcal isolates the proportion of coagulase negative staphylococcus and *S. aureus* were 55.5 and 44.5% respectively. There was an increase in the somatic cell count corresponding to an increase in *S.aureus* and coagulase negative staphylococci (CNS).The most predominant bacterial isolate was *Bacillus* spp., followed by CNS and then *S. aureus.* Teepol test (indirect somatic cell count) was more specific, because it detects the intact cells compared to direct count, which takes into account, cytoplasmic particles.

The final diagnosis must be made by the culture of milk samples. Sub-clinical mastitis in goats may be identified by laboratory culture. Organism coagulase negative or hemolytic on blood agar plate culture can be detected in milk of mastitic goats. *Staphylococcus aureus* and *Corynebacterium pyogenes* are readily identified by laboratory culture of milk samples.

The California Mastitis Test

Mastitis is a very serious ailment and in dealing with problems of the udder the results of the CMT are essential in determining the proper treatment program. Furthermore, the test is very accurate and good at early detection of infection. With a little practice anyone can easily perform the test. It hardly takes any time at all. The mixture of milk and test liquid is converted into gel. Just be sure to mix equal parts milk and liquid. There should have an extra pan or bucket available for this so that you do not have to throw potentially germ-laden milk on the ground or the floor. Somatic Cell Counts are a more accurate measure of udder health. Healthy dairy goat herds can be expected to produce milk with a somatic cell count under 500,000. The presence of mastitis infection in dairy goat herds is reflected in bulk tank milk samples with a CMT of 1 or higher and a somatic cell count exceeding 1,000,000 cells per milliliter. Regular use of the CMT or SCC can give both the owner and the milk consumer confidence that the milk is produced by healthy animals.

Identification of Mastitis by the Nature of the Milk

Slight wateriness with a few flakes - subacute *Staphylococcus aureus* Watery with large yellow clots - streptococcal and staphylococcal Watery brown with fine mealy flakes – *E. coli*

Treatment

Goats which show signs of mastitis should be separated from the rest of the flock and treated with antibiotics. It may be necessary to bottle feed their kids. Treatment usually involves intramammary infusions of antibiotics and systemic antibiotics. It is helpful to collect milk samples from affected does to determine the main bacteria involved and the correct medication to use. Treatment should be continued for several days until the clinical signs have gone away. The udders of goats should be examined after weaning and before breeding. Mastitis can be controlled with good management and sanitation. Bedding in kidding pens should be clean and dry. There should be good drainage around the sheds. Animals should not be overcrowded. The incidence of mastitis is greater in closely confined flocks. Preventing respiratory disease in kids may help to prevent mastitis, as *Pasteurella hemolytica,* the bacteria that causes kid pneumonia is a major cause of goat mastitis. Sore mouth is another contributing factor, as kids with mouth lesions can infect their dams and any other goats they may nurse. Weaning kids from goats whose milk production has not declined sufficiently puts severe stress on the udder; therefore proper management at weaning is also necessary to prevent mastitis. After weaning, it is advisable to restrict the feed and water of goats for 1 to 2 days to rapidly decrease their milk production. Delaying weaning until after milk production has decreased sufficiently will lessen the occurrence of mastitis. Dry mastitis treatment udder infusion formulations are recommended for goats which have had evidence of mastitis infection before drying off and they may be at least as effective in preventing mastitis attack during the dry period. A single dry cow quarter udder infusion dose is recommended for each udder half in the goat. In severe acute attacks of mastitis, systemic administration of antibiotics by intravenous or other parenteral means is indicated. Frequent udder massage with gentle hand milking may be helpful to relieve pressure in the affected gland to aid recovery. Antibiotic udder treatments available are excellent for treatment of infected mammary glands, but success with their use is determined by the level of milking management and sanitation used in milking the herd.

Treatment of Mastitis-Causing Organism

Staphylococcus aureus

This is the most common type of mastitis in sheep and in cattle. Not as common in goats as *Streptococcus agalactiae*. This is one of the most dangerous types of mastitis, also known as gangrenous mastitis. This organism is widespread throughout the environment, with various strains causing life-threatening illnesses in people. The skin of the teat becomes cold, blue and sloughs (hence the old name of "blue bag"). The whole infected quarter may slough. There is a good chance that the doe may go off feed completely. She may stand with her back legs quite far apart. There will be watery, blood-stained, smelly milk which later changes to thick, yellow ropy stuff. Death can be quite sudden. Goat should be isolated from rest of the herd. The kids should probably be pulled and bottle fed or grafted onto other does. If she is close to drying off, this should be done following the appropriate treatment with benzathine cloxacillin. These organisms are penicillin resistant, so use erythromycin or chlortetracycline systemically.

Streptococcus agalactiae

Streptococcus agalactiae is the most common source of mastitis in goats. The organism enters through the teat opening, so it is easy to see the importance of prevention. Fortunately, this is not usually a real dangerous disease in goats. The milk will be watery and possibly have some yellowish clots. There is probably a reduction in milk supply. The udder will feel warm and the doe may or may not run a fever. She will probably show a reduced appetite. The disease may range from mild to serious and is rather hard to differentiate from congestion without using the CMT. An infusion of penicillin (into the udder) will usually take care of the problem. Commercially prepared tubes of penicillin which come with a longer and narrower point for insertion into the teat are available from most suppliers. Penicillin can be given but if Penicillin is not effective after 3 or 4 days, switch to LA200® and if that doesn't work, try Tylan200®.

E. coli

This is usually a fairly serious illness. There is frequently a temperature ranging from 103°C - 108°C. It produces dangerous toxins as part of the inflammatory process. *E. coli* infections frequently occur right after freshening as the udder starts to fill. Milk production ceases. Anorexia (loss of appetite) is fairly complete. Other symptoms include depression, dehydration and weight loss. Secretion from the udder is brownish and watery. This type of mastitis can be accompanied by diarrhea. Treatment is penicillin infusions and injections. Although this is a serious illness, upon recovery the udder usually returns to normal. But even if the affected quarters dry up, the udder will be normal at the next lactation.

Corynebacterium pyogenes

This organism is responsible for a multitude of diseases in farm animals. It is constantly lurking around waiting for an opportunity to strike. Thus, it is known as a secondary invader when it comes to mastitis and this type is fairly common in goats. There is a foul-smelling pus-like discharge that can occur even in does that are not in milk. There may be actual abscesses visible. It is felt that dirty bedding and the presence of flies contributes to its spread. In this type the udder may not return to production after infection. Treatment is penicillin and not very optimistic.

Pseudomonas aeruginosa

This is a persistent disease that shows intermittent episodes of visible signs of mastitis. If often follows infusions that are not done antiseptically. It may lead to toxemia and death. It is very hard to treat and may persist through the drying off period. It occurs mostly in cattle and rarely in goats.

Nocardia asteroides

This is a very dangerous disease. Sudden, acute onset with high fever, complete loss of appetite, rapid wasting away of the doe, marked swelling of the udder are the major symptoms. There will be tissue fibers in the milk and nodules that can be felt. This type may also follow unsanitary infusions.

Yeast Infections

Any infection that survives or occurs after treatment with penicillin can be the result of the development of opportunistic yeast invasion. The doe will have a high temperature. There may be a spontaneous recovery or it may end up as a chronic destructive process. Treatment of this problem is very complicated, frequent use of penicillin can lead to yeast infections.

Mycoplasma

There are a large number of species of mycoplasma organisms which can cause mastitis in goats. Both quarters will be seriously affected and will be hot and swollen. There will be a dramatic loss of milk production followed by serous or purulent (pus) discharge. In goats, the disease is highly transmissible. In one of the more common types painful, arthritic joints will be readily observable. There will usually be fever and loss of appetite. The milk will be thick and yellow and separate on standing, often times with flaky sediments. In another type, the milk may be greenish and there will be symptoms of the central nervous system and pneumonia. In another type, the milk will have a putrid odor. Mycoplasmal mastitis can be spread to the young kids through the milk. The kids may show pneumonia and/or arthritis and may even die. Other symptoms can include limping, hunched back, blindness, obvious pain or discomfort, anorexia, death. If this type of mastitis is suspected, the kids should be pulled and either fed off other does or given pasteurized milk. There is some chance that tetracyclines or Tylan® can cure, but the outcome is not optimistic. Careful consideration should be given to elimination from the breeding herd.

Prevention and Control

Always keep bedding clean, especially if any of the does are leaking milk. Keep a careful watch on any signs of congestion in does. Clean teats and udder before and after milking. Immediately treat any injury to the teats or udder, even if it appears very slight. Use the CMT mastitis test to diagnose the mastitis. Always, after the first five days after kidding, milk the doe completely out with each milking, being careful to massage and strip the teats. Proper use of commercial teat dips is highly recommended. For does which are nursing kids check the udders at least once a day for heat, swellings, injuries and residual milk that needs to be removed (if the doe is producing more milk than the kids can consume otherwise excess milk can lead to congestion and mastitis). Be alert to any unexplained drop in milk production. Start penicillin shots right away (Penicillin G Procaine). Do this twice a day for five days, more if the problem take longer to solve. If it has not been five days since delivery, there is a complexity because of the danger of causing milk fever from intensive milking, which is part of the cure for mastitis. Mastitis right after delivery frequently results from not enough milk being taken out by the kids. If the doe is not a heavy milker, the danger of taking too much is reduced. After milking and with the needle off a syringe gently shoot about 2 ml penicillin up into each teat. Massage it up into the udder a little. Leave it there until milk her the next time, probably about 3 hours. If the udder is hot, apply ice packs. If the kids are nursing, they will help soften the udder as well. Some udders show problems throughout the milking season and have to use a "dry doe" infusion at the end of the milking season.

General Considerations

The most common time for udder problems is the second or third day after delivery. This will usually show up as rather severe hardening anywhere in either or both sides of the udder. It can be a single, relatively small area resembling a flattened golf ball or it can be so large as to include the whole udder. It can even be noticed before delivery. It is not unusual for a doe to quickly develop some hardened areas (perhaps more common in first-time mothers) soon after delivery as the udder starts to fill with milk and grow in size. Fortunately, most of these are not "mastitis." The term for this condition when it is not accompanied by or caused by an infectious organism is congestion. This can be a rather harmless situation which may go away as the kids begin to nurse or milking. On the other hand, it can advance to a more serious, painful and aggravating health hazard with symptoms such as the cessation of milk production, loss (sometimes complete) of appetite, digestive disturbances and a host of other things. Rough hand milking which pulls on the teats and excessively strips after milk-out can be stressful and injurious as bad machine milking. Good hand milking requires full hand milking and no tug and pull on the teats. Both hand and machine milking require good milking preparation - clean dry teats and clean dry hands and/or teat cup inflations. Rough handling, irregular milking times, over milking or inadequate preparation for milking all take their toll in providing stress and injury. These directly affect mastitis resistance and susceptibility.

The use of a bactericidal solution to cleanse the udder and teats also stimulates good milk let-down. Dry the udder and teats with an individual paper towel before milking begins. With hand milking, it is very important that milkers hands be thoroughly washed and dried before milking. Teat dipping procedure has been found useful for preventing spread of mastitis from infected to susceptible glands in dairy cow herds. It is equally effective and useful in goats. However, some teat dipping solutions tolerated by dairy cow teats may be too irritant for goats. Teat dipping solutions should not be used for goats if they produce drying or irritation of the skin of the teats. Regular use of the California Mastitis Test and/or Somatic Cell Counts can successfully monitor the progress of mastitis control and the health status of udders in the herd.

The California Mastitis Test can identify mastitis in the "pre-clinical" stage. This means that it will tell a case of mastitis even before there are any visible symptoms. Got a suspicious udder? Do the CMT. If it's clear, then you probably have congestion. If it's positive, then start treating for mastitis. Congestion and mastitis can usually be distinguished on the basis of other symptoms. Blood in the milk for the first two or three days, if it turns the milk pink and settles to the bottom of a clear glass if left overnight is usually something you can ignore. It is more common in cattle than goats. Although the milk may appear rather unappetizing, the kids will not complain. If the blood persists for several days, does not settle out and appears more like clots or strings, then be sure to do the CMT. If after all of this, still are not certain, always err in the direction of treating mastitis. Not treating mastitis is much more harmful than over-treating any of the more minor ailments.

□□□

6

Viral and Chlamydial Diseases

6.1 Goat Pox

Goat pox is an acute to chronic disease of goats characterized by pyrexia, generalized skin and internal pox lesions, lymphadenitis and a viral pneumonia. Sub-clinical cases also occur. The disease is caused by goat poxvirus (GPV) which is an enveloped, double-stranded DNA virus, classified in the genus *Capripoxvirus* of the family *Poxviridae*.

Etiology

The virus that causes GP is a capripoxvirus, one of the largest viruses (170-260 nm by 300-450 nm). It is closely related to the virus that causes lumpy skin disease; GP virus and lumpy skin disease virus cannot be distinguished serologically. There is only one serotype of GP virus. The genome of pox virus is a double-stranded DNA of 73–91 MDa, with a central, less variable region of about 145 kilobase pairs and more variable termini with cross-links. There is extensive DNA cross-hybridization between species. Two types of particles are observed in negatively stained preparations, namely the mulberry or 'm' form, which is electron-transparent, and the electron-dense capsule or 'e' form. The poxviruses are generally resistant to drying, survive freezing and thawing, and remain viable for months in the lyophilized state. Sensitivity to heat differs among strains. The Iranian and Egyptian strains of GPV are more resistant than the Dushanbe strain as heating at 56°C for 1 h fails to reduce their titer significantly, whereas the Sambalpur isolate of GPV is inactivated within 30 min at 60°C. Moreover, the thermal inactivation of a Turkish strain of SPV leads to the loss of infectivity when it is heated to 55°C for 1 h. In contrast, the infectivity of the Jaipur isolate follows a bimodal curve, with residual infectivity titratable even after 1 h at 50°C. The infectivity is sensitive to ether and chloroform but a few strains are resistant to ether. Replication of viruses is inhibited by Actinomycin D and bromodeoxyuridine.

Propagation and Cultivation of Viruses

Natural hosts are used successfully for cultivation of the capripoxviruses. The virus does not grow in any species of laboratory animal. GPV isolates differ significantly in their growth behavior in embryonated chick eggs. Iranian, Kenyan and Indian strains grow well on chorioallantoic membranes, whereas Russian strains do not. The Iranian Gorgon strain and Egyptian strains grow poorly on chorioallantoic membranes. GPV are adapted to and are able to grow in a number of host systems *in vitro*, most of which are derived from sheep, goats and cattle. Of these, goat kidney cells are very sensitive systems in which the virus produces characteristic cytopathic effects, such as rounding, tract formation, retraction and ballooning of cells, nuclear vacuolation, chromatin fragmentation, loss of continuity of the cell sheet, and intracytoplasmic inclusion bodies. Other cell culture system, e.g. embryonic goat cell culture, is also quite sensitive. Some isolates can grow in bovine fetal muscle cells and calf kidney cells. The virus can also be grown in some established cell lines. GPV can grow in the Vero and MDBK cell lines.

Virus Purification and Characterization

Capripoxviruses are purified by different methods, *viz.*sucrose gradient (40–60%) and potassium tartrate gradient (density range 1.05–1.35 g/ml) centrifugation of partially purified virus through a sucrose gradient (36%). Viruses of scab origin are purified in a similar fashion.The main structural polypeptides of capripoxviruses are identical on analysis by polyacrylamide gel electrophoresis (PAGE), and more than 20 polypeptide bands can be distinguished in a purified virus preparation. The molecular weight of the polypeptides ranges from 14 to 130 kDa, but only a few of them are immunogenic. The major common precipitating antigen, of Goat pox and sheep pox about 67 kDa, induces serum-neutralizing antibodies. Immunization with poxviruses evokes both humoral and cell-mediated immune responses. The relative importance of circulating antibodies and cytotoxic T lymphocytes in the suppression of the infection is not fully understood. However, it is quite clear that infection subsides in the host with the appearance of circulating antiviral antibodies. The serum of the immune host contains a spectrum of antibodies reactive with surface or internal virion antigens as well as with virus-induced antigens.

Antigenic Relationship

Capripoxviruses are antigenically related depending upon the antiserum used. The viruses cross-react in the complement fixation test. The sharing of antigens between the viruses can be further demonstrated by ELISA, sodium dodecyl sulfate–polyacrylamide gel electrophoresis, the leukocyte migration inhibition test, immunoelectrofocusing (IEF) and IEF-gel blotting. The techniques of IEF, IEF-gel blotting and dot immuno binding are useful in the differential diagnosis of goat pox and sheep pox . Though the major portion of the antigens of GPV and SPV are similar, there may be some differentiating antigenic components in each virus. Moreover, within isolates of GPV, the antigenic make-up is largely similar but may not be identical. Furthermore, it seems that GPV is not serologically related to Contagious pustular dermatitis virus, a parapoxvirus; however, the results of cross-reactions often vary, which is probably due to the different methods used by different workers employing different isolates of capripoxvirus.

Molecular Biology

There have been few studies of the molecular biology of GPV. Restriction analysis of genome of the virus isolated from different countries reveals that sheep pox and goat pox are caused by very closely related isolates of capripoxvirus. Such analyses not only demonstrate the genetic relatedness but are also used as molecular epidemiological tools for the characterization of the viral genomes of Capripoxviruses. The extent of nucleotide divergence values within the genomes suggest that the typical sheep and cattle isolates are more closely related to each other than to goat isolates. There is, however, less divergence between Capripoxvirus genomes than that seen in Orthopoxvirus genomes. Furthermore, the Capripoxviruses can be used successfully as viral vectors in the preparation of recombinant vaccines by incorporating foreign viral genes, such as those of the Rinderpest and Bluetongue viruses. A P32 gene (969 bp) of goat-pox virus, Mukteswar isolate; goat-pox virus Uttarkashi isolate and sheep-pox virus Rumanian FANAR Station have been completely sequenced and submitted to NCBI Gene Bank with accession number AY159333, AY382869 and AY38684.

Epidemiology

Goat pox is present in India, and as far East as China and Bangladesh. Hansen first reported goat pox in 1879 from Norway. Goat pox is an ancient disease which has turned endemic in the Middle East, the Indian subcontinent, and Central and Northern Africa. The disease is present mainly in Asia, Africa, the Middle East and parts of Europe, but it has also been reported from many other parts of the world, e.g. southwest Africa, France, Malaysia, Norway, Morocco, Tajikistan, Sweden , Sudan, Iraq, Nigeria , the western USA, China, Bangladesh, and Oman. In India, an outbreak of goat pox was reported for the first time in 1936. Later it was reported from the states of Haryana, Uttar Pradesh, Orissa, Bihar, West Bengal, Madhya Pradesh and Himachal Pradesh.

Contact is the main means of transmission of GPV. Inhalation of aerosols from acutely affected animals, aerosols generated from dust contaminated from pox scabs in barns and night holding areas, and contact through skin abrasions either by fomites or by direct contact are the natural means of transmitting GPV. Insect transmission is possible. The virus can cause infection experimentally by intravenous, intradermal, intranasal, or subcutaneous inoculation. Under field conditions, the incubation of GP is between 4 and 8 days. Experimentally, the first sign (fever) may appear within 3 to 5 days after inoculation. The course of the disease is 4 to 6 weeks with various stages of pox lesions present at the same time. Full recovery may take up to 3 months. The virus commonly enters through the host's respiratory tract, and therefore transmission of disease frequently occurs via an aerosol during direct or close contact between infected and susceptible animals. Transmission also occurs by skin abrasion or mechanically by the action of insects. The disease may be transmitted by *Stomoxys calcitrans* and the tsetse fly. Experimentally, the diseases can be transmitted by intra-dermal, intravenous and subcutaneous inoculation, and by artificially produced virus aerosols. Goat pox virus causes clinical disease in goats. The virus replicates in cattle but does not cause clinical disease. The disease has not been detected in wild ungulate populations. A field isolate of Capri-poxvirus (CPV) from goats was characterized and designated as the Pantnagar isolate. It did not infect sheep, chicken, rabbit, mice and buffalo calf. The virus failed to agglutinate the erythrocytes of sheep, goat, pig, rabbit, dog, chicken and human type O cells. On chorioallantoic membrane (CAM) of embryonated chicken eggs, the virus produced non-haemorrhagic pock lesions of about 2.0 mm in diameter. Under natural conditions, GPV is highly host-specific but the host specificity varies from isolate to isolate. Kenyan and Yemen isolates and an Oman sheep isolate infect sheep and goats equally readily. However, isolates from the Middle East and India are host-specific. GPV does not infect sheep and SPV does not infect goats. It is believed that the virulence of Capri-pox viruses is so pathogenic in goats only, and thus is considered goat poxviruses. Therefore, it is arguable that the malignant pox diseases of sheep and goats should be designated by the single term 'capripox'. Goat pox affects goats of all ages, both sexes and all breeds but are more common and more severe in young and old animals and in lactating females. The virus is present in the vesicular fluid, scab and milk of infected goats and sheep, and blood and spleen emulsions are also infective. Human infections during the handling of infected animals are rarely seen.

The severity of GP varies depending on the strain of the virus and the age and breed of the animals affected. In adult goats, morbidity may range to 80 percent with some subclinical infections. Mortality can approach 50 percent. In susceptible kids under 1 month of age, morbidity may approach 100 percent, and mortality may be as high as 95%. Factors that may complicate the course of the disease and increase the mortality are poor nutrition, heavy parasitism, and severe climatic conditions. Morbidity rate of 50% in Sirohi and 67% in Beetal breeds of goats have been reported. An outbreak of goat pox in the Sirohi, Marwari and Kutchi breeds of goats observed and morbidity rate was higher in the adults than the kids. Mortality in Black Bengal goats in five outbreaks of goat pox was reported and morbidity and mortality rates were 75.85 and 59.97%, respectively. Morbidity and mortality were higher in animals <12 months of age (84.61 and 70.07%, respectively) than in adults (70.7 and 50.48%). To ascertain the extent and economic impact of goat pox (GP), survey of goat producers in the state of Maharashtra was conducted. The survey revealed that GP had a major impact, with average morbidity and mortality rate of 49.5%. Modelling studies suggested it would take about 6 years for a flock or herd to recover from an outbreak, with average annual losses in income of 30-43%, depending on flock type and the owner's actions. Statewide, it was estimated that around 5000 flocks and herds were affected by GP annually in Maharashtra, costing up to 107.5 million rupees.

Clinical Signs

Goat pox virus may cause subclinical infection; clinical cases vary from mild to severe. The first signs may include fever, depression, conjunctivitis, lacrimation, and rhinitis. Within a few days of the prodromal signs, pox lesions develop in the skin. These are more easily observed on the wool-free or hair-free parts of the body such as the perineum, inguinal area, scrotum, udder, axilla, and muzzle. Lesions do occur in wooled or haired skin. Generally, more severe (extensive) skin lesions correlate with more severe illness. The skin lesion first appears as an erythematous area (macula). This lesion progresses to a raised, slightly blanched lesion that presents erythema with edema in the central part of the lesion (papule). Pox lesions with a transudate, representing the vesicular stage of the lesion, may be noted, but rarely is there any gross vesicle in the skin. The center of the lesion then becomes depressed and gray (necrotic) and is surrounded by an area of hyperemia. Late

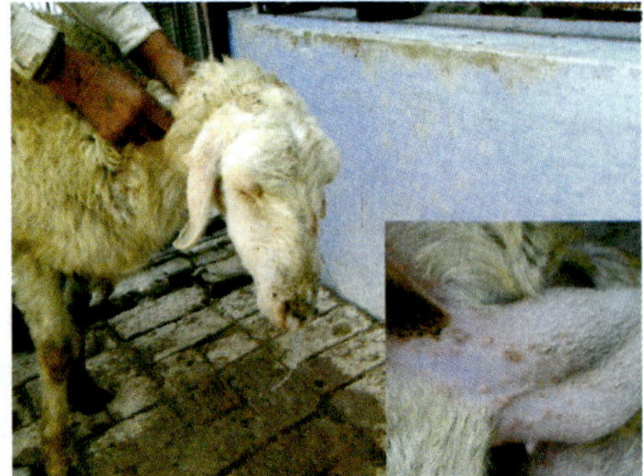

Fig. 20 Pneumonic, nasal discharge, Pock lesions on scrotum

Fig. 21 Pox nodule and scabs (goat)

in the course of the disease (2 to 4 weeks after the first signs), the lesion becomes dry, and a scab forms. A characteristic feature of a pox lesion is that lesions involve the entire epidermis and dermis and penetrate into the subcutaneous tissue; it feels like a nodule. Depending on the severity of the skin lesion, there may be a scar, an area devoid of wool or hair, after the lesion heals. Secondary bacterial infection may complicate the healing process. The muzzle may be swollen, and the nares and oral mucosa may have extensive lesions. In many cases, pneumonia may occur with labored breathing. Depression, anorexia, and emaciation are common and may persist. Kids under 1 month of age may suffer a very severe generalized form of GP. The signs described above for older animals are exaggerated, and there is an increased mortality.

Post Mortem Findings

At necropsy, skin lesions have congestion, hemorrhage, edema, vasculitis, and necrosis and will be seen to involve all layers of the epidermis, dermis, and, in severe cases, extend into the adjacent musculature. Lymph nodes draining affected areas are enlarged up to eight times their normal size owing to extensive lymphoid proliferation, edema, congestion, and hemorrhage.Mucous membranes of the eye, mouth, and nose have pox lesions that, in severe cases, may coalesce. In severe cases of GP, the eyelids may be so seriously affected that the proliferative lesions and inflammation cause the eyes to close. Lesions on the muzzle and nares may coalesce, and proliferative changes and inflammation may be extensive. Pox lesions may occur in the pharynx, epiglottis, and trachea. These usually appear as rounded blanched areas surrounded by an area of hyperemia. Occasionally there may be lesions in the epithelium of the rumen and omasum. Pox lesions in the lungs may be severe and extensive; the lesions are focal and uniformly distributed throughout the lungs as the result of hematogenous infection. Early lesions are congested areas; these then progress to discrete areas of congestion and edema and finally to white nodules. Areas distal to the pox lesions have lobular atelectasis. Mediastinal lymph nodes are often enlarged up to five times their normal size and may be congested,

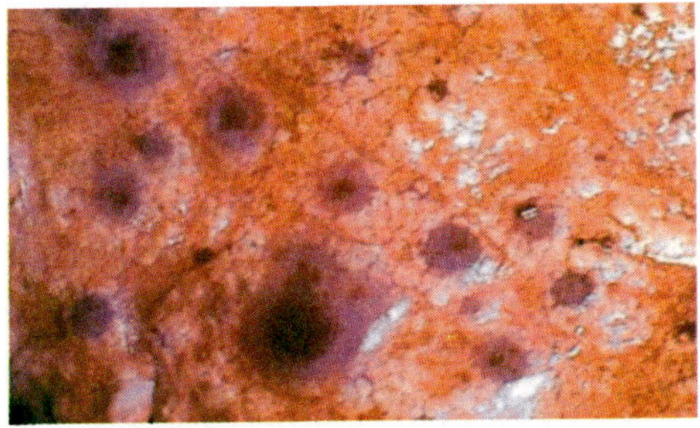

Fig. 22 Pock lesions in lungs

hemorrhagic, and edematous. Pox lesions also may be present on the vulva, prepuce, testicles, udder, and teats. Post-mortem lesions include tracheal congestion, lentil sized, bullet-shaped nodules and white patches on the lungs, an inflamed spleen and lymph nodes with graying white necrotic lesions, and an increased quantity of blood-tinged pleural fluid. In some animals, lesions develop in the lungs as multiple consolidated areas. Microscopically, the affected lung tissue is characterized by congestion, red hepatization and exudation, coagulative necrosis surrounded by a marked zone of inflammatory reaction, and thickening of the interlobular septae. Depletion of the lymphocyte population in the paracortical regions and absence of germinal centers in the spleen and lymph nodes are also observed.

Diagnosis

A tentative diagnosis of GP can be made on the basis of clinical signs consisting of skin lesions, which on, palpation involve the whole thickness of the skin, a persistent fever, lymphadenitis, and often .pneumonia; mortality may approach 50 percent in adults and 95 percent in lambs and kids under 1 month of age. For laboratory diagnosis of GP, skin biopsies of early lesions can be used for virus isolation and histopathologic and electron microscopic studies. Samples aspirated from enlarged lymph nodes can be used for virus isolation. Necropsy samples should include a full set of tissues, but samples of the lungs, trachea, and rumen containing gross lesions are especially valuable for histopathology. Samples for virus isolation should be shipped to the laboratory under wet ice if they will arrive in 2 days and shipped under dry ice if delivery will take longer (send in screw-capped vials with the caps secured with electrical tape). Samples for histopathology should be preserved in 10 percent buffered formalin. Serum samples should be taken from acute and chronic cases. Subsequent serum samples from acute cases may be taken 2 to 3 weeks after the first sample. The laboratory procedures for the diagnosis of GP include virus isolation; observation of the virus by electron microscopy; detection of antibody by virus neutralization, the indirect fluorescent antibody test, or both; and characteristic histopathologic lesions.

Diagnostic Tests

Pox diseases are usually identified according to their clinical signs and gross pathology and the host species affected. It appears that the host preference shown by the virus with respect to either sheep or goats, accompanied by the case history, may be regarded as partially affirmative for either sheep pox or goat pox, but accurate identification requires laboratory studies. It is also known that homologous diagnostic reagents tend to be more efficient than heterologous reagents in the accurate diagnosis of the disease. The non-infectious soluble antigen fraction alone can efficiently replace the use of infected scab suspensions or infectious virus, which pose a considerable disease security problem as the virus is highly infectious. In addition, the diagnosis of goat pox by classical virological or serological techniques that depend on live viruses is not suitable in countries where the virus is exotic and live viruses are not available. There is an extremely useful diagnostic method based on the polymerase chain reaction (PCR) for the detection of the viral nucleic acid of GPV in those countries that are free of these viruses. Besides animal inoculation tests in natural hosts, the following tests can be employed routinely for the diagnosis of goat pox in field samples.

Agar Gel Precipitation Test

A gel diffusion technique for the diagnosis of goat pox was introduced as early as the 1960s, using either homologous or heterologous antiserum. Currently, a more efficient agar gel precipitation test (AGPT) using better diagnostic reagents, such as the soluble antigens, is available. In addition, the use of [35 S] methionine-labeled antigen preparations considerably improves the sensitivity of the AGPT in the detection of capripoxvirus antibody.

Counter-Immunoelectrophoresis Test

Immuno-electroprecipitation is commonly used for the detection of various antigens and antibodies. The counter-immunoelectrophoresis test (CIE) is more sensitive and rapid than the AGPT in the diagnosis of goat pox. Normal saline solution may be used as an alternative to barbitone buffer to carry out the CIE test to make it more efficient.

Latex Agglutination Test

Latex beads provide a convenient carrier for antigens in agglutination tests, and coating the beads with antigen is not difficult. Latex agglutination assays have been used successfully in the detection of various antigen–antibody systems and have proved to be rapid and simple to perform, and they need no expensive equipment. The test is more efficient than the CIE test in the diagnosis of goat pox in field samples.

Reverse-Phase Passive Hemagglutination Test

Capripox viruses are non-hemagglutinating therefore, an indirect hemagglutination test, such as the reverse-phase passive hemagglutination test using sensitized sheep erythrocytes is useful in the laboratory diagnosis of goat pox and is more sensitive than the AGPT and the CIE test. Among the various types of sensitized sheep erythrocytes, glutaraldehyde and tannic acid-treated erythrocytes give the best results for the diagnosis of goat pox. Several agglutination tests, *viz.* coagglutination, passive hemagglutination and spot agglutination are also available for the simple and rapid diagnosis of goat pox. Though all the hemagglutination tests consist basically of an indirect hemagglutination test that has been modified slightly, the coagglutination test is based on the globulin-binding ability of protein A, which is found frequently on the cell wall of certain strains of *Staphylococcus aureus*.

Single Radial Hemolysis Test

This test depends on the fact that complement lyses erythrocytes when antigen–antibody complexes use it. The test gives a simple and quantitative estimate of antigen and antibody and is used successfully to diagnose goat pox.

Enzyme-Linked Immunosorbent Assay

The enzyme-linked immunosorbent assay (ELISA) is now used widely to detect antibodies and antigens in a variety of test systems, and is more sensitive than virus neutralization tests. Various methods of ELISA are available to diagnose goat pox, but problems such as a considerable background reaction and the requirement for special reagents, such as recombinant proteins, often limit their use as routine screening tests. Hence, an immunocapture ELISA has been developed as a relatively simple assay for the detection of GPV antigens in scab suspensions . However, this assay also has a limitation in that it is best used only in combination with the CIE test for accurate and confirmative diagnosis. A dot ELISA, carried out on nitrocellulose strips or paper, is a valuable addition to the battery of diagnostic methods for goat pox and is about three times more sensitive than the single radial hemolysis test. Recently, an avidin–biotin ELISA was introduced for the detection of antibodies to GPV in goat sera; it uses an isolated fraction of the soluble antigens that substantially reduces the background reaction in the assay. This assay is a reliable test for Capripox virus antibody that may be used to assess immunity to goat pox and can also be used in epidemiological studies.

Polymerase Chain Reaction

Although they are sensitive, methods such as ELISA and virus isolation in cell culture fail to detect virus particles that are bound to neutralizing antibody and the sensitivity of precipitation and agglutination tests is relatively low. Hence, highly sensitive molecular biological techniques based on the PCR are now employed for the identification of many viruses, including GPV in skin biopsies and cell cultures. PCR-based diagnostic methods are exceptionally effective for the diagnosis of goat pox in suspected skin samples obtained from the field. However, differential diagnosis of goat pox and sheep pox based solely on the PCR technique is not yet feasible, but may be possible by restriction analysis of PCR-amplified products. Besides, a PCR-ELISA has also been developed for differential diagnosis of Capri pox virus.

Treatment

All infected sheep or goats should be placed in a clean, well-ventilated enclosure and fed a balanced diet at a high level. Animals reluctant to feed should be given 10% glucose saline parenterally. All diseased animals should be treated with antibiotic to restrict secondary bacterial infections. To relieve respiratory-related signs, the nostrils should be cleaned and washed with a weak solution of potassium permanganate (1:10 000). Respiration should be stimulated with eucalyptus oil inhalations or coramine. Antibiotic ointment or powder should be applied topically to the skin lesions.

Prevention and Control

The most likely manner for GP to enter a new area is by introduction of infected animals. Restrictions on the movement of animals and animal products (meat, hair, wool, and hides) are essential to prevent introduction of GP. Wool, hair, and hides must be subjected to suitable decontamination procedures before entry into non-endemic areas. If a new case is confirmed in a new area before extensive spread occurs, the area should be quarantined, infected and exposed animals should be slaughtered, and the premises cleaned and disinfected. Vaccination of susceptible animals on premises surrounding the infected flock(s) should be considered. If the disease has spread over a large area, the most effective means of controlling losses from GP is vaccination; however, consideration should be given to eliminating infected and exposed flocks by slaughter; properly disposing of animals and contaminated material; and cleaning and disinfecting contaminated premises, equipment, and facilities. A carrier state has not been shown for GPV. However, the virus may persist for many months on contaminated premises. The imposition of quarantines on areas and premises containing infected or exposed animals is required to prevent disease spread. Depopulation of infected and exposed flocks should be used if limited spread has occurred. If the disease has spread extensively, massive vaccination followed by cessation of vaccination and control of animal movements from the area represent a strong strategy to control and then eradicate GP.

Inactivated vaccines are available for immunization against GPV infection which gives only short term immunity. The live attenuated vaccines are highly immunogenic but their usefulness is limited because they stimulate a pock reaction and/or lead to the death of some of the vaccinated animals because of generalization of the disease. Usually, homologous vaccination incorporating the locally prevalent strains of GPV is successful in protecting goats against goat pox. Therefore, in different countries and sometimes within a country, various live attenuated vaccines for goat pox have been available from time to time, with varying degrees of protective efficacy. Although it may not always be successful, a single vaccine prepared from a strain of Capri poxvirus that infects sheep and goats equally can be effective in controlling goat pox. A subunit vaccine has proved effective in protecting goats against Capri pox. On the whole, reports of cross-protection of sheep and goats against goat pox and sheep pox and other related diseases, such as contagious pustular dermatitis, are often contradictory and inconclusive because attempts to protect goats with SPV vaccines are usually unsuccessful. Hence, it is recommended that homologous vaccines should be used to protect goats against goat pox.

Vaccine Development

In endemic areas, vaccination could be an effective means of controlling losses from GP. Killed vaccines have not proven to be practical under field conditions because they do not provide solid lasting immunity. The Kenya O 180 strain is possibly the vaccine with the best safety and efficacy. A cell culture vaccine against Capri poxvirus was prepared which was given at 1000 $TCID_{50}$ (0.2 ml) intra-dermally to 6 sheep and 5 goats. After 23 days the sheep were challenged with Sheep pox virus and the goats with Contagious ecthyma virus. Negligible protection was achieved to either challenge virus, and they recommended that homologous vaccines should be used to protect against these diseases. All strains of Capri poxvirus so far examined are antigenically indistinguishable, and recovery from infection with one strain provides immunity against all other strains. Because of this antigenic homology among all strains, there is the potential to use a single vaccine strain to protect cattle, sheep and goats. It should be possible to produce a vaccine which protects against all strains of the virus, and trials of such a vaccine are in progress in Kenya.

The prevalence of two Capri poxvirus species, namely GPV and SPV, with selective virulence for goats and sheep respectively, is obviously due to their specific antigenic properties, and indicates that the genetic constitutions of the viruses may not be the same. This observation is supported by the fact that recombination events do occur naturally between species. Therefore, insights into the genetic organization of these viruses and the identification of the loci that are responsible for their host specificity would lead to the development of a recombinant vaccine assuring protection against both goat pox and sheep pox.Genetic characterization would further allow the Capri poxviruses to be used as viral vectors with the potential for delivering and expressing gene(s) from foreign pathogens, and thus act as vaccine vectors for the stimulation of protection not only against goat pox and sheep pox but also against other specific diseases of goats. In view of the host specificity of Capri poxviruses, at least two different vaccines, incorporating isolates of GPV are necessary to protect goats against goat pox. Alternatively, there is a need to identify the Capri poxvirus isolates that are infective for goats, so that a combined vaccine, as mentioned above, may be formulated to protect the animals against both diseases. Additional knowledge about the differential diagnosis of goat pox and sheep pox, gained from further studies of the antigenic and genetic nature of isolates, would open new vistas in the fields of epidemiology and effective control of the disease.

Public Health Importance

There is no conclusive evidence that GPV infects humans. A report from India implied that goat pox caused human infection was merely based on clinical signs. There was no attempt to isolate the causative virus or perform serology on the convalescent serums of the three patients to differentiate the infection from contagious ecthyma, which is a known zoonotic agent that occurs worldwide. A report from Sweden indicated that human infection occurred during an outbreak of goat pox. Although serological studies seemed to indicate that the apparent causative agent of the outbreak was not vaccinia or contagious ecthyma, no virus was isolated. Therefore, it cannot be said that goat pox virus caused human infection.

6.2 Peste Des Petits Ruminants (PPR)

Peste des petits ruminants (PPR) is also referred as goat plague, has always been noticed as a highly contagious disease of small ruminants (SRs) caused by a Morbillivirus close to Rinderpest virus (RPV). PPR is characterized by classical signs of alimentary tract disorders (ATD's) and respiratory tract disorders (RTD's) characterized by fever, necrotic stomatitis, lacrimation, pneumonia, foetid enteritis and high mortality.

Etiology

Molecular studies of the morbilliviruses (family *Paramyxoviridae*) revealed that type virus of the group is human Measles Virus (MV) but it also includes important animal pathogens such as RPV of cattle, buffalo and wild bovidae, PPR virus of SRs, Canine Distemper (CD) virus of domestic dogs and wild carnivore species, Porcine Distemper Virus (PDV) of seals and two viruses isolated from cetacean species-Dolphin morbillivirus (DMV) and Porpoise Morbillivirus (PMV). The PPR virus has the following characteristics:

- May survive at 60°C for 60 minutes.
- Stable at pH from 4.0 to 10.0.
- Killed by alcohol, ether, and detergents as well as by most disinfectants (eg, phenol, sodium hydroxide) etc.
- Long survival time in chilled and frozen tissues.

Sequencing studies have shown that, despite being only one serotype of PPR virus, there are different genetic lineages of each virus. Studies on lineage of virus allows tracing of disease outbreaks outside their known geographic area. In recent decades, PPR emerged in South India and later spread to North India. Studies on the genetic relationships of the viruses from nineteen independent PPR outbreaks facilitated identification between viruses of distinct geographical origin by nucleotide sequencing followed by PCR amplification. Four separate lineages of the virus were identified and the virus isolates from Asia over the past 2 years were all of one lineage, which had not previously been identified in Africa or Asia. Further, phylogenic analysis of virus isolates from 27 outbreaks in Asian and Middle Eastern countries, reported between1993 to 2000, and two recent outbreaks from the African continent were compared with the prototype African strain. Of the four known lineages of PPR virus the isolates detected to have their lineage. The geographical source of the new lineage 4 viruses is unknown although it is most closely related to African lineage 1. There was some possibility that the circulation of RPV, a related virus of cattle, would have masked its earlier presence of PPR in Northern India.

Epidemiology

Although goat and sheep are the natural hosts of PPR but goat appears to be more susceptible and suffer a more severe clinical disease than sheep. PPR also occur in other species, particularly in captive wild ungulates from three families i.e. Gazellinae (Dorcas gazelle), Caprinae (Nubian ibex) and Hippotraginae (gemsbok). The American white-tailed deer (*Odocoileus virginianus*) has also been infected experimentally. Cattle, buffaloes, camels, and pigs are also susceptible to infection but do not exhibit clinical signs and are unable to transmit the disease to other animals. PPR is not infectious to humans. PPR is endemic in sub-Saharan Africa since several decades; it was first described in Ivory Coast in the 1940s. The virus has been circulating in parts of sub-Saharan Africa for several decades and in the Middle East and southern Asia since 1993. Although, PPR in India, was first confirmed in March 1987 in sheep but suspected of having RP is now believed that many outbreaks in India previously attributed to RP were actually PPR. The virus was isolated four more times by 1992, and major epidemics occurred in the state of Andhra Pradesh. In Rajasthan nearly 50% of susceptible population is all the time under risk. PPR has also been reported in Sudan, Kenya, Uganda and Ethiopia. The Arabian Peninsula, the Middle East, and the rest of the Indian Subcontinent reported PPR incidents during 1993-1995. The disease has remained endemic in these areas. It has also occurred in Turkey, Iraq, Iran, Bangladesh and Nepal. Studies clearly indicate that the disease is also endemic in our neighboring countries viz., Pakistan, Nepal and Bangladesh.

PPR virus survives poorly outside the host. Spread of infection to new areas is largely by close contact between infected and susceptible animals particularly during the movement of infected animals. There are several means of transmission between animals:

- Inhalation of aerosols produced by sneezing and coughing of infected animals.
- Outbreaks are more frequent during the rainy season or the dry, cold season.
- Direct contact with ocular, nasal or oral secretions.
- Direct contact with feces.
- Fomites such as bedding, water and feed troughs.
- No carrier state is known to exist.
- Besides, PPRV is also present in semen and embryos and probably transmited viral infection through gonads.
- Feeding milk from infected animals to kids or lambs can spread infection.

Acute febrile viral infection or sometimes non-febrile, non-clinical infection (in endemic areas) of PPR in goats and sheep elicits humoral antibody response that is easily detected by specific immune assays. A monoclonal antibody-based competitive enzyme-linked Immunosorbent assay has been in use to detect the presence of serum antibodies. Screening of the 7,411 sera samples, revealed 34.2% prevalence of PPRV in small ruminants in India. The prevalence of antibodies to PPRV was noted to differ between species (i.e. sheep versus goats), age groups and geographical regions. The distribution and prevalence of antibodies to PPRV among various age groups of animals indicated that goats were exposed at an earlier age than the sheep, suggesting that goats may be more susceptible to infection with PPRV.

It is a fact that Morbilli viruses are highly pathogenic in their natural hosts and the economic consequences of epizootics of RP and PPR can be devastating. The disease outbreaks were more severe in goats than sheep and that the frequency of disease outbreaks was greater between the months of March and June (51.7%) as compared to other periods of the year. Analysis of another epidemiological study revealed, that most of the outbreaks were reported during the cooler months of the year along with rare reports during dry hot or humid hot months. It was also observed that flocks having mixed population of sheep and goat or goat alone were maximum affected then flocks exclusively of sheep and hence reconfirms established observation of higher susceptibility for PPRV in goats than sheep. Clinical signs and lesions in sheep were comparatively milder than goats. Observations indicated that goat played key role in PPR precipitation and spread in area, where flocks with mixed population of sheep and goats are reared as normal animal husbandry practices.

Morbidity and mortality recorded by various workers in different states and some other countries are as high as 100% and 90%, respectively when associated with other diseases such as capripox. PPR infection is one of the most important constraints to the increased production of small ruminants. Earlier RP use to pose a serious threat to cattle, buffalo and wild ruminants but a global eradication campaign lead to eradication of the RP with possibility to declare zero RP status by 2010. PPR causes death in more than 50% of the affected animals due to high fever, pneumonia, diarrhoea and dehydration. The annual loss due to PPR in the small ruminants in India is approximately Rs.180 crores. In another economic evaluation of PPR in Rajasthan revealed a loss of Rs. 85.13 crore.

Clinical Signs

Although, susceptibility to PPRV infection rises with age but the disease is fatal in the young animals. The clinical signs imitate those of RP, but changes can occur faster. Specific clinical signs are includes sudden onset of the disease with high fever, severe depression and loss of appetite, watery nasal discharge progressing to profuse muco-purulent discharge with catarrhal exudate occluding the nostrils, respiratory distress and sign of bronchopneumonia, conjunctivitis with sometime catarrhal conjunctivitis, necrotic lesions in the mouth (dental pad, hard palate, cheeks), progressing to erosions of epithelium, severe diarrhoea, dehydration, weight loss and abortion. Prognosis of acute PPR is usually poor; especially when lesions do not resolve within 2 to 3 days or when extensive necrosis and bacterial infection give the animal's

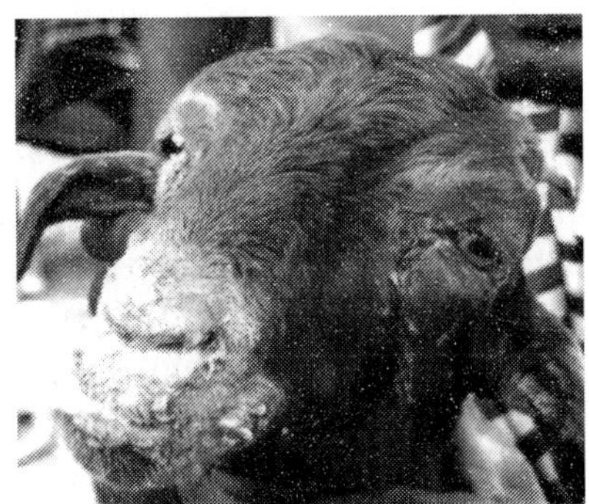

Fig. 23 Pneumonia, nasal discharge & conjunctivitis

breath an unpleasant, fetid odor. Poor nutrition, stress of movement, and concurrent parasitic and bacterial infections worsen clinical signs. The disease was categorized as typical PPR (with ATDs and RTDs) or atypical PPR on the basis of the clinical signs exhibited by the affected sheep and goats. Young animals (4 to 8 months) often have more severe disease.

Typical PPR - (ATD's) -fever, anorexia, grinding of teeth, frothy salivation, dry muzzle, oral lesions inform of erosive stomatitis, glossitis, gingivitis, hemorrhages on hard palate, halitosis, abdominal pain, soiled perineal region with foul smelling pasty feces or blackish diarrhoea and progressive dehydration. Convalescent animals showed poor body condition and debility. (RTD's) - Mucopurulent oculo-nasal discharges with opacity, high fever, and severe congestion of nasal passage, nasal block leading to open mouth breathing in some cases, occasional coughing, abnormal respiration and panting.

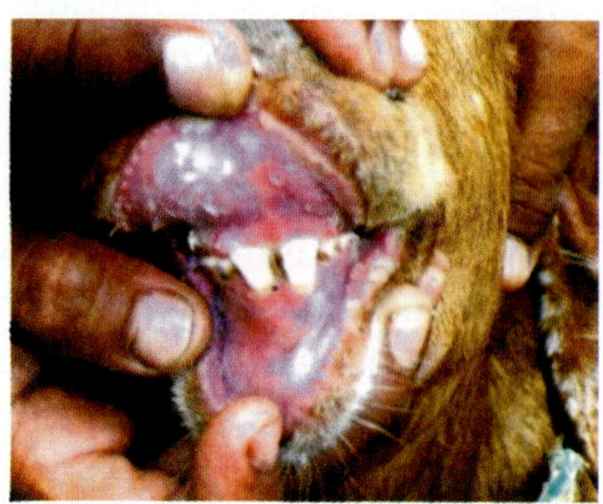

Fig. 24 Buccal congestion & erosion

Atypical PPR- Nasal discharges, coughing some times, soiled perineal region with foul smelling pasty feces or blackish diarrhoea with possibility of reoccurrence. In typical cases of PPR, flocks were having history of one or the other stress born predisposing factors like cold spell, winter/spring rains (particularly after shearing), flocks kept under open sky roof, poor grazing availability during winter/cooler months and physiological stress (gestation & lambing) would also have played significant role in further aggravation of the situation alone or in combination, in favoring disease precipitation in partially undernourished animals. Higher number of clinical cases recorded in winter indicates favorable season for virus circulation through aerosols generated by coughing/sneezing and copious nasal discharge on or around thick fecal bed would have been possible more freely than in dry hot or humid hot months (May to October) of the year when thermo labile nature of the virus would have limited its transmission.

Post Mortem Findings

Lesions are usually seen in the digestive and respiratory tract but can also be seen in other systems as well. In acute cases, the dominating

PM findings includes dehydrated carcass with faecal soiling, necrotic lesions in the mouth/tongue and nose, congestion of the ileo-caecal valve, engorgement and blackening of the folds in the cecum , colon and rectum ('zebra striping'), enlarged spleen, oedematous lymph nodes and pneumonia.

Diagnosis

Differential diagnosis of PPR should be carried out with other infections producing similar clinical signs such as RP, Pasteurellosis, Orf, Pox, CCPP, BTD, Heart water, FMD, NSD, GI Parasite and plant or mineral poisoning. RP, in its classical form, is easy to identify clinically, however, clinical observations for both the diseases should always be confirmed by a laboratory.

Samples should be collected in acute phase of disease and be shipped fresh under cold chain (not frozen) within 12 hours of collection. Samples include blood in EDTA anticoagulant (Blood & anticoagulant should be mixed gently), clotted blood or serum, mesenteric lymph nodes, spleen, lung, tonsils, pieces of the ileum and large intestine, swabs of serous nasal and lachrymal discharges. Ideally, samples should be collected from several animals in an outbreak.

Immuno capture ELISA and counter immuno-electrophoresis (CIEP) are most rapid test for detecting viral antigen. AGID is very simple and inexpensive and gives results within 1 day, but not sensitive to mild forms of PPR. PCR and culture for isolation in lamb kidney or African green monkey cell cultures is confirmatory. Further, indirect fluorescent antibody test have been used to identify PPRV antigen in various tissues of organs of infected animal. Routine serological tests include competitive ELISA (Sensitivity 99.4% Specificity 94.5%) and virus neutralization test, which is a prescribed test for international trade. However, Cross-neutralization with RP virus must be completed. Other serologic tests have been described, but not widely adopted includes CIEP and AGID test.

Treatment

There is no treatment for PPR however, mortality rates may be decreased by the use of drugs that control the bacterial and parasitic complications. Use of specific antibiotics i.e. oxytetracycline and chlortetracycline are recommended to prevent secondary pulmonary infections. However, in absence of specific guidelines lot of animal use to die by the time effective treatment is practiced by the field veterinarians. Palliative ORS for sick animals along with Inj. Belamyl, Inj. Melonex and Mineral mixture powder are to be given. The following managemental practices should be used to minimize the disease occurrence in day to day management.

- Keep the sick animal at home and clean their coral daily.
- Dusting of lime powder in coral once in 15–20 days.
- Protect animals from excessive cold.
- Clean nasal opening with the help of moist cloth or a piece of gunny bag and apply coconut oil containing camphor daily.
- Wash mouth with PP water or water boiled with neem leaves (but after cooling).
- Apply Haldi and sweet oil paste after mouthwash daily.
- First 10 days of sickness is crucial. Take care of starvation and dehydration through palliative treatment.
- Start treatment at evening or after offering the palliative ORS to sick animals.
- Contact area Veterinarian as per need.
- Recovered animals develop life long immunity.

Prevention and Control

In the past, the RP vaccine was used to prevent PPR, however, this practice has now been phased out to avoid confusion during retrospective serologic studies. A homologous PPR vaccine is now available and gives strong immunity. There are a few genetically engineered recombinant vaccines undergoing limited field trials. A combined recombinant vaccine to protect against RP and PPR has also been prepared. Currently a new marker vaccine for RP using reverse genetics technology is also being developed.

In India an effective Vero-cell line based live attenuated indigenous freeze-dried vaccine has been developed and validated using extensive field trials. The vaccine has a shelf life of more than one year at 4°C and provides immunity for three years. The vaccine is safe, potent and acceptable for use even during pregnancy. The vaccine has been developed at Indian Veterinary Research Institute (Mukteshwar campus) and Tamil Nadu Veterinary and Animal Science University,

Chennai. The technology for vaccine production has already been transferred to a few biological product units of the state AH Departments and also to a private organization for commercial production and marketing. The vaccine would help in reducing losses due to morbidity and mortality and thus improve economic condition of poor sheep and goat farmers in the country.

PPR has proved to be difficult in many countries, especially when the disease is first introduced. If we decided to move from RP to PPR it would be very important to discriminate between PPR and RP, because RP has sometimes also been recognized to affect sheep in the Indian subcontinent with signs similar to PPR. Moreover, PPR is an important disease in its own right hence differentiation is essential particularly with RP, at this stage where the RP eradication program represents an international emergency. Unfortunately, unlike the situation with RP, coordinated eradication of PPR is not a likely prospect in the immediate future but progressive control of the disease is possible. Vaccination, combined with common sensible actions such as safeguarding flocks from PDFs and unhygienic management practices may provide good control. However, a few points should be kept in mind while preparing action plan for PPR eradication program.

1) The turnover rate of small ruminant populations is, much faster than that of larger livestock, which means that vaccination has to be used more intensively and more frequently.

2) All sero-surveillance should be carried out using indigenously developed monoclonal antibody-based diagnostic kits of IVRI, Mukteshwar for the detection of PPRV antigen or antibody for uniformity of results.

3) Assessment of disease status through serosurveillance and disease surveillance for the trans-boundary animal diseases would be mandatory for efficient disease reporting, monitoring and forecasting system.

4) Initially a program of progressive control, could be taken up in certain areas to declare them free from PPR. In view of disease profile PPREP can be initiated through a participatory approach among SAARC countries.

5) Financial support should be invited from world agencies to handle PPREP.

6) Differentiation of PPR from a variety of diseases that cause similar ATD's and RTD's for example RP, FMD, BTD, pneumonic mannheimiosis/pasteurellosis and CCPP should be regularly carried out by authentic laboratory following OIE/IVRI guidelines.

6.3 Blue Tongue

Bluetongue is an infectious, a non-contagious, insect-transmitted disease of domestic and wild ruminants. It is caused by blue tongue virus (BTV) belonging to family *reoviridae* (genus *Orbivirus*) is considered as one of the most important diseases of domestic livestock. BT has a broad host range,which naturally infects domestic (sheep, goats, cattle) and wild ruminants (deer,pronghorn, antelope, boghornsheep) ,camelids and some other herbivores such as elephants. It is almost exclusively a disease of sheep, although cattle and goats show sub clinical infection. Cattle are the main mammalian reservoir of the virus and play a very important role in the epidemiology of the disease. It is a cause for serious concern to the livestock industry due to staggering direct and indirect economic losses mainly attributed to high morbidity, mortality, abortions, foetal deaths and fleece losses. BTV is listed under category "A" of the Office Internationale des Epizootics (OIE) and consequently restrictions are imposed on movement of ruminants from bluetongue endemic regions to bluetongue free zones and so are of major importance to international trade. The name "bluetongue" was given to describe the distinctive cyanotic tongue in some severely affected sheep. It was first reported in South Africa after Merino sheep were introduced from Europe in the late eighteenth century.

Etiology

BTV is a member of the Orbivirus genus, currently one of nine genera classified in the family Reoviridae. Within the Orbivirus genus, 14 groups are differentiated on serological basis. There is significant immunological cross-reactivity between members of the BT and Epizootic Hemorrhagic Disease (EHD) serogroups. The BTV is quite stable at 25°C. The slow freezing of virus at -10°C or -20°C results in the loss of infectivity and hence the blood containing virus should be stored at 4°C without any substantial loss of infectivity for a year. The tissue culture virus can be stored in 50% buffered lactose-peptone solution at -20°C without any significant loss of titer after 8 years.

The virus is resistant to ether, chloroform and sodium deoxycholate and is inactivated beyond the pH range of 6-8.The virion is about 86 nm in diameter with icosahedral symmetry and is composed of 80% proteins and 20% nucleic acid of ten segments of double stranded RNA. The base composition of genome has 42.4 % G+C value. Each of these genome segments code for one of the ten viral proteins. Seven of these proteins are 'structural', combining to form the viral particles and three non-structural proteins, NS1, NS2, NS3, are detected within the infected cells. They are non enveloped and composed of three protein layers. The outer layer contains two proteins, VP2 and VP5. VP2 is primarily responsible for the induction of type-specific neutralizing antibodies and its variable sequence resulted in the generation of 24 recognized serotypes of BTV.

In India, so far 21 serotypes (except 19, 22 and 24) have been reported from different states, of which serotype 1, 2 and 23 are found to be most prevalent. The first outbreak of this disease, reported in 1964 in sheep and goat farm from Maharashtra state was based on clinical symptoms. Subsequently infection in sheep was reported in Uttar Pradesh. It was reported the occurrence of disease in this country. Henceforth several outbreaks of BT have been recorded from different states of India in indigenous, crossbred and exotic sheep. The virus was isolated and identified as serotypes 3, 4, 9, 16 and 17 from Himachal Pradesh and serotypes 1 and 4 from Haryana. Subsequently, reported isolation of virus serotypes 9 and 18 from Maharashtra. In southern India, serotypes I, 4, 5, 6, 7, 11, 12, 14, 15, 16, 17, 20 & 23 in Tamilnadu 15 and serotype 1, 2, 12, 16, 17 & 20 in Karnataka were most commonly seen.

Epidemiology

In India, the disease was first reported from Maharashtra state in flock of sheep and goat. Since then, several outbreaks of BT have been reported in sheep from almost all the states of the country. Now, BT has endemic status in India. A serological survey has indicated the presence of bluetongue virus (BTV) antibodies in cattle and buffalo in several states in India. However, clinical BT has not been frequently observing in cattle and or buffaloes. BTV has evolved a life cycle where alternate cycles of virus replication in vertebrate and invertebrate hosts are essential for virus persistence; consequently *Culicoides* spp acts as principal vector in transmission of BTV. There is no evidence of vertical transmission of the virus in the invertebrate host. Observations on the placental transmission of virus in the vertebrate host are contradictory. In addition, there is no convincing scientific evidence that natural BTV infection crosses the placenta to cause fetal infection and production of persistently infected adult animals. There is little evidence of direct or indirect contact transmission in either host, other than rare instances of seminal transmission (here male should be viraemic and semen should contain red and white blood cells) in vertebrates. The virus cannot be spread by meat, milk or dairy products, but can be through mechanical vectoring by means of instruments (blood transfer) during periods of viraemia. Midges of the genus Culicoides are biological vector for BTV. Culicoides transmit BTV to and from susceptible animals, which have become infected by feeding on viraemic vertebrates. After a replication period of 6-8 days, and following its appearance in the salivary gland (extrinsic incubation period), the virus can be transmitted to a vertebrate host during a blood meal. Of approximately 1400 species of Culicoides worldwide, less than 20 are considered actual or possible vectors. There is evidence for the involvement of novel Culicoides vector(s) *(C.obsoletus* and or *C.pulicaris)* and for survival of BTV for significant period in the absence of adult vectors i.e. over wintering.In India, although disease is endemic but detailed study on vector species responsible for disease and virus-vector relationship are scanty. BTV has been isolated from Culicoides in India and China, but the species were not identified. Recent studies on vectors indicated that *C.oxystoma* is found to be most prevalent. Other vectors responsible for transmission include *C. clavipalpis, C. anophelis* and *C. imicola* in West Bengal, *C. anophelis* and *C. actoni* in Uttar Pradesh, *C. simiklis* in Uttaranchal and *C. orientalis* in Gujarat. The virus seems to be maintained in midges-cattle cycle as a primary cycle, and once a certain level of infection is attained, it spills over to initiate a secondary cycle in sheep. This generation occurs in late summer and early autumn when the vector populations are at a maximum. Thus, cattle act as both primary and reservoir vertebrate hosts. Susceptibility of dogs for BTV infection has also been reported on inoculation of inadvertently contaminated canine vaccine with BTV, indeed pregnant bitches that received this contaminated vaccine typically aborted and died.

The presence of BTV year round or seasonal depends on the climatic conditions and therefore, the geographic distribution of BTV is categorized into three ecologic zone types: endemic, epidemic and incursion. Endemic zones are defined as typically tropical regions where BTV transmission occurs throughout the year and sub-clinical infection is common. In endemic zones, clinical disease generally occurs in newly introduced immunologically native susceptible species. Epidemic zones include temperate areas where outbreaks occur seasonally, generally in the late summer when vector population is at peak. Incursion zones are areas that experience outbreaks infrequently when climatic conditions favour disease transmission by vectors as observed in greatest outbreak of Mediterranian basin.

Economic Importance and Impact

Bluetongue is an economically important disease of ruminants for several reasons. It is inflicting great economic losses to livestock industry in India and different parts of the world. These losses are attributed primarily to mortality and high morbidity, including abortion, milk, meat, and fleece losses in sheep. In endemic areas vaccination is a recurring cost. Now in India lot of trials going on to produce effective vaccine by pouring huge amount of money. In the recent years, several countries have shown interest in the germ plasm of Indian breeds of cattle and other animals for genetic and molecular studies and for some animal products such as fetal calf serum. But, inclusion of BTV in the OIE 'List A' has direct consequence that continuingly affecting our global trade, particularly to Canada, New Zealand, Australia and the European Union, which are classified by the OIE as bluetongue free. It has been estimated that in the late 1970s, the ban on US cattle semen exports resulted in an annual loss of $24 million. Worldwide losses due to BTV have been estimated to US$ 3 billion/ yr. India may be the only country which experiencing severe outbreaks due to the prevalence of highly virulent strains.

Pathogenesis

Pathogenesis of BT disease in goat depends on the virulence of the virus strains as well as the nutritional status, immune status, age of the animal and environmental stresses such as high temperature and ultraviolet radiation. On the bite of infected vector, the virus multiplies primarily in the regional lymph node which leads to primary viraemia where all the cells of the blood promiscuously become infected. But, the cell associated viraemia occurs primarily in the RBCs and platelets, although dissemination of the virus occurs mainly through mononuclear cells. Further multiplication of the virus occurs mainly in the endothelial cells including periendothelial and pericytes of capillaries, which is main contributory factor in the development of pathological lesions. This multiplication of virus in the leukocytes leads to panleukopenia and in smaller blood vessels lead to micro-vascular endothelial injury.Endothelial injury leads to increased capillary permeability resulting into oedema of various parts of the body especially in face region, larynx and lungs. It is responsible for generalized vascular thrombosis leading to multi-visceral necrosis, skeletal and cardiac infarction development, and bleeding tendency due to consumption of various clotting factors and platelets with severe hemorrhage, which is especially more so in sheep and white tailed deer. Affected animals frequently suffer pharyngeal or esophageal myo-degeneration leading to dysphagia, or regurgitation, and animal may succumb to aspiration with the development of bronchopneumonia. In severe cases, death of the animal occurs either from pulmonary edema, bronchopneumonia, cardiac insufficiency or any of these combinations.

Clinical Signs

After an incubation period of 4 to 7 days, fever up to 103° - 105°F develops and persists as long as the viraemia is present in the body (approx. up to 6-8 days). Within 24 to 48 hrs onset of fever, animal starts showing nasal discharge, salivation, hyperemia of nasal buccal mucosa and discharge may become mucopurulent and blood stained. In next few days, there will be edema of lips, tongue, face, ears, intermandibular space giving "monkey face appearance", hyperemia may become more intense with petechiation, and cyanosis of tongue in few cases which later may become swollen and protruded. During 5-8 days following onset of fever, there will be necrotic ulcers on cheeks, tongue, gums and they usually heal slowly with diphtheritic membrane formation. Respiration may become laboured and dysentery may occur during this period. When mouth lesions begin to heal, coronitis may occur (at the end of febrile period) which is characterized by the coronary swelling and streaky hemorrhage in the periople owing to affection of underlying sensitive laminae. Loss of body condition, muscle wastage and torticollis may develop as a late sign. Animal

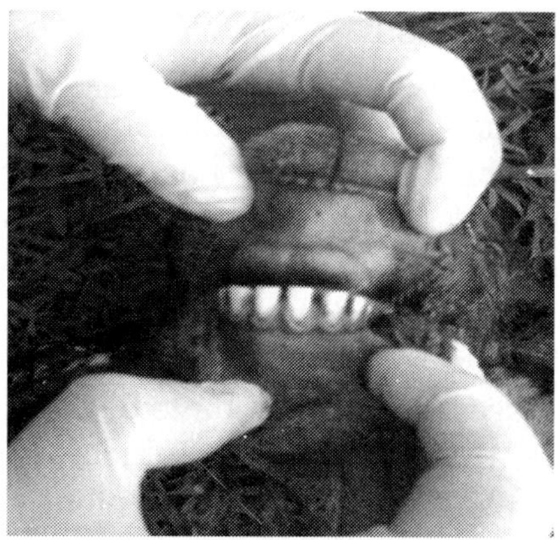

Fig. 25 Buccal lesions (Bluish mucous membrane)

may exhibits stiffness, reluctance to move, arching of back and recumbency. Affected areas of skin show break in the wool or much of the fleece being tender or cast. Death due to pulmonary edema and/ or cardiac insufficiency occurs at any stage up to a month post infection or prolonged convalescence. Mortality may vary from 5 to 70%. *In utero* transmission of the virus in pregnant animals leads to abortion, foetal death or lamb/calf born small, underdeveloped with or without malformations. Various reported malformations include hydroencephaly, cerebral anomalies like cerebral dysplasia, excessive gingival, enlarged tongue, anomalous maxillae, dwarf-like build and rotations and contractures of the distal extremities.

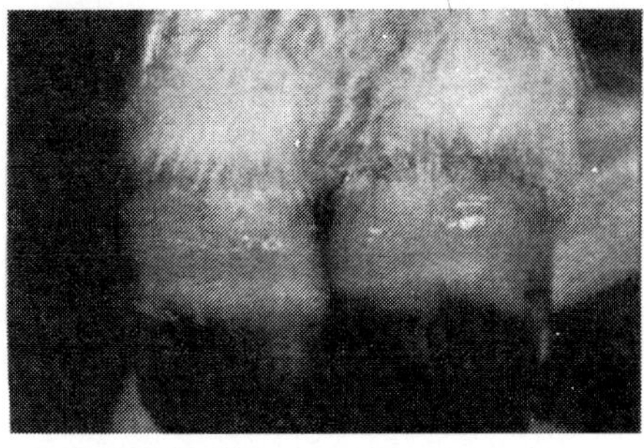

Fig. 26 Bluish congestive ring on coronet

Post Mortem Findings

Postmortem examination of animals revealed all the lesions attributed mainly to micro-vascular endothelial damage. Grossly, there is congestion, oedema, hemorrhage and ulcerations of digestive and respiratory mucosae (mouth, pharynx, larynx, oesophagus, fore stomach, intestine, tracheal mucosa). Principally, necrosis of cardiac muscle involving the papillary muscles of left ventricle and petechial to ecchymotic hemorrhages in the tunica media near the base of the pulmonary artery, are almost considered as pathognomonic findings. Sub-epicardial and sub-endothelial haemorrhages, particularly involving the left ventricle, are common. Hyperaemia and edema of lungs are also consistent findings.

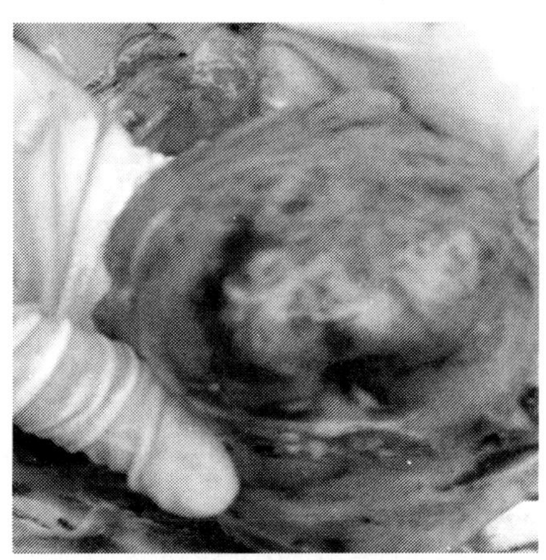

Fig. 27 Hemorrhage in Myocardium

Diagnosis

A. Clinical diagnosis : Diagnosis of BT in affected goat can be made on the bases of clinical signs and lesions.

B. Laboratory diagnosis: BTV infection is usually sub-clinical and laboratory confirmation, either by isolation and identification, serology or by nucleic acid detection.

Samples to be Collected

1. Blood for virus isolation from febrile animal in anticoagulants, such as heparin or EDTA and kept cool (4°C), but not frozen. Since the virus highly cell associated, it coexists with the antibody in the circulatory system, and so upon defreezing, RBCs will undergo lyses with release of virus particles which are then immediately neutralized by the antibodies.
2. Spleen and lymph nodes are preferred tissues for virus isolation or antigen genome detection by ELISA/PCR. They should be chopped finely and preserved in glycerol phosphate buffer with pH 7.6 and stored at 4°C, but should not be frozen.
3. Serum for serological tests. Preferably paired samples should be collected at acute stage and 21 days later.

Virus Isolation in Culture

1. Intravenous inoculation of embryonated chicken eggs: It is the second most sensitive technique for isolation of BTV. It is a prescribed test for international trade. Bluetongue virus may be detected in the inoculated eggs by antigen/ nucleic acid detection methods.
2. Tiss ue culture method: Cell culture BTV isolates are identified and confirmed by tests based on group-specific antibodies using fluorescent or enzyme conjugates. Isolates placed in the BT serogroup are typed by virus neutralization tests using serotype-specific antisera. Commonly, baby hamster kidney (BHK-21) and African green monkey kidney (Vero) cell cultures are used.

Serotyping of BTV by Virus Neutralization

Virus neutralization is the 'gold standard' test for serotype determination of virus isolate using serotype-specific neutralizing antibodies directed against VP2. Neutralization tests are type-specific for the currently recognized 24 BTV serotypes and can be used to serotype a virus isolate, or can be modified to determine the specificity of antibody in sera. The following techniques may be used:

a. Microtitre neutralisation: most widely used test.

b. Plaque reduction

c. Plaque inhibition

d. Fluorescence inhibition

Serogrouping of Viruses

This is based on the reactivity of group-specific BTV protein VP7. Immunocapture ELISA with BT serogroup-specific monoclonal antibody (MAb) is used to avoid cross-reactivity between members of the BT and EHD serogroups. It is rapid, sensitive, most widely used test.

Serological Tests

Competitive or blocking ELISA: It is rapid, sensitive, can be easily standardized and most widely used. It is prescribed tests for international trade.

Agar gel Immunodiffusion Test: This test to detect anti-BTV antibodies is simple to perform.

Nucleic Acid Detection

Polymerase chain reaction (PCR): It is highly sensitive and specific. It is the only priscribed test for the detection of disease of livestock that OIE recommended for international trade.

Prevention and Control

In disease free countries, controlled importation of susceptible animals, serological testing of animals during quarantine period in vector-free area and importation of semen from bulls shown to be serologically BTV negative should be practiced. In disease prevalent countries, restriction of animal movement, slaughtering and burning of infected stock and serological surveillance of area round infected zone constitute main strategies to contain the disease. Typing of the virus isolates prevalent in a particular state or area would facilitate in selection of virus types to be incorporated in the vaccine and thus the disease control.

Of several vaccine options available, namely live attenuated, killed or recombinant, only attenuated virus vaccines are in, current use in several countries. In India, its use is still awaited.

a) Live Attenuated Vaccines

It is the only current licensed vaccines for BT. In South Africa, for example, they have been used for over 40 years and are known to induce an effective and lasting immunity. However, they must be type-specific, and therefore multiple vaccinations may be needed. Most disadvantages of live attenuated vaccines are possible reversion to virulence and transmission through insect bites can occur apart from inducing teratogenesity and congenital malformations in the fetuses.

b) Recombinant Vaccines

There are several options for the development of recombinant BTV vaccines, including live virus delivery of BTV antigens and the virus-like particles generated in infected insect cells by recombinant baculoviruses expressing the four major BTV proteins, VP2, 3, 5 and 7. These vaccines need further-studies on duration of immunity and ability to suppress viraemia following challenge. None of the recombinant vaccines are currently licensed and are not under use.

c) Inactivated Vaccines

Although the efficacy of inactivated virus vaccines has been investigated in some laboratory studies, they do not appear to have been used in the field. However, in view of their being safe, attempts are being made to develop and commercialize such vaccine. Ensuring complete inactivation is often problematic but that can be ensured by testing the safety in proper manner.

Control of Vectors

Spraying nearby animals with long-acting insect repellent, housing animals at night, reducing insect breeding sites by spraying or improved drainage and spraying buildings with insecticide should be undertaken. Disinfection of premises is not appropriate since virus is insect transmitted.

6.4 Foot and Mouth Disease

Foot-and-mouth disease (FMD) is a highly contagious viral infection of goats. The disease is characterized by fever and vesicles with subsequent erosions in the mouth, muzzle, feet or teats.

Etiology

Foot-and-mouth disease is caused by FMDV, an Aphthovirus of the viral family Picornaviridae. The members of this family are small (25-30 nm), nonenveloped icosahedral viruses that contain single-stranded RNA (ribonucleic acid). When such a virus comes in contact with a host cell, it binds to a receptor site and triggers a folding-in of the cell membrane. Once the virus is inside the host cell, its protein coat dissolves. New viral RNA and components of the protein coat are then synthesized in large quantities and assembled to form new viruses. After assembly, the host cell lyses (bursts) and releases the new viruses.There are seven serotypes of FMDV: A, O, C, Asia 1, and Southern African Territories (SAT) 1, 2 and 3. Within these serotypes, over 60 subtypes have been described.

Epidemiology

Foot-and-mouth disease is widely distributed throughout the world. However, most European countries have been recognized free from the disease. The FMD virus gains access into a free area by direct or indirect contact with infected animals, spread of aerosol from infected animals and contact with contaminated objects (hands, footwear, clothing). The primary mode of virus spread is via respiratory aerosols. Other important means of spread is by direct and indirect contact. Most ruminant species can remain as carriers of the virus in their pharyngeal tissues for a long period. When goats get infected with FMDV, the disease usually goes undetected for a considerable period of time because signs and lesions are usually very mild. In susceptible lambs and kids, morbidity can be 100 per cent and mortality can be as high as 95 per cent due to myocardial necrosis. Research on epidemiological features of FMD in goats is largely lacking in India except a few. A continuous monitoring on epidemiology of this disease is desired, in current prevailing agro-climatic conditions and husbandry practices in the country to formulate preventive and bio-security strategies. The disease commonly occur in sub-clinical nature but in one Indian outbreak of caprine FMD due to virus type A_{22}, morbidity was 82.5% and mortality 8.2%. Virus O type is mainly causing FMD in Indian livestock including goats since last 6 to 7 years in India, which has been categorized as pandemic strain (Pan Asia). Moreover, the O strain produces the lowest percentage of animals with lesions in comparison to other strains. This disease usually appears in vaccinated herds as in acute/mild or sub-clinical infections depending upon the immune system of herd, level of challenge efficacy of vaccine used. Significant achievements in molecular epidemiology and antigenic analysis have been made. A large number of FMD outbreaks due to type 'O' followed by type A have been recorded in most of the states in endemic form especially during winter and early summer (Oct-April). In the North Eastern states, type A caused most of the outbreaks followed by type O. Type Asia- 1 accounted for very small number of outbreaks. Reverse transcription, polymerase chain reaction and cycle sequencing are routinely used for obtaining the nucleotide sequence of the FMDV field isolates. In type O, the new strain, which overtook Pan- Asia strain re-established itself as the predominant strain in FMDV type O outbreaks in India. In Asia-1, the previously identified novel subgroup (>10% nucleotide divergence) within the widely circulating lineage was responsible for disease outbreaks in both cattle and buffaloes and was present in six states in the country. Antigenic variation occurs both in the presence and in absence of antibodies. These results have strong implications on efficacy of the vaccine, as the antigenic variants generated in partially immune animals can evade hosts immune response.

Soon after infection, the single stranded positive RNA that constitutes the viral genome is efficiently translated using a cap-independent mechanism driven by the internal ribosome entry site element (IRES). This process occurs concomitantly with the inhibition of cellular protein synthesis, caused by the expression of viral proteases. Processing of the viral polyprotein is achieved by viral encoded proteases, giving rise to the different mature viral proteins. Viral RNA as well as viral proteins interacts with different components of the host cell, acting as key determinants of viral pathogenesis. In depth knowledge of the molecular basis of the viral cycle is needed to control viral pathogenesis and disease spreading.

Clinical Signs

Clinical signs usually develop in 3 to 5 days. The average incubation period of the Foot and Mouth virus varies but is generally around 3-8 days. The course of an FMD infection lasts from 2 to 3 weeks. Clinical signs, if they occur, tend to be very mild and may include dullness, fever and small vesicles or erosions on the dental pad, lips, gums, and tongue. Mild lameness may be the only sign. In lame animals, there may be vesicles or erosions on the coronary band or in the interdigital space. Infected animals may abort. The diagnostic lesions are single or multiple vesicles of about 2 mm in the interdigital space. Foot and mouth disease in goats is inapparent with mild clinical signs and lesions in vaccinated animals. Adult animals may suffer weight loss from which they do not recover for several months as well as swelling in the testicles of mature males, and in goats, milk production can decline significantly. Though most animals eventually recover from FMD, the disease can lead to myocarditis (inflammation of the heart muscle) and death, especially in newborn animals. Some infected animals remain asymptomatic, but they nonetheless carry FMD and can transmit it to others infection with foot-and-mouth disease tends to occur locally - that is, the virus is passed on to susceptible animals through direct contact with infected animals or with contaminated pens or vehicles used to transport livestock. The clothes and skin of animal handlers such as farmers, standing water, and uncooked food scraps and feed supplements containing infected animal products can harbor the virus as well. Goats can also catch FMD from the semen of infected males. In presence of FMD virus, climatic stress may act as predisposing factor.

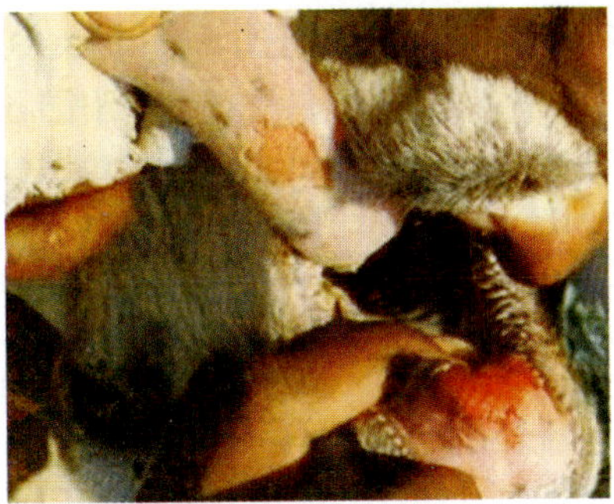

Fig. 28 Tongue erosion in goat

Diagnosis

The major methods available for diagnosis and determination of the strain of virus involved are tissue culture, the complement fixation test and experimental transmission in test animals.A rapid ELISA test is available for examination of fresh epithelial tissues.

Post Mortem Findings

Small vesicles develop at point where virus gain entrance and mostly found on the buccal mucosa and skin of the feet. Gastrointestinal lesions may be found at necropsy particularly of the rumen. In rare cases lesions of the perenium, valva or scrotum are seen. Tiger heart necrosis (grey, white or yellowish myocardial lesion) may be seen. On section the heart muscle appears streaked with patches of yellow tissue interspersed with apparently normal myocardium. Tongue epithelium generally gets sloughed off.

Economic and Ethical Issues

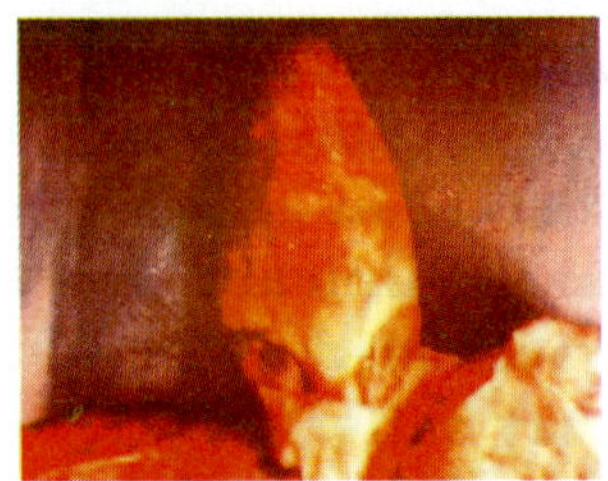

Fig. 29 Tiger heart necrosis in kids

Outbreaks of foot-and-mouth disease have resulted in the slaughter of millions of animals, despite this being a frequently non-fatal disease for adult animals (2-5% mortality), though young animals can have a high mortality. The Taiwan outbreak that only affected pigs also showed a high mortality for adults. The destruction of animals is primarily to halt further spread, as growth and milk production may be permanently affected, even in animals that have recovered. Due to international efforts to eradicate the disease, infection would also lead to trade bans being imposed on affected countries. Critics of current policies to cull infected herds argue that the financial imperative needs to be balanced against the killing of many animals, especially when a significant proportion of infected animals, most notably those producing milk, would recover from infection and live normal lives, albeit with reduced milk production. On the ethical side, one must also consider that FMD is a painful disease for the affected animals. The vesicles/blisters are painful in themselves, and restrict both eating and movement. Through ruptured blisters, the animal is at risk from secondary bacterial infections and, in some cases, permanent invalidisation.

Treatment

Animal should be provided soft, nonirritating food. Mouth and foot should be flushed with mild disinfectant or astringent. A good symptomatic treatment is to be administered. If required, supportive treatment such as glucose solution intravenously may be given in acute cases.

Prevention and Control

Seven main types of Foot and Mouth Virus are believed to exist. Like other viruses, the FMD virus continually evolves and mutates, thus one of the difficulties in vaccinating against FMD is the huge variation between and even within serotypes. There is no cross-protection between serotypes (meaning that a vaccine for one serotype will not protect against any others) and in addition, two strains within a given serotype may have nucleotide sequences that differ by as much as 30% for a given gene. This means that FMD vaccines must be highly specific to the strain involved. Vaccination only provides temporary immunity that lasts from months to years. Sera samples (paired/ post vaccination/post infection/random sera) that were subjected to determination of anti-FMDV antibody titers using liquid phase blocking ELISA, have revealed that compared to previous years more animals had protective titre following vaccination. To differentiate the vaccinated animals from the naturally infected FMD carrier animals, a non-structural protein 3AB of FMD virus was identified, cloned and expressed for production of viral protein, which is having a good potential for differential diagnosis as a serological reagent in an ELISA system.

The essential features of a control and prevention program include restriction of movement of animals, animal products and disinfection of the infected area. Multi component vaccine should invariably be used to control the disease. Formalin inactivated FMDV, as well as cell culture propagated virus has been used to produce FMD vaccine. At present the classical FMD vaccines are prepared using binary-ethyleneimine (BEI) inactivated virus and aluminum hydroxide-saponin or oil as an adjuvant. Double emulsion oil vaccines have been shown to produce an immunity of longer duration than aluminum hydroxide-saponin vaccine. Protection induced by a good aluminum hydroxide vaccine decreases rapidly in 4-6 months. A double emulsion oil vaccine can protect for up to 1 year. Vaccinated animals that are not completely protected can be a source of infection. There are several reasons cited for restricting export from countries using FMD vaccines. The most important is probably that routine blood tests, relying on antibodies, cannot distinguish between an infected and a vaccinated animal. This would severely hamper screening of animals used in export products, risking a spread of FMD to importing countries. A widespread preventive vaccination would also conceal the existence of the virus in a country. From there, it could potentially spread to countries without vaccine programs. Lastly, an animal that is infected shortly after being vaccinated can harbour and spread FMD without showing symptoms itself, hindering containment and culling of sick animals as a remedy.

Currently, the World Organisation for Animal Health recognizes countries to be in one of three disease states: FMD present with or without vaccination, FMD-free with vaccination and FMD-free without vaccination. Countries that are designated FMD-free without vaccination have the greatest access to export markets, and therefore many developed nations, including Canada, the United States, and the UK, work hard to maintain their current FMD-free without vaccination status.

Public Health Importance

Humans can be infected with foot-and-mouth disease through contact with infected animals, but this is extremely rare. Some cases were caused by laboratory accidents. Because the virus that causes FMD is sensitive to stomach acid, it cannot spread to humans via consumption of infected meat, except in the mouth before the meat is swallowed. In the UK, the last confirmed human case occurred in 1966, and only a few other cases have been recorded in countries of continental Europe, Africa, and South America. Symptoms of FMD in humans include malaise, fever, vomiting, red ulcerative lesions (surface-eroding damaged spots) of the oral tissues, and sometimes vesicular lesions (small blisters) of the skin. Because FMD rarely infects humans but spreads rapidly among animals, it is a much greater threat to the livestock industry than to human health. Farmers around the world can lose huge amounts of money during a foot-and-mouth epizootic, when large numbers of animals are destroyed and revenues from milk and meat production go down.

6.5 Contagious Ecthyma

Contagious ecthyma is a viral disease of goats which is highly contagious and zoonotic in nature. It is characterized pustular, scabby and proliferative lesions on the muzzle, lips face, ears, teats or vulva.

Etiology

Contagious ecthyma is caused by an epitheliotropic, Parapox virus that enters the goats through skin abrasions.

Epidemiology

The distribution of the disease is worldwide.The virus replicates in proliferating keratinocytes in the damaged epidermis. Scabs that fall to the ground during resolution of lesions have been incriminated as possible agents of transmission to other animals. Carrier goats may also be an important source of contagion.

Clinical Signs

The lesions (papules, vesicles, pustules and scabs) are found principally on the lips, gums, muzzle and nostrils of kids from where the virus may spread to the teats of the does. In goat herds, adult animals may develop lesions on the mouth. The disease appears in the form of papules which progress rapidly to vesicles, pustules and scabs. Crusty proliferative lesions typically form on the lips but can also affect face, ears, coronary band, scrotum, teats or vulva. Large masses of granulation tissue develop under the scabs. Bronchopneumonia and gastroenteritis have been observed in goat kids. Lesions usually regress in 3-4 weeks. In malignant form lesions are generalized in nature. Recently severe ecthyma lesions extending into the buccal cavity have been reported in the flocks concurrently infected with PPR virus. Morbidity rate is very high 50-60%, but mortality rate may be 5% in goat kids.

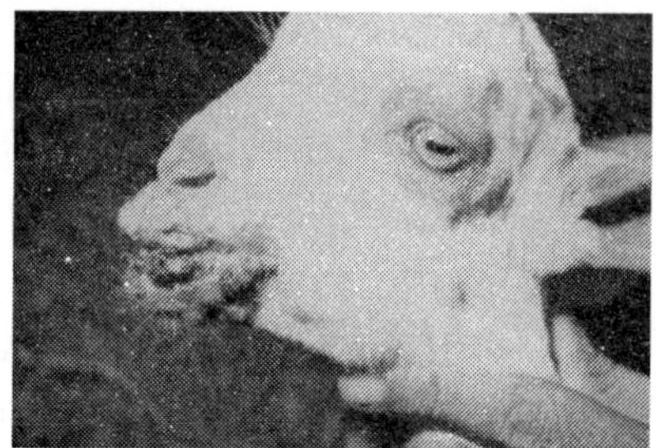

Fig. 30 Scabby lesions on mouth (Kid)

Diagnosis

Field diagnosis is primarily based on clinical signs. Laboratory Diagnosis is by the demonstration of antigen in scabs by immunologic techniques. Electron microscopy and skin biopsies are also used for confirmation of infection.

Treatment

The lesions should be treated with antiseptic drugs. In generalized or systemic infection antibiotic should be given to prevent secondary bacterial complication. Outbreak in kids was effectively managed with potassium permanganate dressing of lesions and administration of antibacterial drug. The lesions should be treated with mild antiseptic drugs. In generalized infections, antibiotics should be administered to prevent secondary bacterial complications.

Prevention and Control

In order to check the spread of infection, infected goats should be isolated and the infected area should be disinfected. Lack of adequate prophylactic measures, is leading to frequent disease outbreaks in sheep and goat farms all over the country. Extreme climatic conditions also aid in the precipitation of the disease. In order to check the spread of infection, infected goats should be kept in isolation and the infected area should be disinfected. Young kids usually die due to their inability to suckle. Therefore, the affected animals should be supplemented with succulent fodder.

Commercially available vaccines are basically unattenuated live virus preparations of ground up scabs or are tissue culture strains, although level of protection afforded appears to vary with strain. The occurrence of colostral immunity in vaccinated animals is disputed.

6.6 Contagious Pustular Dermatitis

Infection in contagious pustular dermatitis occurs classically on the lips on lower legs and oral mucosa. Lesions in other sites of the body are usually trivial and of no consequence.

Etiology

The disease is caused by a parapoxvirus and is mainly seen in lambs and weaners.

Epidemiology

The lesions commence with one or more small focal areas of inflammation which develop into vesicles and pustules then rupture, forming scabs at sites other than the oral mucosa. The disease normally runs a course of 2 to 3 weeks before resolving completely. Infection with the virus of CPD requires a break in the epithelium. Trauma to the skin can be caused by rough feed, thistles or grass seeds. On the legs, lesions occur above the coronary band, behind the pastern or on the interdigital skin. Infections are often preceded by persistent wetting that comes from walking through very wet pasture or in flooded paddocks. The resultant maceration of the skin provides the opportunity for the virus to enter the skin. Infection of the site of application of ear tags can occur, leading to an enlargement of the hole and a loss of the eartag. CPD is seen most commonly in summer months. The virus does not persist on sheep but can survive for years in the environment. Scabs are a rich source of the virus.

Clinical Signs

In lambs, lip lesions can infect ewes teats; the ewes then refuse to allow the lambs to suck. Mastitis and necrosis of the quarter is an occasional consequence. CPD can also occur on the scrotum. The virus has caused severe systemic infection with significant mortality. Occasionally severe forms, but not systemic forms, are seen in Australia.

Economic Consequences of CPD

The results of infection are generally trivial but at times can be serious. The presence of CPD lesions on sheep at particular times may lead to management complications such as delays in shearing or sale of rams. Limb lesions can confuse a diagnosis of benign footrot as well as causing some lameness. Weaners with lip and mouth lesions will graze less and lose weight which, at particular times of the year, may have serious consequences on their survival rates and/or their need for later supplementation. The infection has been transmitted rapidly between sheep in 'Sharlea' sheep sheds and between sheep on ships en route to the Middle-East. The resultant inappetence can lead to increased mortalities but, more importantly, the presence of lesions on arrival has led to the refusal of importing countries to accept delivery. Humans are readily infected but the disease is rarely important.

Diagnosis

The diagnosis is generally made on clinical grounds but can be confirmed by the demonstration of virus in lesion material by electron microscopy. Limb lesions can be mis-diagnosed as strawberry footrot. Severe cases must be differentiated from bluetongue and sheep pox, both of which have high mortalities, unlike CPD. Dermatophilosis occurs in woolled skin only; photosensitisation is diffuse through the face and exposed skin. Sheep and goats rarely contract vesicular stomatitis when exposed.

Treatment

There is no effective treatment. The disease resolves spontaneously after 2 to 3 weeks. Secondary infection of the limbs of valuable animals may warrant antibiotic treatment.

Prevention and Control

A commercial vaccine prepared from scab material is available. It is applied with a small, double-pointed fork which holds a small volume of vaccine suspension and simultaneously scratches the skin and dispenses the fluid. The usual site for vaccination is the inside of the thigh and the operation can be done conveniently at lamb marking. It is advisable to examine the site of inoculation on a number of sheep 3 days after treatment to ensure a successful take. Pustules should be evident along the site. Immunity after one vaccination or natural infection is strong and remains effective for 2 to 3 years. The vaccine is prepared from fully virulent virus so is capable of causing the disease. Consequently, any animals who miss vaccination are likely to develop the disease naturally from the contamination of the environment caused by infected vaccinates.

6.7 Rabies

Rabies is a disease (caused by the rabies virus) primarily of animals, including both wild and domestic animals and human beings. Although people usually associate rabies with dogs, among domesticated animals but Rabies today is more likely to be found in cats. Cats, dogs and cattle account for nearly 90 percent in domestic animals, with horses, mules, sheep, goats and ferrets making up the remaining cases. Among wild animals, the disease is most often reported in skunks and raccoons. Other wild species in this country in which rabies is commonly found include bats, foxes, and rodents.

Epidemiology

The rabies virus, present in the saliva of an infected animal, is usually spread by a bite or scratch that punctures the victim's skin. The virus has a strong affinity for cells of the nervous system. It enters nerve cells at the site of the wound, travels to the brain, and then follows other nerve pathways to muscles and organs that are especially affected by rabies.

There are at least two other ways in which humans have been known to have contracted rabies, both extremely rare. Two people were exposed by breathing the air in caves inhabited by rabid bats. And six people contracted rabies following implants of corneas from donors who had undiagnosed rabies. The virus concentrates in the salivary glands, which explains why it is usually spread by bites. It also invades and damages the muscles involved in drinking and swallowing. Most human victims and apparently lower animals as well, suffer excruciating pain on swallowing liquids. Though they suffer from thirst, animal and human rabies victims can be terrified by the sight of water, hence another name for the disease, hydrophobia.

Clinical Signs

Symptoms usually develop between 20 and 60 days after exposure. Rabid animals may become aggressive, combative, and highly sensitive to touch and other kinds of stimulation. And they can be vicious. This is the "furious" form of rabies, the kind traditionally associated with mad dogs. There is also a "dumb" form of the disease in which the animal is lethargic, weak in one or more limbs, and unable to raise its head or make sounds because its throat and neck muscles are paralyzed. In both kinds of animal rabies, death occurs a few days after symptoms appear, usually from respiratory failure. In humans, the course is similar. After a symptom-free incubation period that ranges from 10 days to a year or longer (the average is 30 to 50 days), the patient complains of malaise, loss of appetite, fatigue, headache, and fever. Over half of all patients have pain (sometimes itching) or numbness at the site of exposure. They may complain of insomnia or depression. Two to 10 days later, signs of nervous system damage appear, hyperactivity and hypersensitivity, disorientation, hallucinations, seizures, and paralysis. Death may be sudden, due to cardiac or respiratory arrest, or follow a period of coma that can last for months with the aid of life-support measures.

Diagnosis

The diagnosis of rabies is one of the most difficult. Since in most cases there is a probability of human exposure, failure to recognize the disease may place human life in jeopardy. If the animal is rabid, it will die and the diagnosis can then be confirmed by laboratory examination. Polioencephalomalacia is accompanied by obvious opisthotonus and convulsions in sheep and goat but salivation, bawling, anesthesia and tenesmus do not occur. Listeriosis in sheep is usually manifested by localizing signs of circling and facial nerve paralysis.

Prevention and Control

Rabies control is possible not by cure but by prevention due to the advent of scientific medicine. Unlike other immunizations, the rabies vaccine is administered after exposure to the virus. This unusual technique is successful because the rabies virus takes a comparatively long time to induce disease, a minimum of 10 days, and in rare cases, up to a year. The length of the incubation period apparently depends on both the location of the wound - the farther from the brain, the longer the incubation - and the dose of virus received. No matter where the wound is there, the first and most valuable preventive measure is thorough cleaning of the site with soap and.water, and immediate medical attention. If rabies vaccine treatment is called for, it should be started as soon as possible after exposure. Counting the first day of vaccine treatment as day 0, injections are administered on days 0, 3, 7, 14, and 28.In addition to vaccine, patients who have not previously been vaccinated for rabies also receive an injection of rabies immune globulin (RIG) on the day they get the first vaccine (day 0).RIG is prepared from the blood of persons who have been immunized against rabies and contains antibodies to the rabies virus. This "passive" immunity helps protect patients during the period before the rabies vaccine causes their own immune system to counter the virus (active immunity).

□□□

7

Mycoplasmal and Rickttsial Diseases

7.1 Contagious Caprine Pleuropneumonia

CCPP is a significant disease of goats and is characterized primarily by its contagious nature. The disease causes interstitial, fibrinous pleuropneumonia, interlobular oedema and hepatization of the lung causing high mortality rates of up to 80%.

Etiology

The causative agent of contagious caprine pleuropneumonia (CCPP) is *Mycoplasma capricolum* subsp. *capripneumoniae (Mccp)*, which was previously known by the strain name of its type species, F38. It is a member of the *Mycoplasma mycoides* cluster which includes *M. mycoides* subsp. *mycoides SC (MmmSC), M. mycoides* subsp. *mycoides LC (MmmLC), M. mycoides* subsp. *capri (Mmc), M. capricolum* subsp. *capricolum (Mcca)* and *Mycoplasma species* bovine group 7, an uncharacterized bovine isolate, which causes other diseases of ruminants. Definite diagnosis is made by culture of the causative agent from lung samples or pleuritic fluid taken at postmortem. Liquid and solid mycoplasma media are inoculated and filtered subcultures from liquid medium may be required if there is evidence of bacterial contamination. Isolates may be identified by biochemical, immunological and molecular tests. Serological tests for the detection of specific antibodies have relied on the complement fixation test which is the test prescribed by the Office Internationale des Epizootics (OIE) for international trade. A latex agglutination test is also available, and is used routinely in some parts of Africa.

Epidemiology

CCPP is an infectious disease which affects only goats, and was first described in the late 19th century. Before the isolation and identification of *Mycoplasma* strain F38 and the subsequent demonstration of its causal relationship with CCPP *M. mycoides* subsp. *capri* was considered to be the aetiological agent of CCPP. So far *M. capricolum* subsp. *capripneumoniae* is the only mycoplasma that fulfils Kochs postulates for CCPP and is believed to be the sole cause of CCPP. *Mycoplasma strain* F38 has recently been reclassified and now all F38-like mycoplasmas are known as *Mycoplasma capricolum* subsp. *capripneumoniae*.

The most important distinguishing features of CCPP, with respect to the other goat respiratory mycoplasmoses, were defined by Hutcheon and are quoted as follows: (1) the disease is readily contagious to susceptible goats; (2) sheep and cattle are not affected; (3) local oedematous reactions do not occur in goats when infective inoculum is given subcutaneously. In natural infections, the organisms are acquired by susceptible goats by inhalation of contaminated droplets from infected goats. The environment as a whole plays an important role in the appearance, evolution and severity of CCPP. Due to the high sensitivity of mycoplasmas to the external environment, close contact is essential between infected and native animals for transmission to take place, and overcrowding and confinement favours close contact and circulation of mycoplasmas. Stress factors due to malnutrition and movement over long distances can predispose the animal to disease.

CCPP has been reported to affect only goats and it does not cause disease in sheep, neither spontaneously nor experimentally. However, there are some reports describing the isolation of *M. capricolum* subsp. *capripneumoniae* from healthy sheep in Kenya that have been in contact with goat herds affected by CCPP and from sick sheep mixed with goats in Uganda suffering from the disease. The isolation of *M. capricolum* subsp. *capripneumoniae* from cattle with mastitis has also been reported and these reports confound the perceived host specificity of *M. capricolum* subsp. *Capripneumoniae.*

Mycoplasmoses is one of the major, causes of goat mortality and is one of the biggest hurdles in development of Goat industry in India. In Africa where extensive and traditional husbandry is practiced, pathogens spread when animals meet at watering points and grazing areas. Breed and sex appear not to affect the epidemiology of CCPP, but age is an important factor. Though all age groups are susceptible, mortality is higher among young animals than adults. Infective *M. capricolum* subsp. *capripneumoniae* may persist in chronic, latent carriers, such as goats or sheep which have recovered from infection without becoming bacteriologically sterile, and are considered to be responsible for the perpetuation of the disease in a herd. This aspect of the epidemiology was described as early as 1881 in the case of CCPP in South Africa. Viruses are important factors that predispose lung tissue to invasion by mycoplasmas. In Africa the virus, peste des petits ruminants (PPR) and capripox viruses are important.

Although a precise description of the distribution of CCPP is not available, the clinical disease has been reported in 30 countries mainly in Africa and Asia. The only African countries where *M. capricolum* subsp. *capripneumoniae* has been isolated are Chad, Eritrea, Ethiopia, Kenya, Niger, Sudan, Tanzania, Tunisia and Uganda. In Asia and the Mediterranean, isolations have been reported in Oman, the United Arab Emirates, and Yemen. In Mali, goats have been suspected of infection based on serological evidence.

Economic Importance and Impact

Goats are important commodities to a large segment of the world population as a source of meat, milk, and skin. CCPP is a disease of major economic importance in Africa and Asia, posing a major constraint to goat production. The direct losses of the disease result from its high mortality, reduced milk and meat yield, cost of treatment, control, disease diagnosis and surveillance. In addition to this, there are indirect losses due to the imposition of trade restrictions.

Pathogenesis

The gross pathological lesions are localized exclusively to lung and pleura and are often unilateral. Affected lungs can be totally hepatized, and have a port wine colour. A lung section shows a fine granular texture with various colours, but usually without any thickening of the interlobular septa. There is often an abundant pleural exudates and conspicuous pleuritis. The pleural exudates can solidify and form a gelatinous covering sometimes over the whole lung. In acute cases, the pleural cavity contains an excess of straw-coloured fluid with fibrin flocculations. In chronic cases there is a black discolouration of the lung tissue and sequestration of the necrotic lung areas. Adhesions between the lung and the pleura are very common and often very thick. Histological examination of the lung tissues may show acute serofibrinous to chronic fibrino-necrotic pleuropneumonia with infiltrates of serofibrinous fluid and inflammatory cells, mainly neutrophils, in the alveoli, bronchioles, interstitial septae and subpleural connective tissue. Intralobular oedema is more prominent but interlobular oedema has also been reported. Peribronchial and perbronchiolar lymphoid hyperplasia with mononuclear cell infiltration is also present.

Diagnosis

In the field, diagnosis of mycoplasma pneumonia can not be established on clinical signs or on postmortem examinations alone. In outbreaks of classical acute CCPP, the high mortality and typical early thoracic lesions in goats are highly indicative of *M. capricolum* subsp. *capripneumoniae* infection, but all cases of caprine mycoplasmosis need additional laboratory tests to establish a presumptive diagnosis. It may be difficult to distinguish CCPP from an infection by *M. mycoides* subsp. *mycoides LC or M. mycoides* subsp. *mycoides SC*, which have a pulmonary location. In the case of *M. mycoides* subsp. *mycoides LC* infection, thickening of the interlobular septa may be evident. These lesions are similar to those observed in the case of CBPP. Sometimes the thickening is absent or inconspicuous and laboratory confirmation is needed. Recently, sequestra in the lungs of goats infected with *M. mycoides* subsp. *mycoides SC* have been described.

Isolation

Definite diagnosis is made by the isolation of *M. capricolum* subsp. *capripneumoniae* from clinical samples, usually lung tissue and may be a long and difficult process. The success of isolation depends primarily on the attention that is given to sample collection. The main difficulties in isolating *M. capricolum* subsp. *capripneumoniae* is that it grows very poorly *in vitro* and samples are often contaminated by other mycoplasmas which are generally faster growing and overgrow *M. capricolum* subsp. *capripneumoniae*. In addition to this the frequent use of antibiotic therapy has impaired the cultivation of these mycoplasmas from clinical material. An immunobinding assay that detects *M. capricolum* subsp. *capripneumoniae* in pleural fluids that overcomes some of these problems has been described. Liquid medium and a solid agar medium which allows the presumptive identification of *M. capricolum* subsp. *capripneumoniae* by the production of coloured colonies are available commercially.

Biochemical Tests

Only a limited number of biochemical tests perform a useful function as a preliminary screening system and are based on specific enzyme activities or nutritional capabilities. For instance, digitonin sensitivity distinguishes mycoplasmas from acholeplasmas, and serum digestion differentiates members of the *M. mycoides* cluster from all other small ruminant mycoplasmas. Also, phosphatase production separates *M.capricolum* subsp. *capricolum* from other members of the cluster. Substantial metabolic differences between *M.capricolum* subsp. *capricolum* and *M. capricolum* subsp. *capripneumoniae* exist, but differences in glucose metabolism were described between strains of *M. capricolum* subsp. *capripneumoniae*. These tests can not differentiate *M. capricolum* subsp. *capripneumoniae* from all members of *M. mycoides* cluster. The interspecies variation in some biochemical reactions is often considerable, rendering their application valueless.

Growth Inhibition (GI) Test

The growth inhibition (GI) test is the simplest and most specific, but the least sensitive of the tests available. It depends on the direct inhibition of mycoplasma growth on solid media by specific hyperimmune serum, and detects primary surface antigens. The GI test is particularly useful in identifying *M. capricolum* subsp. *capripneumoniae* because they appear to be serologically homogeneous, and antiserum to the type strain produces wide inhibition zones free of break through colonies against field isolates from diverse sources. *M. capricolum* subsp. *capripneumoniae* cross-reacts with *M. equi genitalium and M. primatum* in the GI test, but since these cross-reactive species do not occur in goats, they present no difficulties when identifying field isolates. However, a small proportion of *M. capricolum* subsp. *capripneumoniae* isolates also cross-react in the GI test with antiserum to *M. capricolum* subsp. *capricolum* which may confuse the identification of field isolates. A monoclonal antibody has been produced which specifically inhibits the growth of *M. capricolum* subsp. *capripneumoniae* but not of other members of the *M. mycoides* cluster. It was later demonstrated that this monoclonal antibody was not specific. Cross-reactions with some strains of Bg7 were observed with the GI test and with a strain of *M. capricolum* subsp. *capricolum* in the immunofluorescence test.

Fluorescent Antibody Test

The direct and indirect fluorescent antibody tests are one of the most effective, simple and rapid serological methods of identification for most mycoplasma. Several forms have been described, the most commonly used one is the indirect fluorescent antibody (IFA) test which is applied to unfixed colonies on agar.

Serological Test

The complement fixation test (CFT) was used for the detection of CCPP infection and it was found to be more specific, though less sensitive, than the indirect haemagglutination test. The latex agglutination test uses latex beads sensitized with the polysaccharide produced by *M. capricolum* subsp. *capripneumoniae* in culture supernatant in a slide agglutination test. The use of the more defined antigen such as the polysaccharide provides greater sensitivity without cross-reactivity with sera against the other three principal caprine mycoplasmas. The indirect haemagglutination test (IHA) has been used for the diagnosis of CCPP. The specificity of IHA test for the *M. mycoides* cluster has been evaluated and results were found to show cross-reactivity between these organisms. An indirect immunosorbent assay (ELISA) was developed to screen goat sera at a single dilution of antibody to *M. capricolum* subsp. *capripneumoniae*. Some problems due to cross-reactions from other members of the *M. mycoides* cluster were encountered, but in spite of these, ELISA was more sensitive than CFT in detecting antibodies in serum. More recently, a competition ELISA (c-ELISA) was developed which permitted the specific detection of antibodies in sera from animals affected by CCPP. Analysis of field sera showed that seroconversion did not occur in all animals, whatever test was used. The percentage positive animals in affected herds varied between 30 and 60% with this test. The test was therefore unsuitable as an individual screening test.

Molecular Diagnosis

Until recently, isolation was the only way to confirm the presence of CCPP. A DNA probe which differentiates *M. capricolum* subsp. *capripneumoniae* from other members of the *M. mycoides* cluster was developed. Diagnostic systems based on PCR have been developed for the rapid detection, identification and differentiation of members of the *M. mycoides* cluster and the specific identification of *M. capricolum* subsp. *capripneumoniae*. The sequence of the gene for 16S ribosomal RNA has also been used to develop a PCR-based test where the final identification of *M. capricolum* subsp. *capripneumoniae* is made dependant on the pattern of the products after digestion of the PCR product with the restriction enzyme Pst1.

Treatment

The macrolides (erythromycin, spiramycin, and tylosin), tetracyclines, quinolones and chloramphenicol are active against *M. capricolum* subsp. *capripneumoniae*. Tylosin, tetracyclin, tiamulin or streptomycin are recommended but their success depends on early intervention and treatment.

Prevention and Control

An experimental vaccine inactivated with saponin that protects goats for approximately a year has been produced in Kenya. In the field of immunoprophylaxis/vaccine development, inspite of lot of experimental work done, no vaccine has been developed till date for CCPP in India which can give 100% and long lasting protection in the host. An attenuated as well as killed Aluminium Hydroxide gel vaccine was developed. The vaccine gave immunity up to 7 month duration and 75% protection under field condition. Fractionated *M. capri* vaccine and sonicated suspension with IFA to be 100% protective up to 5 month duration. Several immunogenicity trial of *M.capri* organism with various adjuvants was conducted and found saponified vaccine to be most affective as compared to Potash Alum, CFA & Vit. E. Killed *M.capri* strain vaccine were single shot- freeze-dried and were found to be 100% protective up to 6 months duration. Another vaccine prepared by using saponified *M.capri* also found to be protective up to 8 month duration. The disease Caprine pleuropneumonia (CCPP) in India still need a suitable vaccine which can give 100% protection in goats for longer duration/life long immunity. Movement restrictions and slaughtering infected animals is recommended for countries that are newly infected.

7.2 Mycoplasma Infections

Mycoplasma infections account for serious morbidity and mortality in goats throughout the world. Several distinct disease entities are recognized, notably contagious caprine pleuropneumonia, contagious agalactia, and infectious keratoconjunctivitis. However, mycoplasma infections are usually septicemic, and polyarthritis frequently occurs in these and other *Mycoplasma spp.* infections of goats.

Etiology

There are three genera of organisms commonly referred to as mycoplasmas that infect goats; *Mycoplasma, Ureaplasma,* and *Acholeplasma.* There are two forms of the species, *M. mycoides* subspecies *mycoides* that share similar immunologic and serologic characteristics but differ on the basis of colony morphology and specific biochemical tests. The small colony type is the cause of contagious bovine pleuropneumonia and has been isolated from goats. The large colony type is principally a pathogen of goats and has a much wider geographic distribution. *M. bovigenitalium,* has been isolated from the genital tracts of does with repeat breeding problems in India.

Mycoplasmas are the most primitive prokaryotes capable of *in vitro* growth on culture medium and have numerous special growth requirements including serum factors and yeast extracts. They do not have a rigid wall. Many species produce characteristic spreading colonies with a "fried-egg" appearance. In general, these organisms are fragile away from the host and are rapidly inactivated by sunlight, dehydration, and heat but are more resistant to cold. Effective disinfectants include formalin, phenols, cresols, peracetic acid and iodophors.

Epidemiology

Arthritis caused by Mycoplasma in goats has been associated with numerous species. Major cause of importance include *M. agalactiae, M. capricolum, M. mycoides* subspecies *capri, M. mycoides* subspecies *mycoides* (large colony type), and *M. putrefaciens.* Minor causes include *M. arginini, M. conjunctivae* and *M.mycoides* subspecies *mycoides* (small colony type). *M. agalactiae* is the cause of contagious agalactia in goats and sheep. Currently the disease is reported from many countries in Africa, Europe, and the Middle East and is enzootic in much of the Mediterranean region. Goats are considered more susceptible than sheep. The highest incidence of disease is recorded in late spring and summer when does have kidded and are lactating. The morbidity approach 100% and mortality ranges from 10% to 30 %. The disease is characterized by the concurrent appearance of mastitis, arthritis, and keratoconjunctivitis. Respiratory involvement is minimal or absent. Transmission is by direct contact with infected animals or their discharges including milk, urine, and lacrimal and nasal secretions. Introduction into the udder during milking is important in the spread of infection. Chronically ill or convalescent goats can sporadically shed the organism.

M. capricolum is considered to be highly pathogenic yet disease due to this organism is reported only sporadically from diverse geographic locations, including USA, France, Spain, Australia, Egypt, India and Morocco. Morbidity and mortality in these reports are quite variable. Arthritis is the predominant sign in *M. capricolum* infection, but neonatal septicemia, pneumonia, keratoconjunctivitis, mastitis, and agalactia can also be seen. Transmission of *M. capricolum* occurs by direct contact with infected goats, inhalation of the organism from feces, urine or respiratory discharges and especially by ingestion of infected milk by sucking kids. *M. mycoides* subspecies *capri* until recently was considered the primary cause of contagious caprine pleuropneumonia (CCPP). Though the organism does indeed produce pleuropneumonia, the causative agent of the classically described disease CCPP is now considered to be the *Mycoplasma* F38 strain. While the F38 strain infects only goats, M. *mycoides* subspecies *capri* can also infect sheep and cattle experimentally. Pleuropneumonia is the predominant clinical feature of *M. mycoides* subspecies *capri* infection, but arthritis is also reported. Morbidity and mortality rates can approach 100%. Goats are the natural reservoir of infection. Transmission is primarily through direct contact with sick goats via inhalation of expiratory aerosols from sick animals; however, organisms may also be present in urine, milk, and genital secretions and contact with inanimate vectors may lead to infection. Recovered and treated animals can act as sub-clinical shedders. Stress predisposes to disease and though the condition occurs year round, morbidity is highest in cold, wet winter months.

M. mycoides subspecies *mycoides* (large colony type) has a wide geographic distribution. It is the most frequently reported caprine mycoplasma infection in the United States, particularly from eastern and western coastal states and is currently known on all continents except South America and Antarctica. Kids are more frequently and severely affected than adults. Though variable, mastitis may affect 25% to 33% of does while morbidity and mortality rates may exceed 90% in kids on the same premises. The most common clinical findings in adults are fever, mastitis, pleuropneumonia, and arthritis. In kids, arthritis, septicemia, and meningitis are more common. In addition, polyserositis, osteomyelitis, keratoconjunctivitis, abscesses, and abortion are also reported. Reports from France, California, and Israel suggest that morbidity and mortality are highest in intensive commercial goat dairy operations. Economic losses caused by decreased production, diagnosis and treatment costs, and death of replacement stock can be devastating in severe outbreaks. Infected, lactating does can be asymptomatic carriers with high numbers of organisms being shed in the milk. These does may become clinically ill themselves as a result of management, nutritional, or climatic stresses. The most explosive disease outbreaks occur among kids after the onset of kidding season. The principal mode of transmission to kids is the oral route via daily ingestion of infected colostrum and milk. The mode of transmission among adults is not as clear. Direct contact is possible but not very efficient. Transfer of infection during the milking process by introduction into the teat seems to play a larger role, particularly because improvement of sanitary procedures slows infection rates in known-infected herds. Infection may be introduced into a herd by the introduction of subclinically infected milking does.

M. putrefaciens was first isolated from mastitic goats. The major mode of transmission of this organism among adults is by introduction through the teat due to poor hygiene during milking. For kids, the disease is spread by feeding of infected colostrum or milk. Stresses such as crowding, poor nutrition, and lack of shelter from wind and rain may predispose kids to clinical disease. All five of the major mycoplasma species just discussed have been found in the ear canals and ear mites of normal goats. This represents a previously unrecognized method for maintaining a carrier state and transmitting new infections to susceptible goats. In addition, all the species except *M. mycoides* subspecies *capri* have been isolated from the nasal and oral cavities and tonsils of clinically normal goats at necropsy.

Among the minor causes of caprine mycoplasma arthritis, *M. arginini* is probably of the least consequence. This ubiquitous organism causes minimal disease alone and is often isolated from clinically normal goats. It is frequently isolated in conjunction with other bacterial pathogens such as *Pasteurella* spp. from pneumonic lungs and may serve as an initiator of bacterial pneumonia. There are occasional reports of *M. arginini* being isolated from the joints of arthritic goats but the etiologic significance in arthritis is doubtful. *M. conjunctivae* is found worldwide and associated primarily with keratoconjunctivitis in sheep and goats. However, it has the potential to become septicemic and concurrent pneumonia and arthritis have been observed in at least one enzootic of keratoconjunctivitis caused by *M. conjunctivae* in a goat herd.

Pathogenesis

Mycoplasmal arthritis in goats always occurs as a result of septicemia. Therefore, it frequently involves multiple joints and is accompanied by generalized malaise, fever, and evidence of infection at other organ sites. Septicemic goats may die without manifestation of localizing signs. This can be attributed to circulatory collapse caused by the action *of* mycoplasmal galactan and an endotoxin-like activity *of* mycoplasma. Experimental infections with *M. mycoides* subspecies *mycoides* (large colony type) in goats suggest that vasculitis, coagulopathy, and thrombosis play significant roles in the pathogenesis of mycoplasma septicemia, leading to infarction and necrosis in numerous organs. Complement activation by *M. mycoides* subspecies *mycoides* (large colony type) may also be an important initiator of inflammation.

Clinical Signs

Mycoplasma arthritis may be seen in all ages and breeds of goats but is most common in young kids and yearlings in dairy goat breeds. In many instances, there is a history of purchased goats introduced into the herd some months before onset of disease. Arthritis rarely occurs alone in an outbreak of mycoplasmosis. Mastitis, abortion, agalactia, pneumonia, or keratoconjunctivitis may occur concurrently or precede the onset of arthritis. The presentation may vary with age or production group. For example, mastitis may occur in lactating does several months before arthritis is noted in kids. This underscores the importance of obtaining a thorough history. There is a tendency for disease incidence to increase around the kidding season.

Affected animals usually first show evidence of septicemia including fever of 104 to 108.5°F (40.0 to 42.5°C), anorexia, depression, weakness, and rough hair coat. One exception is in *M. putrefaciens* infection, where fever does not regularly occur. Some animals may die of septicemia without developing additional localizing signs. In most cases involving arthritis, lameness develops concurrently or within several days of the onset of fever but may precede the onset of joint swelling. Though a single joint may be affected, polyarthritis is the rule. Any diarthrodial joint may be involved but the carpi, tarsi, and stifle joints are most frequently affected, followed by the elbow and fetlock joints. The joints become swollen, hot, and painful. The pain is frequently so severe that affected goats refuse to bear weight on a single affected leg, or become persistently recumbent when multiple joints are involved. The clinical course is usually 4 to 10 days and commonly ends in death if not treated. Occasional spontaneous recoveries do occur, sometimes with complete resolution of joint swelling and lameness noted over several weeks. Morbidity and mortality are almost always higher in kids and yearlings than adults. Goats with hot swollen joints should be carefully examined for indications of other organ system involvement, especially nasal discharges, increased lung sounds, corneal opacity or conjunctivitis, and mastitis or agalactia, all of which can occur in mycoplasmosis. Diarrhea is not usually associated with mycoplasmosis but has been reported in affected kids concurrently experiencing coccidiosis.

Clinical Pathology

Leukopenia with neutropenia is likely in the early stage of septicemia and neutrophilic leukocytosis with hyperfibrinogenemia in the later stages. Synovial fluid analysis can be useful in diagnosing arthritic mycoplasmosis. While the volume of fluid may not be increased, the color is frequently yellow to red brown, and cytologic examination reveals increased cellularity, particularly with increased numbers of neutrophils.

Serologic tests are of little value in the diagnosis of individual cases of mycoplasmosis, but may be of some value on a herd basis when acute and convalescent samples are obtained 3 to 8 weeks apart. Serologic cross reactions occur, particularly between the *M. mycoides* subspecies. Tests commonly in use for caprine mycoplasmosis include the complement fixation test, the indirect hemagglutination test, plate and tube agglutination tests, and more recently the ELISA test. In one follow-up study *of* a goat herd with confirmed *M. mycoides* subspecies *mycoides* (large colony type) infection, goats that remained healthy all had indirect hemagglutination test titers less than 1:32, while sick goats had convalescent titers ranging from 1:8 to 1:1024. Culture and identification of pathogenic strains *of Mycoplasma* spp. should be attempted. Samples of affected tissues and body fluids from clinically ill goats are suitable for culture. For survey and control activities, it has been demonstrated that swabs of the ear canal from clinically normal goats are a reliable and convenient means of isolating pathogenic *Mycoplasma* spp. Swabs should be inoculated into suitable *Mycoplasma* transport media and kept cool for shipment to laboratories.

Post Mortem Findings

A variety of gross necropsy findings may occur in goats with septicemic mycoplasmosis. These include patchy or diffuse pneumonia, pulmonary congestion or edema, a watery, yellow or red-tinged hydrothorax, fibrinous pleuritis with adhesions to the chest wall, fibrinous pericarditis, serous or fibrinous peritonitis, meningitis, mastitis with firm, hyperemic glandular tissue, swollen, edematous lymph nodes, and fibrinopurulent arthritis. In advanced or severe cases, the affected joint is filled with a white to yellow fibrinopurulent exudate. There may be erosions of the articular cartilage, the joint capsule and periarticular tissues are hyperemic, thickened, and edematous, and fibrin tags may be noted extending up tendon sheaths.

Diagnosis

Definitive diagnosis of mycoplasmosis of goats depends on isolation of a known pathogenic *Mycoplasma* spp. from affected tissues. Mycoplasma infection can be suspected when arthritis occurs in conjunction with fever and general malaise and when other localized infections such as pneumonia, mastitis, abortion, and keratoconjunctivitis are occurring in the herd or flock, cither concurrently or historically. Mycoplasma infection should be considered even when a single case of arthritis is observed in a herd or flock as morbidity is not always high, and sporadic cases may precede a generalized outbreak.

Treatment

The prognosis for recovery is generally poor in clinical mycoplasmosis involving arthritis, pneumonia and mastitis, even with aggressive therapy. An extended course of parenteral antibiotic therapy is required, usually ranging from 5 to 14 days. Complications of lameness from repeated injections are common. Because mycoplasmosis can spread rapidly through a herd flock, all contact animals should be treated at the same time when the disease is identified in individual animals. Treatment, no matter how successful in controlling clinical disease, may not eliminate the carrier state. In general, *Mycoplasma* spp. are most often sensitive to tetracyclines, the macrolide antibiotics (tylosin, erythromycin, oleandomycin, spiramycin), and tiamulin. The aminoglycosides, including spectinomycin, are sometimes effective, as are nitrofurans and chloramphenicol. Lincomycin has also been used successfully Susceptibility of the various *Mycoplasma* spp. to these different antibiotics is variable, even within the same drug class, so that *in vitro* culture and sensitivity should be carried out whenever possible.

A wide range of doses has been suggested for treatment of caprine mycoplasmosis and antibiotic therapy is often met with less than satisfactory results. All the doses given are for once daily intramuscular injection administered for at least 5 days; oxytetracycline at 15 mg/kg b.w.; streptomycin at 30 mg/kg, b.w.; tiamulin at 20 mg/kg; and tylosin at doses of 5 to 44 mg/kg b.w., but most commonly at 20 mg/kg. Spiramycin has been given at a loading dose of 50 mg/kg followed by daily doses of 25 mg/kg. Tiamulin was reported to be severely irritating at the injection site, producing general excitation in young goats. A combination of lincomycin at a dose rate of 5 mg/kg and spectinomycin at 10 mg/kg given intramuscularly once daily for 3 days produced recovery in 55% to 80% of treated goats and sheep with contagious agalactia in a field trial. Lincomycin used orally however, has been associated with severe toxic reactions and high mortality in a large sheep flock.

Prevention and Control

Specific mycoplasmal diseases such as contagious agalactia and contagious caprine pleuropneumonia are of great regulatory concern to countries free of disease and importation of goats from known infected countries is prohibited. Outbreaks that occur are usually controlled by stamping out procedures. In enzootic regions, restrictions on animal movement, testing and culling of infected animals, and in some cases vaccination are carried out. Vaccines are available in some countries against *M. agalactiae, M. mycoides* subspecies *capri,* and

M. mycoides subspecies *my coides* (large colony type). Vaccines against *Mycoplasma* F38 are currently being developed. Several recommendations can be made to control the introduction and spread of mycoplasmas in individual goat herds. Where there is no history of mycoplasmosis, especially in commercial goat farms, closed herds should be maintained. No animals should be purchased and animals from the herd should not be brought to shows because of the risk of exposure to carrier animals. When mycoplasmosis has occurred on the premises, management procedures should be evaluated, particularly in the areas of milking and kid rearing, and several recommendations can be made to reduce the risk of transmission. When mastitis has occurred, milk samples should be cultured from all lactating does and infected, carrier does should be culled or at least managed as a separate string, housed and milked separately. During milking strict hygiene should be observed. Kids should be separated from dams at birth and fed heat treated colostrums. Kids should then be housed separately from adults and fed pasteurized milk or milk replacer. Herd-wide milk cultures should be repeated on an annual or semiannual basis until all carrier animals are identified and culled. Regular herd-wide treatment of ear mite infestations is also advisable because of the potential role of these parasites in mycoplasma transmission. Ivermectin is effective against *Psoroptes cuniculi,* but not against *Raillietia capra.* It should not be used in lactating does.

7.3 Chlamydial Infection

Chlamydia psittacosis, a gram-negative intracellular organism is a well known cause of abortions in goats world over. Elementary bodies are the infectious particles that carry the infection to other cells. This abortion disease in sheep and goats is commonly referred to as enzootic abortion.

Etiology

Chlamydia psittaci is one of the four species of the genus *Chlamydia,* which also includes *Chlamydia trachomatis, Chlamydia pneumoniae* and *Chlamydia pecorum.* In ewes and goats *C. pecorum* causes pneumonia, conjunctivitis and arthritis but is very often isolated from asymptomatic intestinal infections. Except in very rare instances, strains inducing abortion in goats belong to *C. psittaci* serotype-1, which is also responsible for pneumonia, conjunctivitis, arthritis and sometimes, intestinal infections without any clinical signs. *Chlamydia trachomatis* and *Chlamydia pneumoniae* are both human pathogens. *C. psittaci* infects a wide variety of birds, mammals and occasionally humans and *C. pecorum* contaminates ruminants and swine. Phylogenetic analyses of 16S and 23S rRNA genes suggest the existence of nine differentiated species in the *Chlamydiaceae*. The genus *Chlamydia*, which corresponds to the old *Chlamydia trachomatis* denomination, includes 3 species: *C. trachomatis* (human strains), *C suis* (porcine strains related to *C. trachomatis* isolated from spontaneous abortions, vaginal infections, and pneumoniae) and *C. muridarum* (mouse-hamster strains). The genus *Chlamydophila* regroups 6 species: *C. pneumoniae, C. pecorum, C. psittaci* (previous *Chlamydia psittaci* avian strains), *C. caviae*, the agent of guinea pig inclusion conjunctivitis (GPIC), *C. felis* (*C. psittaci* strains that infect cats) and *C. abortus* (classical serotype-1 *Chlamydia psittaci* strains).

Epidemiology

Chlamydiosis has an economic and public health impact in numerous goat farms throughout the world. Chlamydial abortions were reported for the first time in Germany in 1959.After that the disease was diagnosed in Bulgaria, Spain, USA, France, India, Japan, United Kingdom, Chad, Greece, and Tunisia. In many areas, chlamydial abortion is the second cause of infectious abortions after brucellosis, and the main cause in most of the countries where brucellosis is controlled.

Pathogenesis

The intestinal tract is the natural foci of infection. *Chlamydia* multiplies in epithelium of genital tract. Intestines and conjunctiva may also offer multiplication niches. It reaches placenta and fetus after chlamydemia. Routine transplacental transfer of nutrients is prevented by inflammation and necrosis. This results in death of fetus and abortion results. Carrier does or males in endemic herds continue the spread of infection. Kid born alive also carries infection. Ingestion of placenta and uterine discharges causes the infection to spread in other animals resulting in future abortions. Infection of other pregnant goats from previous abortion may result in abortions if sufficient duration (6 weeks) of pregnancy is left.

Chlamydia psittaci, which grows in the cytoplasm of eucaryotic cells in a unique cycle of development in which a resistant infectious form, the elementary bodies (EB) alternate with a metabolically active non infectious form the rèticulate bodies (RB). The EB attaches to the membrane of the host cell and promotes its own endocytose in a membrane limited vacuole called the inclusion, which does not fuse with lysosomes. Then the EB transforms in RB, which replicates by binary fission. After several divisions, the RBs, which fill all the inclusion, transform back into infectious EBs. These EBs are released through host cell lyses or extrusion of the inclusion out of the host cell.

Clinical Signs

Chlamydiosis is clinically characterized by abortion during the last months of pregnancy, stillbirths or premature births of weak kids with low birth weight. Abortions occur without previous clinical specific signs even if some goats may develop persistent cough without breathlessness, or arthritis and keratoconjunctivitis. In experimental infections, slight vaginal discharge was observed the day before abortion on some goats. Retained placentas and metritis are not usual, even if they are more frequent than in ewes. Infected does excreted large numbers of *Chlamydiae* in placenta and fetal fluids at the time of kidding and at the time of abortion. Some goats may shed *Chlamydia* in vaginal fluids from more than two weeks before abortion to more than two weeks after abortion. This may explain the higher incidence of abortion in newly infected herds of goats, since the susceptibility to infection varies in relation to the physiological status of the animal. Goats that are less than 100 days pregnant are more susceptible than those at the end of gestation or those that are barren. Smaller amounts of *Chlamydiae* can also be shed in urine, milk and feces during several days after abortion. Young goats born from infected mothers may retain the infection in the herd or transmit it to other herds.

Vaginal shedding of organism starts a week before abortion and lasts at least two weeks after it. *Chlamydia* has also been isolated from semen and genital of rams but vertical transmission has not been confirmed. The abortion in goats in final eight weeks of pregnancy and particularly in last 14 days is more common in endemic herds. Primiparous does are more prone to abort although; in a native herd abortion may occur in goats of all ages and parities. Chlamydial infection in goats exhibit in different clinical form, viz. pneumonic form, reproductive form in which animal may abort at early stage of pregnancy, arthritic form in which animal suffers swelling of different joints, ocular form with severe conjunctivitis leading to various deformities of eyes; and in nervous form, where animals show bizarre symptoms of nervous disorders.

Diagnosis

Demonstration of Chlamydial bodies in samples from diseased animals provides confirmatory diagnosis.

Public Health Importance

The disease is highly contagious to humans. Pregnant ladies assisting in parturition and cleaning work are at higher risk of infection and abortion. Consumption of unpasteurised milk is a high risk factor for disease. Men may suffer from flu-like illness. Preventive clothing like hand gloves and face masks should be worn to prevent infection. Pregnant women avoid contact with herd during parturition season.

Treatment

Chloramphanicol and Enrofloxacin are the drugs of choice in the treatment of pneumonic animals. Long acting oxytetracycline at the rate of 20 mg/kg every 3 days by intramuscular route is recommended to check storm of abortion.Ocular infections are generally self limiting in nature, require no treatment but antibacterial eye drops may be advocated to prevent secondary contamination.

Prevention and Control

As a preventive measure, long-acting Tetracyclines are advised in threatened abortion caused by chlamydial infections. However, no vaccine is available; therefore, management should be given highest priority by identifying the disease at an early stage and providing treatment. The development of an efficient vaccine would be very useful for the control of the disease. Depletion of the "diagnosis specific antigen" from the live vaccine is not currently an option since suitable genetic and molecular methodologies are not yet available and it seems difficult to delete this protein family. For these reasons, a cellular vaccine, which protects against abortion and excretion, would be of interest. Previous studies have shown that both T and B lymphocytes are involved in protective immunity. Enzovax (Intervet) is a live vaccine licensed in UK for use in sheep only.

7.4 Anaplasmosis

Anaplasmosis is a disease of ruminants caused by an intracellular parasite. Cattle, sheep, goats, deer, antelopes may be infected. Anaplasmosis is an infectious disease that is characterized by progressive anemia due to destruction of red blood cells. It is found in sheep and cattle but no natural infection has been reported in goats. In countries where it does occur it presents as a sub-clinical infection and may contribute to abortions in pregnant does.

Etiology

Anaplasmosis is a hemoparasitic, infectious, and transmissible disease characterized by progressive anemia with intraerythrocytic Anaplasma bodies. *Anaplasma marginale* infects cattle, while *Anaplasma ovis* infects sheep and goats. The disease can be divided into four stages. The first stage, or incubation stage, ranges from the time of introduction of the organism into the susceptible animal until the time that 1% of the erythrocytes are infected. The first stage typically varies from one to three months in length. No clinical signs are seen during this time. The second stage, or developmental stage, is the stage during which anemia develops and lasts until reticulocytes appear in the peripheral circulation. The second stage typically varies from four to nine days in length. Most of the clinical signs characteristic of anaplasmosis appear during this stage. The third stage, or convalescent stage, marks the resolution of the anemia. The duration of this stage varies greatly. During the fourth stage, or carrier stage, the animal remains infected, but Anaplasma bodies cannot be detected in the peripheral blood. The fourth stage can last indefinitely.

Epidemiology

Anaplasmosis is typically transmitted by ticks or biting flies. Iatrogenic transmission can occur when instruments are re-used without proper sanitation, including instruments used for dehorning, ear tagging, castrating and vaccinating. *In utero* transmission has been reported.

Clinical Signs

The symptoms that occur with anaplasmosis are usually subclinical, but clinical manifestations may be triggered by malnutrition and other stress conditions such as parasitism. The only way to accurately diagnose anaplasmosis requires laboratory confirmation.

Clinical signs increase in severity as the animal ages. The first clinical sign is typically fever, ranging from 103° F to 106° F and lasting 12-24 hours. Most other clinical signs are manifestations of acute anemia, including mucosal pallor, muscle weakness, tachycardia, tachypnea, exercise intolerance, and behavioral changes. Additional signs that may be present include depression, anorexia, ptyalism, dehydration, constipation, and frequent urination with dark yellow urine. Hemoglobinuria does not occur because the anemia results from the destruction of parasitized erythrocytes in the spleen, not from intravascular hemolysis. Jaundice and weight loss may occur later in the disease. Milk production declines rapidly in dairy goats. *A. ovis* infection in sheep and goats is typically asymptomatic.

Post Mortem Findings

There are no pathognomonic gross lesions for anaplasmosis, but lesions can be suggestive of the disease. In acute cases, the blood is thin and watery and fails to readily clot. Gross postmortem findings typically include severe anemia with pallor of tissues and occasionally icterus. The spleen is generally enlarged with reddish-brown pulp and enlarged splenic follicles. The liver may be enlarged with rounded edges, and may be yellow in cases of icterus. The gall bladder is typically distended with bile. Histologic findings include bone marrow hyperplasia and extramedullary hematopoiesis in the spleen and other organs. Anaplasma organisms may be found in erythrocytes with smears of the peripheral blood.

Diagnosis

A positive diagnosis of anaplasmosis depends upon positive transmission and complement-fixation tests. The history of outbreak occurrence of the disease in the area and the presence of insect vectors may suggest the possible presence of anaplasmosis. Babesiosis is clinically much more acute, is accompanied by hemoglobinuria and can be distinguished on examination of smears of peripheral blood.

Treatment

Early and aggressive intervention is the best treatment for anaplasmosis. The drug of choice is oxytetracycline and tetracycline hydrochloride given intramuscularly for 3 to 5 days. Tetracyclines have been shown to reduce the severity of the disease, and blood transfusions have been used as supportive therapy. After recovery, additional treatment with tetracyclines is necessary to prevent animals from becoming asymptomatic carriers. Control of vectors seems to be of limited value in controlling the spread of anaplasmosis. There may be some value to dipping or spraying for insect vectors. There is no vaccine available, but prophylactic administration of tetracycline is regarded as the measure of choice in areas where there is an outbreak.

□□□

8

Fungal Diseases

8.1 Ringworm

Etiology

A variety of dermatophytes have been cultured from ringworm in goats. These include *M. gypseum, Trichophyton mentagrophytes, T. schoenleinii* and *T. verrucosum.*

Epidemiology

The higher incidence of dermatophytes was reported during the months of November to March in India, may be probably due to prolonged confinement of goats during the cold season. It may, however, at least partly also be due to poor nutritional status during winter months. Dark and damp sheds are more conducive for the spread of the disease. Further, during the winter months most of the animals are huddled together providing maximum chances of contact between healthy and infected animals.

Clinical Signs and Diagnosis

Lesions in goats consist of alopecia, scaling, erythema, and crusts. They typically involve the face, external ears, neck or limbs. Wide spread grayish white crusted lesions which were typically circular in the beginning and became confluent to form extensive irregular patches later on ,were seen on the head, ears, nose and other body parts. Pruritus is not typical, but has been reported. Microscopic examination of hairs and keratin from the periphery of an active lesion may reveal ectothrix invasion of hair shafts. Species identification requires culture and examination of both colony and microscopic morphology.

Treatment and Prevention

Young animals or those living in a dark, damp, dirty environment, or those with debilitating nutritional or infectious diseases are most at risk for developing ringworm. Management changes may be required to control an outbreak in goats. Most cases of dermatophytosis in large animals regress spontaneously in 1 to 4 months. Treatment does reduce contamination of the environment and the risk of spread to other animals or man. People handling infected goats should take precautions to avoid contracting the infection themselves. Lime sulfur (2% to 5%), Iodophors, and 0.5% Sodium hypochlorite as total body sprays daily for 5 days and then weekly are recommended for ringworm. Topical thiabendazole paste or iodine ointments can be used on small lesions. All in contact animals should also be treated and the environment disinfected with sodium hypochlorite if possible. Pens that have previously housed young cattle with ringworm should be thoroughly disinfected before goats are introduced.

8.2 Dermatophilosis

Dermatophilosis, also known as streptothricosis, is a common skin infection in goats worldwide. Cattle, sheep, horses, various wildlife species, and humans are also affected.

Etiology

Dermatophilus congolensis is a Gram-positive, pleomorphic, facultative anaerobic actinomycete.

Pathogenesis

Dermatophilus congolensis may survive in soil or in dust on an animal's hair coat during dry weather. It is introduced into the epidermis following injuries of any sort, and its life cycle is activated by moisture. Outbreaks often occur during periods of heavy rain or high humidity.

Clinical Signs

Several localizations of the disease have been reported. Ears are commonly involved, especially in young kids. Tiny wart-like scabs first appear on the inner hairless surface of the ear pinna. They are easily rubbed off, exposing dry circular light-colored areas beneath. Raised scabs with matted hairs form on the external portions of the ears and are more tightly attached. The lesions are nonpruritic and benign and last 2 to 3 months in kids if not treated.

Other areas of the body affected include the nose, muzzle, feet, scrotum, and underside of the tail. These areas of skin frequently are exposed to moisture or mild abrasion from vegetation. Thick crusts may be mistaken for lesions of contagious ecthyma (sore mouth). The dry crusts, scaling, and alopecia of healing or chronic lesions resemble ringworm. Secondary bacterial infections (*Staphylococci, Corynebacteria, Fusobacterium necrophorum*) are to be expected and may lead to pruritus or pain. Sometimes the entire dorsum of the goat is involved with lesions clinically resembling rain scald in horses; continuous exposure to wet weather presumably is an important etiologic factor. Damage to hides can be extensive. Suppurative lymphadenitis has been reported in Beetal goats from which *D. congolensis* was demonstrated by smear and culture.

Diagnosis

The diagnosis can be confirmed in several ways. When lesions are moist, an impression smear underside of scab reveals Gram-positive branching filaments either unsegmented or in railroad track arrangements of cocci. Giemsa stain may also be used. The organisms in smears also fluoresce under ultraviolet light after staining with acridine orange. Fluorescent antibody technique has been used for rapid identification of the organism in smears of exudates. In dry lesions, skin biopsy used to demonstrate the organism.

Culture of the organism by a diagnostic laboratory will also confirm the diagnosis, but this is best done under microaerophilic conditions with increased carbon dioxide. A medium that selects for Gram-positive organisms (such as colistin-nalidixic acid medium) is helpful. Tiny adherent colonies composed of branched mycelia be visible after

48 hours and should then be subcultured: it is common for the original plate to be rapidly grown by contaminants. Several serologic tests have been used to identify bodies against *D. congolensis.*

Treatment and Prevention

The treatment most commonly recommended for dermatophilosis in individual animals is penicillin strepmycin. Tetracycline is also effective. Shelter from the rain and bathing (iodophors, 2% to 5 lime sulfur) and grooming to remove crusts should be recommended. Improved nutrition and control of external parasites (especially ticks) are also desirable for treatment and for prevention. Cleansing thick crusts with hydrogen peroxide will help to control secondary anaerobic infections. Brushes should be disinfected before being used on other animals. Goat handlers should also be warned that people are occasionally infected with this organism. Carrier animals appear to be the reservoir for the agent. Vaccination against dermatophilosis has not been successful and recovery does not appear to provide immunity.

□□□

9

Parasitic Diseases

A. Endoparasites

9.1 Coccidiosis

Coccidiosis is the most common cause of diarrhea in young goats between one to six months of age, particularly when goats are group housed in confinement. Infection of goats with coccidia is extremely common and eimerian oocysts may be found in 100% of goat faecal samples. In older goats, the intensity of infection is low but these serve as the source of environmental contamination for the kids.

Etiology

Coccidiosis is caused by protozoal parasites of the genus *Eimeria.* In the past, the *Eimeria* species infecting goats and sheep were presumed to be the same. But in well-controlled experiments proved that each small ruminant host has its own host-specific *Eimeria* species that are not readily cross-infective. This led to a re-examination of *Eimeria* morphology and a renaming of species thought to be cross-infective. The *Eimeria* species infecting goats (Capra hircus) are *E. alijevi, E. apsheronica, E. arlongi, E. capralis, E. caprina, E. caprovina,*

E. charlestoni, E. christenseni, E. hirci, E. jolchijevi, E. ninakohlyakimova, E. pallida and *E. punctata.*

Life Cycle

Oocysts passed in the feces sporulate in the goats environment to produce infective sporocysts containing sporozoites. These sporocysts are ingested by other goats. Ingested sporocysts release their sporozoites which then enter host cells to form schizonts. Schizonts undergo asexual reproduction to produce a first generation of daughter merozoites. Depending on the species of *Eimeria,* anywhere from a few dozen to one hundred thousand merozoites may be formed by each schizont. The merozoites then break from the disrupted host cell and each is capable of invading a new host cell to form a second generation of schizonts. The number of cycles of schizogony varies among *Eimeria* spp. The final cycle of schizogony leads to the differentiation of merozoites into male and female gametes. Male gametes (microgametes) then are released from host cells to fertilize female gametes (macrogametes) within their host cells. Zygote formation results in oocyst formation with release of oocysts from disrupted host cells and subsequent passage in the feces.

Epidemiology

Eimeria spp. has been isolated from goats in all continents. Prevalence studies throughout the world demonstrate that *Eimeria* oocysts are widely present in the feces of both normal and diseased goats, with a reported range of 38% to 100% of all goats infected. Concurrent infection with multiple *Eimeria* spp. is reported. It is safe to assume that where there are goats, there are coccidia; however, it is important to discriminate between infection with coccidia and disease caused by coccidiosis. All ages of goats may be infected with *Eimeria,* but numerous factors contribute to the highest incidence of clinical disease occurring in young goats between 1 and 6 months of age. These include host, parasite, management, and environmental factors. An important epidemiological factor was that kids of different ages were frequently housed in groups on deep-litter straw bedding in pens previously used to house older animals.

Age related resistance to clinical coccidiosis is reported in all ruminant species. The resistance is immunologic in nature and is maintained by continuous exposure to coccidial infection. Immunity is relative and not absolute because it does not eliminate infection, but effectively checks rate of coccidial reproduction in the host intestinal tract. This is reflected by a good correlation of oocysts shed in feces with age of infected goats. There is a steady decline in oocyst numbers from 6 months of age. Complete elimination of coccidial infection may lead to a failure of immunity and the development of clinical disease on re-exposure to pathogenic *Eimeria.* The immunity that develops is specific for a given *Eimeria* species and animals of any age may develop clinical disease, if exposed to a population of *Eimeria* spp. if not experienced in the past. In most management situations, goats become resistant to clinical disease at approximately 6 months of age. Resistance may be impaired, however, by stresses such as concurrent illness, lactation, transport, feed changes, weather changes, increased levels of exposure, or exposure to new species of *Eimeria.* Therefore, sporadic outbreaks of clinical coccidiosis may occur in a variety of unexpected settings besides the predictable occurrence in confinement. Breed variation may exist in resistance to coccidia. In India, Barbari goats are considered more susceptible to clinical coccidiosis than other breeds. However, a study of fecal oocyst counts in three Indian breeds showed approximately equal counts for Barbari, Jamunapari and Jhakrana goats.

Oocysts are quite resistant to environmental degradation, and are even more resistant when sporulation occurs. Over wintering of sporocysts is not uncommon, and many disinfectants, including 5% formalin, will not destroy them. Sporulation depends on a combination of oxygen availability, temperature and moisture conditions. In general, with adequate moisture and oxygen, sporulation of oocysts occurs optimally within 2 to 5 days at temperatures between (24°to 32°C) and readily at temperatures down to (12°C). Synchronous sporulation can occur among oocysts accumulating in a contaminated environment when optimal conditions prevail. This means that susceptible goats may be challenged by massive numbers of infective sporocysts under appropriate conditions of warmth and humidity. Some of the more pathogenic *Eimeria* spp. such as *E. ninakohlyakimovae* produce thousands of Merozoites per schizont during asexual reproduction inside the goat intestine. The various species may also differ in the number of cycles of schizogony involved in reproduction. The more merozoites produced, the more intestinal epithelial cells are disrupted because every merozoite potentially can invade a target host cell. Pre-patent periods also vary among goat *Eimeria* spp., and oocysts may be shed by kids as young as 2 weeks of age when kids are exposed to sporocysts at birth.

Clinical coccidiosis occurs more often in intensive management situations than in extensive ones because of the concentrating effects of confinement on both host and parasite. Most outbreaks of clinical coccidiosis occur around the time of weaning, particularly if kids are weaned abruptly and not offered solid feed *ad libitum* before termination of milk feeding. Several factors may increase exposure to infective sporocysts at this time. Feeding young goats on the ground promotes ingestion of sporocysts. Poorly designed feeders in which goats can stand, climb, or defecate lead to contamination of feed and water sources. Watering systems that leak or overflow easily contribute to environmental moisture, which promotes sporulation. Overcrowding of pens or failure to segregate kids by age augments exposure to sporocysts and increases disease risk. When goats are kept on solid floors, failure to maintain clean, dry bedding is a major factor in outbreaks of clinical disease. Even when floors are scraped clean, well-intentioned spraying or hosing of the pens may actually aggravate the problem because of the amount of moisture added to the atmosphere. Buildings that do not allow in adequate sunlight also contribute to the persistence of oocysts, particularly in the winter months in temperate regions when days are short. Hot, humid weather is particularly conducive to sporocyst development and outbreaks of clinical coccidiosis are common in temperate regions during summer, especially because spring-born kids are being weaned during the hot, humid months. Even when goats are managed extensively and the risk of coccidiosis is ostensibly less, it is important to scrutinize management practices for evidence of temporary mobbing that can predispose to outbreaks of coccidiosis. In general, most exposures to coccidia cause sub-clinical infection and the acquisition of a protective, though relative immunity. Most clinically affected individuals will recover from clinical disease after a period of diarrhea as long as 2 weeks, and mortality rates do not usually exceed 10% when exposure to sporocysts has been gradual and continual before disease. In situations where exposure to coccidial sporocysts is abrupt and intense, however, mortality rates in young goats can reach 50%. While the costs of treatment and death loss can be high, the major economic impact of coccidiosis in production animals may be reduced growth rates and weight gains after clinical or subclinical infection.

Pathogenesis

The adverse effects of *Eimeria* infection in goats result from destruction of the gastrointestinal epithelium as these intracellular parasites play out their complex life cycle in the digestive tract of the host. Because of the multiple cycles of asexual and then sexual reproduction that can occur, successive waves of host cell disruption can occur. There are species specific differences in the host cells invaded by various *Eimeria* spp. during their development cycles. For example, sporozoites of *E. ninakohlyakimovae,* considered the most pathogenic of the goat species, enter basal epithelial cells in the crypts of Lieberkuhn in villi of the small intestine. Many thousands of merozoites are produced in each schizont. These merozooites then invade epithelial cells in the large intestine. The subsequent gametogonous stages invade the ileum, caecum, and large intestine where oocysts are produced. Since multiple infections are common in goats, disruption of gastrointestinal epithelium can be very widespread. Diarrhea results from disruption and inflammation of the intestinal mucosa. In massive infections, severe hemorrhage into the intestinal lumen can result in death caused by blood loss. In the more typical acute form of the disease, fluid and electrolyte loss result from a compromise of the normal restorative potential of the intestinal epithelium as well as leakage of plasma and lacteal constituents from the inflammed and disrupted mucosa. When extensive, these losses can lead to fatal systemic sequelae of dehydration, acidosis, and serum electrolyte derangement. Disruption of the mucosal integrity may also lead to increased susceptibility to secondary bacterial infection and resultant septicemia, thus increasing mortality rates. The ill thrift and poor growth commonly seen in the aftermath of clinical coccidiosis are manifestations of prolonged malabsorption and maldigestion associated with permanent damage to the intestinal mucosa. In many cases regeneration of a normal epithelial lining is not complete and evidence of mucosal scarring or villous atrophy may be observed at necropsy in poorly doing animals.

Clinical Signs

Subclinical coccidiosis should be suspected when complaints of poor growth, weight loss, or loss of fecal pellet formation are reported for young, susceptible animals in management situations conducive to the persistence and multiplication of coccidia. Clinical coccidiosis may develop within 1 to 2 weeks of ingestion of a large dose of infective sporocysts. Peracute cases, caused

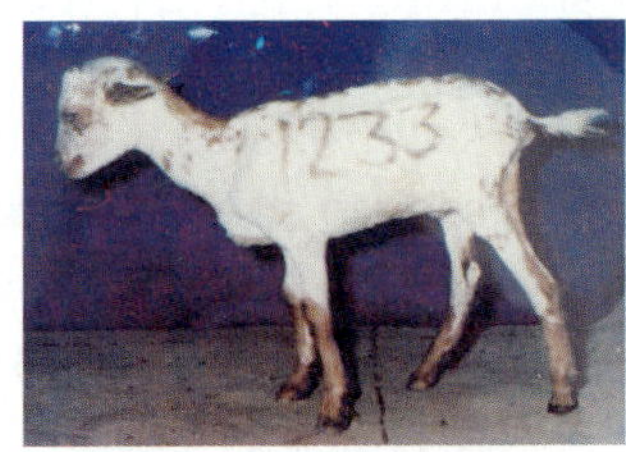

Fig. 31 Debilitated, shrunken eyes in kids

by severe blood loss in the intestinal lumen, may present as sudden death before signs of diarrhea or abdominal discomfort are seen. The intestinal lumen may be filled with blood due to massive destruction of the intestinal mucosa. In acute cases, early signs will include decreased appetite, listlessness, weakness, and abdominal pain that may be manifested by crying and frequent rising up and lying down. The feces may first be unpelleted, then pasty, and then, a watery yellowish-green to brown diarrhea develops.

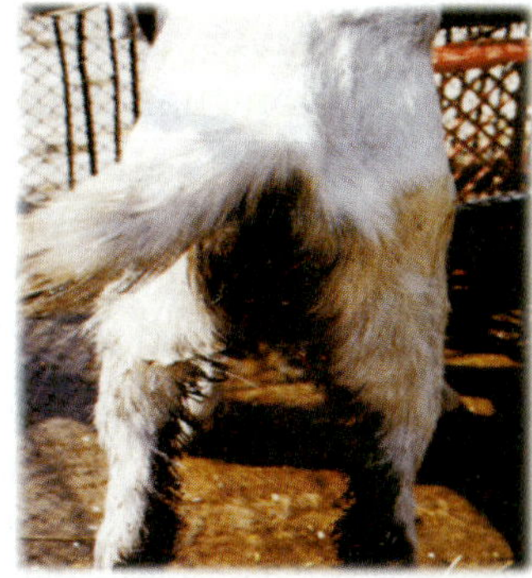

Fig. 32 Yellowish green diarrhea, tenesmus in goat kid

Fresh blood or melena may be seen. Tenesmus is less common than reported in cattle or sheep. The hind quarters and tail become stained and feces coated as diarrhea persists. If dehydration becomes severe, animals will become recumbent with cold extremities and subnormal temperature. They will eventually become moribund and die. Polioencephalomalacia is reported to be a common sequel to coccidiosis in goats. Very young, highly susceptible animals may die of acute coccidiosis within 1 or 2 days. Older or more resistant animals may exhibit diarrhea and weakness with resultant weight loss for as long as 2 weeks before spontaneously recovering. Animals that have recovered from clinical coccidiosis may develop ill thrift. They may be stunted, show a poor hair coat and develop a pot bellied appearance.

Clinical Pathology

The hemogram in coccidiosis may be normal, but a marked leukocytosis is possible when intestinal mucosal damage is extensive. Anemia is variable depending on severity of disease, and may be masked by an increased packed cell volume caused by hemoconcentration associated with dehydration. Varying degrees of hyponatremia, hypocalcemia, hypophosphatemia, and metabolic acidosis can occur. Hyperkalemia has been reported in goats with coccidosis and this may be caused by compensatory efforts to counter acidosis. The use of fecal floatation methods to detect oocysts is not a wholly satisfactory approach to diagnosis of clinical coccidiosis. When oocysts are not found in the presence of active diarrhea, the affected individual may be in an early stage of infection. Intestinal disruption with resulting diarrhea can occur during the schizogony that precedes oocyst formation. In these cases, direct fecal smear may be employed to identify merozoites. Conversely, oocysts may be found routinely in the feces of all normal goats older than 4 to 6 weeks of age. Therefore, finding oocysts in the feces of diarrheic goats does not confirm coccidiosis as the cause of illness. Performing oocyst counts is of limited value, unless the oocysts of pathogenic species are morphologically identified and quantified, as some of the less pathogenic *Eimeria* spp. is prolific egg producers. If oocyst counts are undertaken, a modified McMaster technique should be used, with a floatation solution made of magnesium sulfate at a density of 1.18.

Post Mortem Findings

Necropsy examination offers more reliable indications of clinical coccidiosis. Grossly, signs of enteritis may range from mildly catarrhal to hemorrhagic or necrotic in nature. In peracute cases, the intestinal lumen may contain fresh blood. Thickening of the intestinal wall caused by edema is not uncommon. The most consistent and characteristic lesion is the occurrence of multiple raised, white nodules measuring between 1 and 6 mm in diameter on the intestinal mucosa, which may be apparent even when the intestine is viewed from the serosal side. These nodules represent sites of active gametogony and on smears or histological examination contain numerous macrogametes and oocysts. Lesions of enteritis and nodules may be seen in subclinical, peracute clinical, and acute clinical cases. The extent and severity of the lesions in conjunction with the history must be considered in establishing coccidiosis as the cause of death.

Diagnosis

The diagnosis of coccidiosis is presumptive, based on an appropriate history, clinical signs, and necropsy findings. In very young kids between 3 week and 1 month of age, other causes of abdominal pain and acute diarrhea must be differentiated. These would include cryptosporidiosis, colibacillosis, enterotoxemia, salmonellosis, viral enteritis, and dietary diarrhea. Abdominal pain alone would suggest abomasal bloat, mesenteric torsion, or other intestinal accidents if diarrhea does not develop. In cases of peracute death due to coccidiosis, the differential diagnosis includes enterotoxemia, bacterial septicemias, and plant and chemical toxins. In weanlings 2 to 5 months of age, particularly those with access to pasture, kid sheds or poorly drained dry lots, helminthiasis is the major differential diagnosis for diarrhea.

Gastrointestinal nematodiasis and paramphistomiasis should be considered. In fact, multiple parasitic infestations are common and it may be difficult to attribute diarrhea or ill thrift to coccidiosis alone. Other causes of diarrhea in this age group would include simple indigestion and acute carbohydrate engorgement due to sudden feed changes or excessive grain feeding and enterotoxemia. In subclinical coccidiosis, or after severe clinical coccidiosis, ill thrift or poor growth may be the primary clinical complaint in growing kids. Selenium deficiency or cobalt deficiency must be considered in these cases, particularly where soil deficiencies of these minerals are known to exist. Chronic helminthiasis must also be ruled out.

Treatment

Supportive care is the principal therapeutic intervention in active cases of coccidiosis. Diarrheic goats should be removed from the group, and given oral or parenteral balanced electrolyte solutions depending on the degree of dehydration. In preweaned kids, milk should be fed only in small amounts, as disruption of the intestinal mucosa will produce maldigestion and may promote osmotic diarrhea from undigested lactose. Weaned kids should be offered good quality grass hay and brought back on full feed gradually. Severely anemic kids with acute intestinal blood loss may require blood transfusions. Broad spectrum antibiotics are indicated in severe cases to prevent bacterial septicemia secondary to disruption of the intestinal mucosal barrier.

The use of anti-coccidial drugs in individual, active cases may have limited value. Most anti-coccidial drugs are coccidiostats that inhibit but do not eliminate coccidial reproduction. These drugs usually act on early stages of the reproductive cycle. Animals already showing diarrhea have often progressed beyond the stages of infection where coccidiostats may prove beneficial. Nevertheless, goats are commonly infected with multiple species of *Eimeria* and some of these may still be in the early stages of development. In these cases, the duration of clinical disease may be shortened by treatment. The main goal in using coccidiostats is to reduce the number of additional cases developing in group of animals at risk, rather than curing existing cases. Drugs reported for use in the treatment of caprine coccidiosis include sulfonamides, nitrofurazone, ionophores, and Amprollium. The following drugs can be used aş treatment and prevention of Caprine Coccidiosis.

Sulfonamides

Sulfadimethoxine: 75 mg/kg bw orally for 4-5 days

Sulfadimidine: 135 mg/kg bw orally for 4-5 days

Sulfaguanidine: 280 mg/kg bw orally for 4 days

Sulfamethazine: 50 grams per ton of feed

Antibiotics

Nitrofurazone: 10-20 mg/kg bw orally for 5-7 days

Amprollium: 10-20 mg/kg bw orally for 3-5 days,50 mg/kg bw orally for 5 days,25-50 mg/kg bw orally in feed or water continuously from 2 weeks to several months of age as prophylaxis regimes

Ionophores

Monensin: 15 to 20 grams per ton of feed

Lasalocid: 20-30 grams per ton of feed (1.1-2.2 mg/kg bw/day) continuously

Salinomycin: 100 ppm in concentrate fed for 3 weeks after weaning

Quinolones

Decoquinate: 0.5-1.0 mg/kg bw orally in feed; may be fed continuously.

Other Drugs

Toltrazuril: 20 mg/kg bw orally once, 20 mg/kg bw orally once every 3-4 weeks as prophylaxis regimes.

Clopidol and Methyl benzoquate: 12 mg/kg bw clopidol and 1 mg/kg bw methylbenzoquate per day for 5 weeks as prophylaxis regimes.

The sulfonamides and nitrofurazone may offer the added benefit of control of secondary bacterial infections. Because goats with clinical coccidiosis are often dehydrated, the sulfonamides must be used with care as nephrotoxicity can result from poor renal perfusion and decreased urine output. Sulfonamides are the oldest class of drugs used as coccidiostats and it is generally considered that resistance to sulfonamides is widespread among *Eimeria* spp. However, efficacy continues to be reported in some goats treated with sulfonamides. Sulfa drugs are available for oral and parenteral use. A variety of oral preparations can be given as boluses, or added to feed or water. Nitrofurazones can produce neurotoxicity when overdosed, as may occur if the medicine is not properly mixed into milk, or solid feeds. This drug is rarely used anymore in coccidiosis control. Amprolium is a thiamine antagonist and works as a coccidiostat by blocking utilization of thiamine by coccidia. If Amprollium is used for too long or at excessive doses, the potential for producing polioencephalomalacia in goats exists, though this should not discourage its proper usage. Polioencephalomalacia of ruminants is believed to result from thiamine deficiency. Amprollium is for oral use and is available as a liquid for addition to milk, milk replacer, or water, and in solid forms for addition to feeds. The ionophores include monensin, lasalocid, and salinomycin; all have been used in goats. They are used mainly in solid feed in weaned animals. A new coccidiocidal agent used in poultry, toltrazuril has been evaluated in goats and a single oral dose of 20 mg/kg produced a rapid, significant reduction in oocyst shedding that remained low for 2 to 3 weeks suggesting that all developmental stages of the coccidia present were killed.

Prevention and Control

In addition to the use of coccidiostats, implementation of management changes is important in controlling outbreaks of coccidiosis. Attempts should be made to reduce exposure to infective sporocysts by removal of contaminated bedding and feed, reducing the stocking rate, and moving animals to a new, uncontaminated environment. Control in situations where outbreaks of coccidiosis can be reliably anticipated, measures should be taken to control coccidiosis before the observation of clinical disease. Proper control is a balancing act. The goals are to reduce oocyst numbers in the environment sufficiently to avoid massive challenge to susceptible animals while allowing sufficient exposure so that immunity and resistance can properly develop. This is accomplished by combining sound management and hygiene practices with the use of coccidiostats. Unfortunately, in production systems where group housing of weanlings is unavoidable, control of coccidiosis without the use of coccidiostats is rarely achieved, even when management is excellent. In attempting to maximize the immunologic resistance of the host, it is important to consider the mode of action of the various coccidiostats.

The host immune response appears to be most stimulated by active schizogony, while most coccidiostats inhibit reproduction earlier in the infective cycle. Sulfonamides, amprolium, and ionophores have been used as prophylactics and as therapeutics in goat rearing programs with mixed success. The sulfonamides are folic acid antagonists, and are most inhibitory during the second schizogony. This late action allows for development of host immunity. However, as already mentioned, resistance to sulfas may be widespread. The ionophores primarily inhibit the early, asexual stages of coccidial development. However, they offer an added benefit of improved weight gains above and beyond their action as coccidiostats. Unfortunately, outbreaks of clinical coccidiosis have been observed in young goats being fed lasolacid in their ration. Monensin can be used in goats as a feed additive at 20 grams per ton. Decoquinate, a quinolone not used for treatment, has been found to be effective in prevention of caprine coccidiosis. It works early in the *Eimeria* life cycle by inhibiting sporozoite development. Dose ranges from 0.3 to 4.0 mg/kg have been evaluated. All doses prevented clinical coccidiosis and allowed equal rates of weight gain. Doses at the higher end however, resulted in a more rapid and pronounced decrease in the number of oocysts shed. Coccidiostats are most often used between the ages of 1 and 4 months, usually in conjunction with grain feeding. Good hygiene and thoughtful management are essential to effective coccidiosis control. Weaning should be as unstressful as possible. To minimize weaning shock, kids should have access to grain well in advance of weaning. Whenever possible, kids should be raised away from adults and be housed in small groups, segregated by age, and provided feed and water in devices that minimize contamination with feces. They should never be fed on the ground, and waterers must not leak or be liable to spillage. Housing that allows shelter with ready access to sunlight is preferred. No matter what type of housing is used, ground, floors, and bedding should not be allowed to become excessively wet. Adequate ground drainage, frequent floor scraping, and regular provision of clean bedding are necessary. Scraping floors and allowing them to dry is probably more effective in coccidia control than scraping followed by scrubbing or hosing because disinfectants are not effective, and excess moisture promotes sporulation. In general, when proper control methods are implemented and goats are not exposed to new species of *Eimeria* not previously experienced, immunity to reinfection is usually achieved by 5 months of age. Coccidiostat feeding may be terminated and, if possible, kids should be moved to uncontaminated housing at this time.

9.2 Cryptosporidiosis

Cryptosporidiosis is a protozoan disease of the intestinal tract of the neonatal kids characterized by varying degree of diarrhea in young kids younger than 1 month of age, particularly in intensive management conditions.

Etiology

Cryptosporidiosis is caused by a protozoal parasite of the subclass Coccidia and the suborder Eimeriina. It comprises a separate genus from the *Eimeria* and *Isospora* genera commonly associated with coccidiosis. Cryptosporidia infect a wide range of vertebrate hosts including all the mammalian domestic livestock species, and poultry. It was initially believed that different species of Cryptosporidia infected different hosts. More recently, cross-infection of different hosts has been widely reported, and the pathogen is now known as *Cryptosporidium parvum*. Initially, these parasites of the gastrointestinal tract were considered to be extracellular, in close association with the brush border of the intestinal epithelial cells. Electron microscopic studies have confirmed that they are in fact intracellular, but extra cytoplasmic, residing underneath the brush border of intestinal epithelial cells in a parasitophorous vacuole created from the host cell microvillous and plasma membranes.

Life Cycle

Life cycle is similar to the *Eimeriidae*, with some notable differences. The entire life cycle can be completed within 1 week. Oocysts can sporulate after ingestion by the host to release 4 sporozoites. Two cycles of schizogony and one cycle of gametogony leading to zygote formation occur within the brush border of the intestinal epithelial cells. Subsequent sporogony leads to mature oocysts that can sporulate and excyst within the same host animal leading to reinfection, or oocysts can be passed in the feces to infect new animals.

Epidemiology

Cryptosporidiosis probably occurs worldwide. The first reported field case of diarrhea attributed to cryptosporidiosis in a goat involved a four week old kid. All kids examined were shedding cryptosporidial oocysts and tests for enteric viruses and *K99+ E. coli* were negative. In some studies on the causes of diarrhea in kids younger than 1 month of age in large commercial goat herds, cryptosporidia were isolated more frequently than either rota virus, corona virus, adenovirus or enterotoxigenic *E. coli.* In these studies, cryptosporidia were not identified in the feces of any normal goats while all of the other agents mentioned were identifiable one or more times in non-diarrhoeic kids. A more extensive kid survey on intensively managed goat farms demonstrated an overall mortality rate from diarrhea of 10.3%, mostly during the first 2 weeks of life. Cryptosporidia were identified in feces of virtually all fatal diarrhea cases, but not in any non-diarrhoric animals save those recovering from diarrhea or those that subsequently developed diarrhea. Cryptosporidia also have been reported in association with diarrhea in 4 to 25 days old kids in conjunction with *Clostridium perfringens.* Cryptosporidiosis is a common cause of diarrhea in young, intensively managed kids in a large goat herds. Transmission is by the fecal-oral route from oocysts shed in the feces. Colostrum-deprived kids develop more severe disease and a higher rate of mortality than colostrum-fed kids. It is not yet clear if this is directly caused by Cryptosporidium-specific immunoglobulins or other nonspecific nutritional or immune factors available from colostrum. The severity of disease does not appear to be affected by the state of the intestinal flora . While cryptosporidia are found infrequently in non-diarrheic kids during field investigations, experimental challenge can result in oocyst shedding in diarrhea-free kids indicating that subclinical infections may occur. It is known that cryptosporidial oocysts are resistant to environmental degradation and many common disinfectants are reported to be ineffective in destroying them. Oocysts remained infective after 18 hours of contact with iodophors, cresylic acid, sodium hypochlorite, benzalkonium chloride, and sodium hydroxide. Only ammonia (5%) and formal saline (10%) destroyed infectivity of oocysts.

Pathogenesis

The pathogenic effects of cryptosporidial infection are most likely due to the impairment of enterocyte function and integrity resulting from the rapid, cyclic reproduction of cryptosporidia that occurs within the brush border of enterocytes. In goats, infection leads to increased senescence of epithelial cells and villous atrophy in the ileum. The severity of these effects is related to the degree of infection. This can lead to both malabsorption and maldigestion, resulting in the clinical diarrhea. That diarrhea has been observed in kids as young as 3 to 4 days of age probably reflects the short life cycle of cryptosporidia. That diarrhea has been observed to persist or recur for several weeks may be due to possible re-infection resulting from the ability of cryptosporidial oocysts to sporulate. Prolonged weight loss may be associated with malabsorption. Age-related immunity probably occurs, as clinical disease is far less common in older animals.

Clinical Signs

The most common presentation of cryptosporidiosis is an acute, white to yellow, watery diarrhea in kids younger than 2 weeks of age. The diarrhea may last from several days to 2 weeks, and may be mild to severe, depending presumably on the magnitude of the initial exposure to oocysts. Diarrhea may be persistent or recurrent. In addition to diarrhea, kids may show depression, inappetance, and a rough hair coat. Sequelae of dehydration, electrolyte imbalance, acidosis, and death may occur, depending on the severity of diarrhea and the time of intervention. Spontaneous recovery can also occur. Explosive outbreaks of diarrhea with mortality rates of 10% to 15% have been observed in intensively managed goat kids housed in groups in intensively managed herds. Some of these outbreaks have begun toward the end of the kidding season, suggesting that environmental buildup of oocysts occurs.

Clinical Pathology

Current evidence suggests that cryptosporidial oocysts are not commonly shed in the feces of healthy goats. Detection of oocysts in feces of diarrheic goats may then be useful in confirming a diagnosis of cryptosporidiosis. Numerous staining techniques have been used for identifying cryptosporidia oocysts in feces. Currently, acid fast staining techniques are generally favored. Oocysts stain pink to red with interior blue granules. Direct fecal smears have been routinely used for examination because they are quick and easy to prepare. Oocysts, however, may be sparse, difficult to discover in fecal debris, or hard to distinguish from yeasts. Fecal flatation and sedimentation methods are preferable. Centrifugal flatation techniques including the use of potassium dichromate/saturated sucrose and potassium iodomercurate solutions have been used. Yeasts will not float in the potassium dichromate/saturated sucrose solution. Concentrated oocysts can be viewed stained or unstained, using direct light or phase-contrast microscopy. A semi quantitative flotation technique for counting oocysts in feces has been reported for use in calves and applied to goats.

Post Mortem Findings

Gross lesions are nonspecific and unremarkable and may be limited to a mild to moderate enteritis most commonly located in the ileum. Smears should be made of the ileal mucosa for microscopic examination to detect oocysts. Histological sections of intestine will reveal numerous small round basophilic bodies of various sizes along the mucosal surface. These are the different endogenous stages of developing cryptosporidia in the brush border of the enterocytes. Free oocysts may also be seen in the intestinal lumen. Infection is accompanied by marked villous atrophy and villous fusion in the ileum and possibly the jejunum and large intestine. For accurate diagnosis, it is essential to fix fresh tissues in formalin immediately after death as autolysis of developing cryptosporidia and the related enterocytes is rapid. Oocysts in feces are hardier, although freezing will destroy them. If necessary, they may be preserved as long as 120 days by mixing one part feces with 2 parts of 2.5% potassium dichromate solution.

Diagnosis

A presumptive diagnosis of cryptosporidiosis is based on identification of oocysts in feces or various endogenous stages of the organism in intestinal smears, or histological sections. Since cryptosporidia are frequently observed in association with other etiologic agents capable of producing diarrhea, laboratory evaluation for the various other bacterial, viral, and protozoal causes of kid diarrhea is indicated before a definitive diagnosis can be made. Careful history taking is necessary to rule out dietary causes of diarrhea.

Treatment

Currently, treatment of cryptosporidiosis is limited to supportive care, particularly fluid therapy. Because the diarrhea may last as long as 2 weeks, it is not possible to withhold milk for that period. Because the organism undoubtedly produces a maldigestion, however, it is recommended to reduce the volume of milk fed per feeding while increasing the frequency of feeding to reduce the risk of osmotic diarrhea from undigested lactose. Many anticoccidial, antibacterial, and antiparasitic drugs have been tested for therapeutic efficacy against cryptosporidiosis in numerous species and *in vitro,* and none have been found effective. In goats specifically, sulphadimethoxine and sulfaquinoxalone were found to be ineffective. Preliminary results from field trial suggest that the anticoccidial drug Halofuginone lactate may be effective in arresting diarrhea and eliminating shedding of oocysts in goat kids with cryptosporidiosis. The drug was given orally at a dose of 0.5 mg/kg daily for 3 to 5 days. This drug has a very narrow margin of safety. It can be fatal when administered at three times the dose used. It is recommended that the drug be extensively diluted before administration and not be used longer than 3 days consecutively.

Prevention and Control

Because effective chemotherapeutic control agents are presently unavailable, the thrust of cryptosporidiosis control is in improved management and hygiene. Kids should be separated from does at birth and fed colostrum sufficiently. Kids should be housed away from the adult herd and housed individually or in small groups. When kids are housed in groups, affected kids should be isolated at the first sign of diarrhea to reduce oocyst contamination of the environment. Excellent sanitation in housing, bedding, and feeding utensils is probably most critical in controlling the spread of cryptosporidiosis among susceptible kids. Personnel handling diarrheic kids should not handle healthy kids. Hands should be washed thoroughly after handling sick kids because of the zoonotic potential of cryptosporidiosis. No vaccine is available and the role of maternal antibody in protection of neonates in this disease is not well clarified.

9.3 Babesiosis (Piroplasmosis, Tick Fever)

Babesiosis is a tick-borne protozoan hemoparasitic disease of goats, characterized with fever, anaemia, parasitemia and haemoglobinuria. The economic losses associated with caprine babesiosis may be significant, particularly in the Middle East and India.

Etiology

Two main species of *Babesia* infect sheep, *B. motasi* and *B. avis,* but in goats *B. motasi* infection predominates. *B. motasi* is classified as a large *Babesia* sp with a length of 2.5 to 4.0 microns while *B. ovis* is a small *Babesia* spp. with a length of 1.0 to 2.5 microns. Another small species, *B. taylori,* infective for goats has been reported from India.

The current status of this species is unclear. The agents causing bovine babesiosis are not pathogenic for goats although inapparent infections of goats with *B. bigemina* have been observed. Ticks serve as biological vectors but mechanical transmission is also possible.

Epidemiology

B. motasi infection of small ruminants has been reported from much of Europe, the Middle East, the former USSR, India, southeast Asia, and parts of Africa. The primary biological vector for *B. motasi* is the tick *Haemaphysalis punctata,* although it is likely to be also transmitted by other species in the genus *Haemaphysalis.*

Several strains of *B. motasi* exist, with variable infectivity and pathogenicity for goats. An outbreak of naturally occurring disease in both sheep and goats caused by *B. motasi* has been reported from India. A strain of *B. motasi* isolated from sheep in Wales produced fever and anemia in a splenectomized goat, but pathogenicity for intact goats was not evaluated. Babesiosis was the cause of debilitating disease in sheep and goats in southwest Nigeria but caused death only in goats. *B. ovis* is transmitted primarily by *Rhipicephalus bursa.* The geographic distribution of *B. ovis* coincides in part with that of *B. motasi,* but is associated primarily with disease in sheep where it occurs. Infection of goats with *B. ovis* has been reported only from Somalia, and the infection was subclinical. The pathogenicity of *Babesia* spp. in goats and sheep may be intensified by concurrent infections, particularly other hemoparasites. No cross immunity develops after infection with either *B. motasi* or *B. avis.*

Pathogenesis

Merozoites introduced by feeding ticks invade erythrocytes. Asexual reproduction of the protozoa occurs within the red blood cells to form a pair of trophozoites that are released and re-invade other red blood cells. The release of trophozoites from parasitized red cells results in intravascular hemolysis. The degree of anemia does not always correlate with the degree of parasitemia and immune-mediated hemolysis may also be occurring. In other species, infection may lead to the development of neurologic signs from cerebral thrombosis, hypotension from activation of plasma kallikrein, and disseminated intravascular coagulation. These syndromes have not yet been reported in the goat. Antibodies are produced in response to infection. A protective immunity does develop in nonfatal infections. It does not completely prevent reinfection, but does inhibit recurrence of clinical disease with the same-infecting strain.

Clinical Signs

Affected goats may have fevers as high as 107°F (41.7°C) and show anorexia, weakness, and signs of anemia, including dyspnea, and tachycardia. Sheep consistently show icterus and hemoglobinuria, with coffee-colored urine. These signs occur less consistently in goats, but aid in diagnosis when present. Additional signs of coughing and diarrhea have been reported but may be because of other concurrent infections. Affected goats may harbor ticks, though infective ticks may have dropped off before clinical signs develop. Morbidity and mortality rates can be high. Death can occur within 48 hours of the onset of signs. Chronic infections may occur, with anemia and ill thrift the prominent findings.

Clinical Pathology and Necropsy

Giemsa-stained blood smears should be examined for evidence of piroplasms. Thin smears are usually adequate at the height of parasitemia, but in chronic cases or when parasitemia is low, thick blood smears prepared from ear vein blood are preferred. Blood from several animals in a suspect group should be examined to avoid missing the diagnosis. *B. motasi* is a large piroplasm and occurs in single and double pear-shaped forms. *B. ovis* is a smaller piroplasm and often assumes a round form near the periphery of the erythrocyte. The *Babesia* piroplasms must be distinguished from *Theileria* piroplasms that can also affect goats and have a similar geographic distribution. An indirect fluorescent antibody test can be used for the serodiagnosis of *B. motasi* infection, with titers of 1:640 or more considered as evidence of infection.

At time of necropsy, the predominant findings are splenomegaly, lymphadenopathy, and enlarged liver. Icterus and hemoglobinuria are variable. Enlarged, yellow liver and distended gall bladder containing thick dark bile is important finding. The bladder mucosa is edematous and yellow. Subcutaneous tissue and connective tissue in the muscles are edematous and jaundiced. Thin watery blood and red urine in the bladder.

Diagnosis

The diagnosis of caprine babesiosis is suggested by signs of anemia, hemoglobinuria, and acute collapse in animals where babesiosis is known to occur. Definitive diagnosis depends on the identification of the organism in blood smears. Theileriasis can be differentiated by the morphology of piroplasms and a more prominent lymph node enlargement. Anaplasmosis does not normally produce hemoglobinuria. When hemoglobinuria is present, plant poisonings, copper toxicity, anthrax, and leptospirosis must also be considered.

Treatment

A single dose of prescribed treatments is often satisfactory. The diamidine derivatives diminazene (Berenil) and imidocarb diproprionate are effective at doses of 3 mg/kg and 1-2 mg/kg, respectively. Diminazene in intramuscular doses as high as 12 mg/kg have been used in goats without adverse effect. Quinuronium sulfate compounds (Pirevan, Acaprin) at a dose of 2 m1 of a 0.5% solution per 10 kg b.w. are also effective.

Prevention and Control

No specific recommendations for control of caprine babesiosis have been made. In cattle and sheep, control efforts are directed at vaccination and reduction of tick infestations using acaricides. Vaccination protocols include vaccination with live protozoa in conjunction with imidocarb chemoprophylaxis; or chemoprophylaxis alone, leading to a controlled natural exposure and self-immunization. Inoculation with indigenous strains of *Babesia* in conjunction with chemoprophylaxis may be essential for survival of exotic breeds of goats imported into areas where babesiosis is enzootic.

9.4 Toxoplasmosis

Toxoplasmosis is a contagious disease of animals and man caused by protozoon *Toxoplasma gondii*. Toxoplasma in sheep and goats is manifested with abortion and stillbirths.

Etiology

The protozoan *Toxoplasma gondii* is a very important cause of abortion, mummification, stillbirth and birth of weak young in goats.

Epidemiology

Cats serve as the definitive host for this parasite; they become infected by consuming uncooked meat scraps, placentas, and small rodents. Recently infected cats then shed oocysts in their feces. Oocyst shedding typically lasts from 3 to 19 days, but oocysts may persist in moist and shaded soil for as long as 18 months.

Pathogenesis

Goats become infected by eating grass, hay or grain contaminated by cat feces. After initial invasion of the goat's small intestine and associated lymph nodes, the *Toxoplasma* organisms spread via the blood stream to other tissues, including muscle, brain, and liver. Here the parasite may remain encysted for months or even the life of the goat. If the goat is pregnant at the time of initial infection, *Toxoplasma* commonly invade the placenta and fetus approximately 2 weeks after initial infection of the doe. Fetuses infected in the first half of pregnancy are more apt to die than fetuses infected in the second half. Sometimes abortion is repeated in the next gestation, but previously infected goats are usually resistant to abortion or other clinical signs when challenged with *T. gondii.*

Clinical Signs

Caprine toxoplasmosis is manifested by Perinatal deaths, including abortion and stillbirths. However, the systemic disease, with a high case fatality rate, is more common, especially in young goats. In sheep although a syndrome of fever, dyspnea, generalized tremor, abortions and stillbirth can occur, the systemic disease is rare. Abortions occur during the last 4 weeks of pregnancy and the rate may be as high as 50%. Full-term lambs from infected ewes may be born dead or alive but weak, with death occurring within 3-4 days of birth.

Diagnosis

Fetal serology is a very specific test for abortive toxoplasmosis. The ovine or caprine fetus begins to develop immunocompetency at 60 to 70 days of gestation. First immunoglobin M is produced, but by 90 days immunoglobin G is being synthesized. Thus if the fetus is infected after mid-gestation, it will usually produce antibody detectable by a variety of immunologic tests. A modified direct agglutination test

(MAT) is considered very sensitive and can be used on any species, including goats, because species specific conjugates are not used. Heart blood or thoracic fluid can be harvested. Fetal autolysis does not always preclude identification of toxoplasma antibodies. Absence of antibodies in the fetus does not rule out toxoplasmosis because the fetus may have been infected too young or antibodies may have decomposed. Screening at both high and low dilutions is recommended to avoid false negatives when the antibody concentration is so high as to occupy all binding sites.

Serologic testing of the doe at time of abortion is useful. Absence of antibodies is considered to be conclusive evidence that toxoplasmosis was not the cause of abortion. An increasing titer would at least indicate recent infection, but a stable, even high titer can persist for months or even years after initial infection of the doe. Thus a high titer in the dam is not proof of abortion caused by toxoplasmosis. Large surveys of goats in the western United States have revealed a sero-prevalence of 20% or more in yearlings or older goats, while 54 of 99 does (55%) on 9 farms in or near Tennessee were positive by indirect hemagglutination. In a study in Ontario, Canada, 63% of 399 sera tested positive (1:16 or greater) by the Sabin-Feldman dye test. The Sabin Feldman dye test uses intact parasites as antigen and requires maintenance of live organisms. Enzyme-linked immunoabsorbent assay tests, which are easily automated and use soluble antigens from disrupted toxoplasma organisms, may become more useful.

If the diagnosis is to be made by histology, it is very important that placenta be submitted. Small yellowish white foci of mineralization confined to cotyledons are apparent grossly if abortion is delayed until 45 days or more after infection. Washing the cotyledons thoroughly in isotonic saline solution makes deeper foci easier to visualize. Microscopic foci of necrosis can be identified after 30 days; tachyzoites are sparse in these lesions and difficult to locate. Nonsuppurative encephalomyelitis is found more consistently than myocarditis, but even under ideal experimental conditions it is difficult to find *T. gondii* organisms in tissue sections.

Post Mortem Findings

Multiple granulomatous lesion in the lungs, hydrothorax, ascites, intestinal ulceration, necrosis in the liver, spleen and kidneys, necrosis of placenta, brain haemorrhage, edema and ventricular dilatation and inflammation of the brain are the major post mortem findings. *Neospora caninum* is a newly recognized parasite that can cause abortion in goats. Its definitive host is unknown. Very careful histological examination or immunohistochemistry will distinguish *Toxoplasma* from *Neospora* infection.

Prevention and Control

The control of toxoplasmosis can be approached in several ways. The first is to prevent exposure of susceptible goats to the oocysts in cat feces during the period of danger, which is pregnancy. In particular, grain should be stored in covered containers and the mangers kept clean. Contamination of the hay supply has been implicated in several outbreaks. Because exclusion of all cats from a farm is very difficult, it is usually suggested to maintain a stable population of adult neutered cats; kittens younger than 6 months of age are far more apt to shed oocysts than are adults.

A live toxoplasmosis vaccine available for sheep in the U.K., would be effective. Repeated administration of killed vaccine or vaccination with a related but nonpathogenic organism *(Hammondia)* might be effective in preventing abortion but would certainly be expensive; such vaccines are unlikely to be marketed in the near future. Chemoprophylaxis, possibly by feeding ionophores such as lasalocid or Monensin during pregnancy at the normal anticoccidial rate, holds promise. When abortions caused by toxoplasmosis are diagnosed, emphasis should be put on proper disposal of fetuses and placentas, the wearing of protective gloves when handling these items, and proper pasteurization of milk and cooking of meat. Pregnant women should be especially careful.

9.5 Fascioliasis

Fascioliasis is the most common liver fluke disease of goats worldwide. In certain regions, chronic fascioliasis represents a major constraint on goat production.

Etiology

There are two *Fasciola* species that infect goats. *F. hepatica*is smaller, with an adult length of 18 to 32 mm and a width of 7 to 14 mm.

F. gigantica is longer with an adult length of 24 to 76 mm and a width of 5 to 13 mm. *Fasciola* spp. has an indirect life cycle, with water snails of the family *Lymnaeidae* acting as the intermediate host. More than 15 species of these snails serve as hosts for *F. hepatica* world wide. When *Fasciola* are introduced into new areas by importation of livestock, local species of *Lymnaeidae* may serve as suitable adaptive intermediate hosts. Infected snails introduced into a new area may also act as the source of new infections. Adult flukes reside within the biliary tract of definitive hosts such as goats and lay eggs that are passed via the bile duct to the feces. Egg production is substantial with a mean of 303,000 eggs found in the gallbladders of infected, slaughtered goats. Eggs passed in feces hatch within 10 to 12 days, releasing miracidia that invade intermediate host snails. Eggs hatch optimally at a temperature of 79°F, while temperatures must be at least 50°F for miracidia to survive, invade snails, and develop further. Within the snail, the miracidium undergoes subsequent multiplications through sporocyst, rediae, daughter rediae, and cercariae stages. The average time from infection of snails to shedding of cercariae is 5 to 8 weeks but can be as short as 21 days, or prolonged as long as 10 months under adverse conditions.When cercariae are shed by snails, they swim to a nearby plant, attach to herbage, encyst, and become infective metacercariae that are then ingested by grazing ruminants. The ingested metacercariae excyst, penetrate the intestinal wall, and migrate through the abdominal cavity to the liver. They then penetrate the liver capsule and migrate through the liver parenchyma to the bile ducts where they mature. They reach the bile ducts about 6 to 7 weeks after infection. Maturation within the bile ducts to egg laying adults requires an additional 2 to 4 weeks. Adult flukes may remain *in situ* for months to years. Occasionally, flukes are found in aberrant tissue locations such as kidney and lungs caused by inaccurate migration out of the intestine.

Epidemiology

Fascioliasis occurs principally in domestic ruminants, but other non ruminant livestock, numerous wild mammals also can be infected and may serve, as reservoir hosts. Man can contract fascioliasis from ingesting fresh plants contaminated with metacercariae. The distribution of *F. hepatica* is cosmopolitan, and reports of fascioliasis in goats come from around the world. Prevalence varies considerably. Approximately 4% of sheep and goat livers were infected in a Mexican study; while as many as 79% of sheep and goat livers inspected in Pakistan had *F. hepatica.*

The distribution of *F. gigantica* is more limited, primarily to the tropical regions of Asia and Africa, and the Middle East. Caprine infection with *F. gigantica* has been reported from all these regions. In a year-long Indian abattoir survey, 14.7% of goat livers were infected with *F. gigantica,* and 22.3% of potentially marketable liver was condemned. Besides a suitable snail host, the key element that regulates the occurrence of fascioliasis is a suitable combination of moisture and temperature that allows persistent surface wetness on pasture for the snail and the free-living stages of the parasite to thrive. The relationship of infection rate to weather conditions is sufficiently well known that meteorological calculation of increased fascioliasis risk is done in Europe to determine the need for prophylaxis. Low-lying, poorly drained, persistently marshy, or swampy areas in warm climates grazed by ruminants are ideal for completion of the life cycle of *Fasciola* spp. Water sources, however, do not need to be permanent. Intermittent flooding of fields by overflow of streams or frequent irrigation can facilitate infection because snails can burrow into mud and survive temporary dry periods. In temperate regions, infection of snails typically peaks in late spring and early summer, leading to peak pasture contamination with infective metacercariae in late summer and early fall. Peak incidence of clinical disease in ruminants is then seen in late fall and winter. Where winters are not extremely cold or excessively dry, some snails and eggs will survive winter and recommence development in early spring so that a less significant cycle of infection with metacercariae occurs on pastures in late spring and early summer as well. In tropical regions, metacercariae can be present all year long, and there is no well-defined season for clinical outbreaks in ruminants. Grazing is not a prerequisite for infection if stabled animals are fed fresh green herbage contaminated with metacercariae. Feeding fresh cut grass to penned goats is a typical husbandry practice in much of Asia. Goats, like sheep, do not develop as strong an immunity to *Fasciola* infection as cattle, and remain largely susceptible to reinfection. However, some immunity does develop as evidenced by reduced size and numbers of flukes recovered from repeated challenge infections.

Life Cycle

The adult flukes of *Fasciola hepatica* are found in the bile ducts and gall bladder. The eggs are shed into the bile duct from which they pass to the intestine. With animal faeces, the eggs are expelled out on the pasture.The larvae *(miracidia)* enter aquatic snails (*Limnea truncatula*) which are the intermediate hosts and develop into sporocysts and later into rediae. The rediae will further develop into the final larval stage (cercaria). Cercaria will transform in the external environment to metacercaria. If ingested by herbivorous animals, metacercaria will penetrate the small intestinal wall, cross the peritoneal space and reach the liver. In the bile ducts, metacercaria will mature into an adult fluke. The metacercariae which do not reach the bile ducts will encapsulate in the liver parenchyma.

Fascioloides magna is a large liver fluke which is prevalent in elk, deer and moose. Sheep and goats are susceptible to infection if they share the pasture with those wild animals. *F. magna* in sheep continuously migrate through the liver parenchyma and may cause death in less than six months.Lancet flukes (*Dicrocoelium dendriticum*) in sheep cause little damage to the liver parenchyma except for a moderate to marked thickening of the bile ducts.

Pathogenesis

There is little or no damage produced during migration of metacercariae through the intestinal wall or the peritoneal cavity. The potential for damage begins when metecercariae penetrate the liver capsule and begin migration through the parenchyma. When large numbers invade simultaneously, as may happen in the early fall, a traumatic hepatitis ensues and is recognized clinically as acute fascioliasis. When moderate numbers invade over an extended period of time, then subacute fascioliasis can occur. One thousand infective metacercariae can produce fatal acute fascioliasis in goats and 200 can produce severe subacute fascioliasis. *Fasciola gigantica* is more pathogenic to goats than *F. hepatica* and goats may be more susceptible to *F. gigantica* than sheep. In acute fascioliasis, there is marked disruption of liver parenchyma with extensive hemorrhage. The liver capsule may rupture, leading to severe intraabdominal bleeding and death. Even without fatal hemorrhage, animals may die within a few days as a result of hepatic insufficiency from parenchymal damage.

In the subacute form, invading metacercariae still produce migratory tracts, hemorrhage, and necrosis within the liver and lead to a more prolonged clinical picture of hepatic insufficiency. Extensive fibrosis occurs in the liver during subsequent attempts at healing. Acute fascioliasis is not commonly seen in goats and the reason is not clear. It may be related to feeding behavior because goats are not intensive grazers, and may not ingest sufficient numbers of metacercariae in short periods to produce severe acute reactions. Chronic fascioliasis occurs after arrival and maturation of flukes in the bile ducts. Persistence of adult flukes in the biliary tree causes hyperplastic cholangitis with ongoing loss of plasma proteins caused by increased permeability of the biliary mucosa. Feeding activity and movement of flukes within the biliary tree produces chronic blood loss. Under natural circumstances of exposure, individual animals may simultaneously experience acute, subacute, and chronic fascioliasis. In sheep, infectious necrotic hepatitis, or black disease, is a common complication of fascioliasis. In these cases, *Clostridium novyi* organisms already present in the liver suddenly proliferate in the anaerobic environment of necrotic tissue created by migrating fluke larvae. The release of toxins by proliferating bacteria produces an acute fatal toxemia. It is presumed that goats are less frequently affected than sheep because of their browsing habits. The prevalence in goats increases when they are forced to graze in wet areas during drought because C. *novyi* is a soil-borne organism.

Clinical signs

Acute fascioliasis, though uncommon, may present as sudden death or as progressive weakness, depression, anorexia, and pallor lasting as long as 3 days and resulting in death. Before death, pain may be elicited by deep palpation over the liver. Sub acute fascioliasis may show similar signs for as long as several weeks. Black disease usually presents as sudden death with no other signs except some possible frothing at the mouth and nostrils. Chronic fascioliasis is the most common presentation. The history may include a period of depression, poor appetite, lethargy and weight loss for 1 month or longer. In prolonged cases, some animals will have diarrhea. There may also be a complaint of decreased milk production in milking does. Physical findings include poor body condition, rough hair coat, pale mucous membranes, and tachycardia. Inter mandibular edema frequently occurs in long standing cases.

Clinical Pathology

Acute fascioliasis is most often diagnosed at necropsy. In subacute disease, the hemogram may document normochromic anemia, eosinophilia, and possibly neutrophilia. Hypoalbuminemia and increases in serum concentration of liver related enzymes may be noted in the chemistry profile. In chronic fascioliasis, normocytic or macrocytic anemia, hemoglobinemia, hypoalbuminemia and eosinophilia can be expected. Total protein may be in the normal range as a result of accompanying hypergammaglobulinemia. Increases in liver specific enzymes in the blood occur less reliably in chronic fascioliasis, although increases in GDH and OCT have been reported to occur even when flukes are present only in the biliary tree. In chronic fascioliasis, fluke eggs can be found in the feces of goats. Sedimentation methods for fecal examination are preferred to flotation methods. Eggs are yellow colored, ovoid, and operculated at one end. Eggs of *F. gigantica* are slightly larger than eggs of *F. hepatica.* Where rumen flukes occur, their eggs must be differentiated from those of liver flukes. Serologic tests have been used as an aid to diagnosis. An enzyme-linked immunoabsorbent assay test detected 13 of 14 infected goats in a natural outbreak of chronic fascioliasis.

Post Mortem Findings

Necropsy in acute fascioliasis demonstrates a swollen, traumatized, friable liver with multiple perforations of the capsule and sub capsular hemorrhage. Fibrin tags may be present on the liver surface and blood accumulated in the peritoneal cavity. Migrating flukes are not easily seen on cut surface of the liver parenchyma, but may be visualized by shaking a piece of liver in a pan of water and looking for the larvae to settle out at the bottom. When black disease complicates acute fascioliasis, signs of clostridial toxemia may be evident. These include blackening of the subcutaneous blood vessels observed as the skin is reflected, serosanguinous pericardial and peritoneal effusions, and patchy areas of liver necrosis with reddish borders. These reddened edges are likely

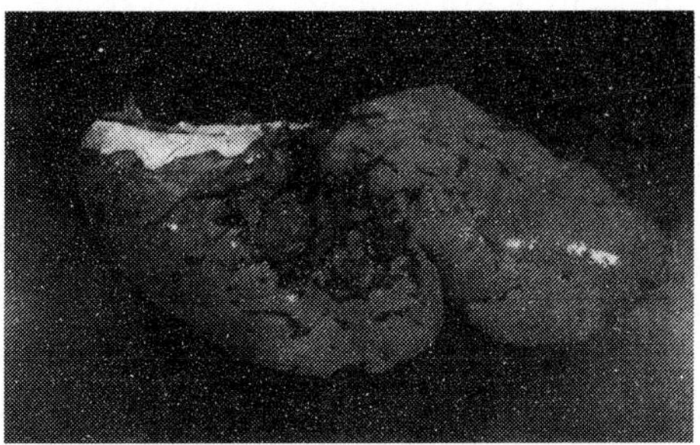

Fig. 33 Black parasitic debris in liver

sites for isolation of *Clostridium novyi.* In chronic fascioliasis the carcass is emaciated, subcutaneous edema may be present, and organs are pale. The liver may be cirrhotic and irregularly nodular and the capsule opaque in areas. On cut surfaces, there is extensive fibrosis of the parenchyma. Bile ducts are thickened and fibrosed, but mineralization is uncommon. Flukes have also been found in the pancreatic duct of goats at necropsy.

Diagnosis

Acute fascioliasis is one of many potential causes of sudden death in goats. Definitive diagnosis is by necropsy. Chronic fascioliasis is most easily confused with gastrointestinal nematodiasis, and, in practice, concurrent nematode and trematode infections are common in goats. When chronic weight loss is the major complaint or presenting sign, a wide variety of other diseases needs to be considered.

Treatment

The choice of therapy is largely dependent on the stage of infection requiring treatment. Carbon tetrachloride, hexachloroethane, bromsalans, oxyclozanide, niclofolan and albendazole are only effective on mature flukes, 10 weeks of age or older. Nitroxynil and clioxanide are moderately effective in flukes from 8 to 9 weeks of age and highly effective against mature flukes. Brotianide and rafoxanide are most effective against flukes 8 weeks and older but only have 50% to 90% effectiveness against flukes 6 to 7 weeks of age. Clorsulon is effective against all flukes 2 weeks of age and older. Triclabendazole and diamphenethide are effective against flukes from 1 day of age through maturity. Other considerations in drug selection include safety, cost, regulatory concerns, and lactation and pregnancy status. The earliest available treatments, namely carbon tetrachloride and hexachloroethane, are still used in some areas, but these compounds are themselves hepatotoxic and can cause mortality in goats. Though inexpensive, their use should be discouraged. Bromsalans, which are substituted salicylanilides, can be toxic when combined with benzimidazoles, and should be used cautiously in goats. Diamphenethide is expensive and may be cost prohibitive for widespread use in commercial flocks. Only oxyclozanide and albendazole can be used in lactating animals. Oxyclozanide is mostly used in goats with no statutory with-holding time. Specific doses for goats are established for a number of these drugs for use against *F. hepatica*. Albendazole orally at 15 mg/kg was 95.9% effective against adult flukes and produced no signs of toxicity even when given at 75 mg/kg. Diamphenethide given orally at 150 mg/kg removed 87.5% of juvenile and adult flukes. Treated animals included pregnant does and no adverse reactions were observed. Oral triclabendazole completely eliminated ova shedding in chronically infected goats at a dose of 5 mg/kg. The drug also has been recommended in goats at 15 mg/kg. In sheep, it has been shown that the lower dose kills flukes only as young as 2 weeks of age whereas the higher dose is effective on flukes as young as 1 day of age. Niclofolan was 90% effective against *F. hepatica* when given subcutaneously at a dose of 0.8 mg/kg in goats, with no adverse effects. Rafoxanide was reported 100% effective against *F. hepatica* in goats at a dose of 7.5 mg/kg orally. Nitroxynil was 89% effective against 6-week-old larvae at a subcutaneous dose of 15 mg/kg. Both rafoxanide and nitroxynil are also effective against *Haemonchus contortus* in goats. Oral bromopheiophos at 16.5 mg/kg eliminated all adult *F. gigantica* in goats but was only 50% effective against immature forms. Doubling the dose had little additional effect. In general, drugs and dosages effective against *F. hepatica* are also effective against *F. gigantica.* Acute and subacute fascioliasis should be treated with diamphenethide or triclabendazole. All animals at risk should be treated whether they are showing signs at the time of first diagnosis or not. The prognoses in acute and sub acute fascioliasis are poor.The prospects are much better in treating chronic fascioliasis. All drugs effective against mature flukes can be used in these cases. However, if drugs not effective against immature forms are used, then animals should be retreated at an interval appropriate for that drug to eliminate subsequently maturing larvae. Early response to treatment can be demonstrated by elimination of ova shedding. It may take 1 month or longer for erythrocyte and protein parameters to return to normal in treated goats.

Prevention and Control

Control is aimed primarily at reducing pasture contamination with infective larvae. Several approaches can be taken. First, adult fluke burdens in goats should be kept to a minimum. This involves maintaining good nutrition, reducing exposure to high-risk grazing areas with large snail populations, and the strategic use of anthelmintics to control concurrent nematode infections.Pasture contamination by metacercariae is a dynamic process largely dependent on weather conditions. When calculated, the meteorological index can be helpful in determining the amount of tactical control necessary to avoid clinical disease and production loss on a seasonal basis. At the minimum, in temperate regions or arid zones, flukicide treatments should be administered strategically once in the late winter/early spring before turnout to pasture using an adult flukicide. Goats should be retreated in the autumn to kill developing and adult flukes picked up over the grazing season. An anthelmintic that kills immature and mature flukes should be used. In tropical condition, particularly when year-round grazing occurs, treatment every 8 to 10 weeks with triclabendazole or diamphenethide may be indicated. This facilitates destruction of all ages of immature flukes within the pre patent period to eliminate egg production while controlling both acute and chronic fascioliasis in the goat. Egg production can also be controlled by use of rafoxanide or closantel at 5 or 6 week intervals, but the risk of acute fascioliasis is still high because immature larvae are not killed. Such intensive control measures will be effective in reducing pasture burdens and may not have to be repeated in subsequent years. When goats are part of mixed grazing systems, all potential hosts need to be treated simultaneously. Limiting access of goats to snail populations is an adjunct to the use of flukicides. High-risk areas such as low-lying, marshy portions of grazing land, drainage ditches, irrigation channels, or stagnant ponds should be identified and drained or fenced off when possible. Cutting back vegetation at the edges of standing water makes the environment less amenable to snails. High risk situations should also be identified such as intensified grazing by animals near water after drought or at the end of the grazing season when forage is in short supply. Housed goats should not be offered freshly cut herbage likely to be contaminated with metacercariae. When mild winters and warm, moist summers promote high snail populations, the application of molluscicides to snail habitat should be considered. Effective compounds include copper pentachlorphenate at a dose of 11.2 kg/ hectare diluted in 4500 liters of water for use as a spray , copper sulfate in solutions of 1:100,000 to 1:5,000,000 or as a powder at 10 to 35 kg/ hectare, and N-tritylmorpholene at a dose of 0.45 kg diluted in 680 liters/hectare. Applications in spring kill over wintered snails and applications in summer and fall will kill newly developed snails. Goats are less susceptible to copper toxicity than sheep, but caution should still be taken in turning goats out on copper-treated pastures before a dilution by rain has occurred. These compounds are toxic to fish. The use of ducks and competitive nonhost snail populations as means of biological control has also been attempted with limited success.

9.6 Lancet Fluke Infection

Dicrocoelium dendriticum, the lancet fluke, generally produces a chronic form of liver fluke disease in goats that is less severe than that produced by *Fasciola* spp. It has a cosmopolitan distribution.

Etiology

Dicrocoelium dendriticum is a small liver fluke measuring 6 to 10 mm in length and 1.5 to 2.5 mm in width. It has an indirect life cycle involving two intermediate hosts, terrestrial snails, and ants. Forty species of snails from 9 families may serve as primary intermediate hosts but all secondary host ant species are of the subfamily *Formicinae.* Goats, sheep, and cattle are the common definitive hosts, but other livestock species and wild mammals and man can be infected. Adult flukes reside in the bile ducts and excrete eggs to the outside in feces. Eggs are ingested by snails where they develop into cercariae. Cercariae are expelled by snails in slime balls and are ingested by ants. Infectious metacarcariae develop in the ants and are neuropathogenic, causing ants to remain paralyzed on herbage during peak times of livestock grazing. Infected ants are ingested by grazing livestock and the young flukes excyst from metacercariae in the small intestine of the definitive host. In contrast with the *Fasciolidae,* these immature flukes do not migrate through the peritoneal cavity and liver parenchyma. Instead, they enter the hepatic biliary system directly through the common bile duct via the small intestinal lumen. The prepatent period for *D. dendriticum* is 8 to 12 weeks.

Epidemiology

Dicrocoelium dendriticum infection is common in North America, Europe, Asia, North Africa, and the Middle East. Prevalence in goat populations is reported as high as 45% in endemic areas. It is less common in South America and is not found in Australia, New Zealand, and much of Africa.Numerous habitats are potentially contaminated with *D. dentriticum* because snail hosts are numerous, terrestrial ant populations are widespread, and many wild mammals serve as reservoir hosts. Grazing areas near forests are particularly high-risk sites. The fluke eggs are also resistant to desiccation and freezing. Hibernating ants can maintain cercarial infections so that grazing stock are at risk from the beginning of spring grazing.

Pathogenesis

These flukes are less pathogenic than the *Fascioloidae* because they do not migrate through liver parenchyma. *Dicrocoelium dendriticum* infection, which may involve thousands of flukes per goat, can cause progressive inflammation of the entire biliary tree and potentially lead to fibrosis, biliary cirrhosis and hepatic insufficiency.

Clinical Signs

Dicrocoelium dendriticum may be a subclinical infection or produce a disease syndrome similar to chronic fascioliasis. Affected goats have a history of weight loss. They appear thin and depressed and may show signs of anemia and hypoproteinemia, including pallor and intermandibular edema.

Clinical Pathology

Identification of lancet fluke eggs in the feces of goats is the most reliable means of diagnosis. The dark brown, operculated eggs are much smaller than those of *F. hepatica* or *F. gigantica* and can be found by both fecal flotation and sedimentation methods. *Dicrocoelium dendriticum* infection may produce elevated serum levels of GGT and AP caused by biliary inflammation. Anemia and hypoproteinemia are evidenced in the hemogram.

Post Mortem Findings

At necropsy, adult lancet flukes are found in the bile ducts. In early cases there is thickening of the biliary ducts. In severe, chronic infections, biliary fibrosis, cirrhosis and scarring of the liver surface may be noted.

Diagnosis

Lancet fluke disease must be differentiated from chronic fascioliasis, gastrointestinal helminthiasis, paratuberculosis, and other causes of chronic wasting diseases of goats.

Treatment

Albendazole is the drug of choice against the lancet fluke at dose rate of 15 mg/kg orally with a precaution against use during the first 3 months of pregnancy. Thiophanate is used at 50 mg/kg orally with a milk withdrawal time of 3 days. Diamphenethide at 220 to 330 mg/kg orally is highly effective, with no toxic side effects. Drugs reported to have an efficacy of 90% or greater in sheep include thiabendazole at 200 mg/kg, fenbendazole at 100 mg/kg, praziquantel at 50 mg/kg.

9.7 Paramphistomiasis

Paramphistomes are trematode parasites commonly referred to as rumen flukes, stomach flukes, or conical flukes. Adult flukes may be present in the rumen of goats in large numbers, but are essentially nonpathogenic. Clinical paramphistomiasis is associated primarily with the immature forms that feed voraciously in the small intestine before moving to the rumen to mature.

Etiology

Numerous genera in the family Paramphistomatidae infect domestic ruminants. In goats *Cotylophoron cotylophorum, Paramphistomum cervi, P. explanatum, Fischoederius elongatus,* and *Gastrothylax crumenifer* are found. These trematodes all have a similar, indirect life cycle involving water snails as intermediate hosts . The size of mature flukes ranges from 5 to 20 mm making them visible to the unaided eye. They are pink to red. One paramphistome of Asia, *Gigantocotyle explanatum,* migrates to and matures in the bile ducts. It is essentially nonpathogenic, and should not be confused with the pathogenic liver fluke, *Dicrocoelium dendriticum* adult paramphistomes inhabit the rumen and lay clear, operculated eggs that are passed in the feces. When feces are passed into water, miracidia develop from these eggs over a period of 12 to 21 days depending on temperature. Swimming miracidia enter water snails of various genera including *Planorbis, Bulinus, Lymnaea,* and others. Mature sporocysts containing rediae develop in snails over a period of 11 days. Rediae are released within the snail and after an additional maturation of 10 days, contain multiple cercariae. In turn, cercariae are released from rediae and mature within the snail for approximately 13 more days. Mature cercariae are released from snails into water under stimulation of strong sunlight. Liberated cercariae attach to herbage and encyst as metacercariae, remaining viable for as long as 3 months. When ruminants ingest contaminated herbage, metacercariae excyst in the small intestine where they feed aggressively and mature for a period of 6 to 8 weeks. Young flukes then migrate anteriorly to the rumen where they undergo additional maturation after attachment to the rumen mucosa, begin laying eggs, and complete the life cycle. Migration to the rumen and subsequent development in that organ can be delayed or prolonged up to several additional months when fluke infestations are heavy. Development of flukes varies in goats, cattle, and sheep. After experimental infection of the three ruminants with metacercariae of *P. microbothrium,* migration of maturing flukes from the intestine to the rumen was virtually complete in sheep and cattle by 34 days after challenge, but was just beginning in goats at that point, and subsequent egg laying began 2 weeks later in goats than cattle or sheep. This prolonged residence of maturing paramphistomes in the small intestine may contribute to increased pathogenicity of the fluke in the goat.

Epidemiology

Though paramphistome infection of domestic ruminants occurs throughout the world, clinical disease is confined to Africa, Asia, Australia, Eastern Europe, Russia, and the Mediterranean countries. While most reports involve cattle and sheep, reports of caprine paramphistomiasis from India and Pakistan suggest that goats are at risk wherever sheep and cattle are affected. In general, outbreaks of paramphistomiasis are increased during dry seasons when snail and livestock populations become concentrated around shrinking natural water sources with restricted grazing and browsing. Animals grazing in low-lying, poorly drained, swampy areas or with access to irrigation ditches and other standing water supplies are at high risk of infection caused by increased contact with metacercariae produced by water snails. In north central India, most outbreaks are seen in small ruminants between late September and January after the rains that promote increases in snail populations. During subsequent dry periods of decreased forage availability, goats may be fed rice paddy straw and other grasses grown underwater that contain large numbers of metacercariae. Morbidity rates in goats in India during such outbreaks have been reported in the range of 35% to 79% and mortality rates in the range of 45% to 88%. Slaughterhouse surveys in the region show that mature rumen fluke burdens are highest from March to October and lowest from November to February, while immature fluke burdens are highest from September to April, the time period correlating to clinical outbreaks.

Pathogenesis

The immature fluke developing in the upper small intestine is most responsible for the pathologic effects of paramphistome infection. These immature flukes imbed deeply in the intestinal mucosa via suckers and feed by drawing a plug of mucosa into the sucker. This mucosal plug becomes necrotic and sloughs, leaving an erosion and petechiation. The pathogenicity of the paramphistomes is correlated directly with the immature fluke burden. It is estimated that attachment of 50,000 immature flukes would completely denude the upper small intestine where most paramphistome infections are confined during the developmental stage. Worm burdens of this intensity do occur under natural conditions. Large numbers of simultaneously feeding, immature flukes cause marked intestinal irritation and mucosal disruption leading to hypoproteinemia, diarrhea, and general debilitation.

Clinical signs

The clinical signs of paramphistomiasis are similar to those of nematode gastroenteritis. Young goats are more frequently and seriously affected than older goats. Affected goats are listless and have diminished appetite. They may be polydipsic, and stand for long periods with their muzzles in the water. They will show a fluid diarrhea that may be projectile at first but later may just drip from the rectum, staining the hindquarters. The diarrhea may contain mucus, epithelial shreds, and immature flukes. It has a characteristic and pronounced fetid odor. Intermandibular edema is pronounced and may extend over the face and brisket. Anemia, if present, is usually mild or moderate. The course of the disease is approximately 5 to 10 days in goats. During that time the animals lose weight, get progressively weaker, and eventually die.

Fig. 34 Bottle Jaw in goat

Clinical Pathology

Marked hypoproteinemia and hypoalbuminemia occur. Anemia may also be present. Because clinical disease is caused by immature non-egg laying flukes, fecal sedimentation techniques for egg identification have little value because false-negative results would be common. As an alternative to checking for ova, diarrheic feces can be sieved through a sieve and the residue examined either microscopically or macroscopically against a black background for immature flukes that are commonly passed in the feces. The immature flukes are pinkish-white with a large prominent sucker.

Post Mortem Findings

At necropsy, the carcass is thin and the hindquarters are soiled with feces. Subcutaneous edema, ascites, hydrothorax, hydropericardium, and lung edema may be present. Significant gross lesions of the alimentary tract are limited to the pyloric portion of the abomasums and the first 2 to 3 m of the small intestine. The abomasal folds are enlarged and edematous and immature flukes may be seen attached to the mucosa, which also contains numerous erosions and petechia. The affected intestine is thickened and edematous. The pitted mucosal surface is covered with catarrhal exudate. Careful inspection may reveal flukes deeply imbedded in the mucosa, with only the external ends present in the lumen. The mucosa in this affected region appears corrugated with many elevated ridges, and multiple foci of hemorrhage. Immature flukes may also be found free in the gut lumen. Mature flukes may be present in the rumen and omasum. Histologically, affected areas of intestine show hypertrophy, edema, inflammatory cellular infiltrates, and fibrosis in the mucosa and submucosa.

Diagnosis

Presumptive diagnosis is difficult because of the clinical similarity of paramphistomiasis to nematode gastroenteritis. Haemonchosis is more likely to produce anemia and less likely to produce diarrhea. However, other nematode infections may mimic paramphistomiasis exactly. Definitive diagnosis depends on ruling out nematode gastroenteritis by fecal flotation or necropsy and ruling in paramphistomiasis by identification of immature flukes in feces or at necropsy. Chronic fascioliasis must also be considered in the differential diagnosis. It occurs under similar environmental circumstances and may appear similar to the chronic form of paramphistomiasis.

Treatment

Elimination of the immature flukes with appropriate anthelmintic therapy is the major therapeutic objective and is life saving if treatment is started early in the course of disease. Morantel citrate at a dose of 6 mg of morantel base/kg b.w. has recently been shown to be 99.5% effective against immature paramphistomes. Other drugs with a reported efficacy of 95% against immature flukes in sheep and goats include Bithionol (25-100 mg/kg), Niclofolan (6 mg/kg), Niclosamide (50-100 mg/kg), and Resorantel (65 mg/kg). Oxyclozanide (15 mg/kg) has a slightly less consistent efficacy range of 85% to 100% against immature flukes, but is also 100% effective against adult flukes. A new tertiary butylbenzathiole has been shown to be 99.7% effective against immature flukes and 100% effective against mature ones. In general, the benzimidazoles are minimally effective against paramphistomes despite their relative effectiveness against liver flukes. Bithional may be toxic to goats at the increased effective dose ranges. Directing treatment toward adult flukes is controversial. It has no direct benefit in the face of clinical outbreaks. Elimination of adult flukes will reduce fecal egg counts and subsequent infection of snails, thereby reducing the general environmental burden of paramphistomes. Elimination of adult flukes, however, may also reduce the level of immunity that they induce in infected hosts thereby increasing the risk of clinical disease in subsequent exposures.

Prevention and Control

The major thrust in the prevention of paramphistomiasis is to keep grazing animals away from areas with heavy concentrations of infected snails and from herbage contaminated with metacercariae. This means avoidance of low-lying, poorly drained or swampy areas, ponds, ditches, and paddy fields. Small ruminants should not be grazed with cattle under such circumstances, as the latter are known to shed large numbers of eggs and promote increased infection of snails. Strategic use of anthelmintics can be helpful in controlling paramphistomiasis. In India it is recommended that livestock be treated in June and August to reduce mature fluke burdens and thereby reduce egg laying, and in November, January, and March to control immature flukes. Though not yet available for widespread practical application, immunization may be a major future tool in paramphistomiasis control. Oral vaccination with irradiated metacercariae significantly reduced paramphistome burdens in goats on subsequent challenge compared with unvaccinated control goats.

9.8 Coenurus Cerebralis Infection (Gid, Sturdy)

Coenurosis is a disease of the brain and spinal cord caused by the intermediate stage of *Taenia multiceps* which inhabits in the intestine of dogs, cats and wild carnivores. The clinical disease occurs in goats and sheep and rarely in cattle, causing primarily a focal encephalopathy.

Etiology

The metacestode involved in this disease, is the intermediate stage of the tapeworm of carnivores,*Taenia multiceps,* formerly known as *Multiceps multiceps.* The metacestode is identified in sheep and goats as *Coenurus cerebralis. T. multiceps* is a large intestinal tapeworm of dogs, foxes, coyotes, and jackals.

Pathogenesis

Eggs or gravid tapeworm segments are passed in the feces of the host, contaminate herbage, and are ingested by grazing ruminants. Embryos are released after ingestion and penetrate the small intestine, enter portal vessels, and are distributed widely throughout the tissues via the circulation. In sheep, only larvae that reach the central nervous system develop into metacestodes; all others die. In goats, metacestode development occurs primarily in the central nervous system, but frequently in the skeletal musculature and the heart as well. Cysts also have been found in mesenteric lymph nodes. There is a period of migration through nervous tissue that may result in clinical signs of acute meningoencephalitis if large numbers of larvae are present. In most cases, however, this migratory phase is subclinical. After migration, the stationary phase of metacestode maturation lasts from 2 to 7 months. The mature metacestodes are cystic structures up to 7 cm in diameter, transparent, thinwalled, and filled with clear fluid. They contain up to several hundred scolices that are visible as white plaques on the clear cyst wall in excised cysts. The majority of mature cysts are located superficially in the parietal or frontal region of the cerebral cortex, but location is variable throughout the central nervous system and musculature. The life cycle of the tape worm is completed when carnivores ingest mature cysts while feeding on tissues of infected ruminants.

During early growth of cysts, irritation of adjacent nervous tissue occurs and may be reflected in some cases by early excitatory signs.

As the metacestode enlarges, pressure on adjacent tissues increases and neuronal death occurs, leading to signs of neurologic deficit. The pressure that develops is considerable and is reflected in a high incidence of papilledema and rarefication of cranial bone adjacent to superficial cysts. Cerebrospinal fluid pressure also can be markedly elevated. The clinical signs that develop reflect the location of the cyst or cysts present.

Life Cycle

Eggs expelled with dog faeces are ingested by the intermediate host (goats). The larvae hatch in the intestine and pass with the blood stream towards different organs. The larvae which reach the brain and spinal cord grow to the coenurid stage. Coenurus cerebralis will further mature in the brain and spinal cord.

Epidemiology

T. multiceps is present in carnivores worldwide, but coenurosis is not common in goats in most countries. There are numerous clinical reports originating from the Indian subcontinent. Thirty two caprine cases were recorded in 1 year at a single veterinary institution in India. A slaughterhouse survey in Bangladesh demonstrated a prevalence range of 1.9% to 13.3% for cerebral coenurosis. Caprine coenurosis is sporadic in Europe and Africa. The disease is rare or absent in goats and sheep in North America, Australia, and New Zealand. The disease is most commonly reported in young female goats, but this may reflect their disproportionately large presence in goat herds rather than a predilection to disease. Morbidity rates are low but mortality rates are high in untreated cases. The disease is infrequently reported in man, with most documented cases found in Africa. The goat is not directly involved in the zoonotic aspect of the disease.

Clinical Signs

During migration of larval stage, there is a blindness, muscular tremor and incoordination, excitability and collapse. Infection with the fully developed larval stage, salivation, wild expressions, frenzied running and convulsion, deviation of eye and head, loss of function, dullness, incomplete mastication, head pressing, incomplete paralysis and, in spinal cord involvement, inability to rise are the main symptoms. Because of the prolonged development phase for coenurosis, cases may be seen anytime of the year. Affected animals may have a history of ill-defined, low grade neurologic abnormalities for 1 month or longer. Animals with the neurologic form of the disease may show a variety of signs. The clinical presentation is variable, but it has been suggested that there are three phases of development. In the first stage, animals may be depressed, and show intermittent periods of abnormal head carriage, especially head tilt. There may be short, intermittent convulsions, or intermittent circling. This phase lasts 4 to 8 days and may be overlooked. In the second phase, which usually begins 8 to 16 days after onset of signs, signs become more consistent. The most common findings are head tilt, unilateral blindness, circling, and loss of balance with a staggering gait. Some animals may show determined forward walking with the head held high and a hypermetric gait followed by head pressing against obstacles instead of circling. Other signs include grinding of the teeth, salivation, exophthalmos, nystagmus, and irregular rumen contractions. The third or terminal phase usually begins 12 to 25 days after onset of signs. Animals become recumbent, exhibit extensor rigidity, and paddle frequently with the hind limbs. These stages overlap considerably and the clinical course is frequently 1 month or longer. Papilledema and hemorrhage of the optic papilla are common in coenurosis cases, so a fundic examination should be performed. In addition, firm, digital palpation of the head will frequently identify dyssymmetry or softened areas of skull suggestive of underlying superficial cysts. Digital pressure on such spots may exacerbate clinical signs. Cysts deep in the parenchyma will not be associated with bony changes.

Clinical Pathology

Total protein, cell counts, and pressure are often elevated in the CSF. Mean total protein in affected goats has been reported at 80.3 mg/ dl, mean cell counts at 85/mm^3, and mean pressure at 262.4 mm water. Eosinophils and degenerative cells may be common on cytologic examination. Allergic skin testing has also been used with positive goats having a reaction of more than 1.2 cm.Radiographs can reveal meta cestode-associated rarefaction of the skull when bone softening has not yet reached the point of detection by digital palpation. For identification of deep parenchymal cysts, cerebral angiography has been used in goats. The angiograph reveals stagnation of contrast material around the avascular zone of the cyst surrounded by concentric arches of displaced vessels.

Post Mortem Findings

At the time of a necropsy, cysts ranging in size from 1 to 7 cm in diameter can be identified in predilection sites such as brain, spinal cord, and skeletal muscle.

Diagnosis

Though the clinical course is prolonged, it may not always be recognized as such, and animals may present as acute onset of circling, falling or head pressing. The differential diagnosis should include listeriosis, polio encephalomalacia, brain abscesses, sinusitis, cerebral hematomas or tumors, cerebrospinal nematodiasis, and rabies. The nose bot, *Oestrus ovis* may occasionally penetrate the frontal sinuses and produce a clinical syndrome.

Treatment

There are no known effective anthelmintic therapies for elimination of the metacestode. When bone softening is pronounced so the site of the cyst is obvious, aspiration of cyst contents is accomplished via an aseptic needle passed through the softened or perforated bone. After this, a mild antiseptic is introduced, such as 0.1% acriflavine solution. When bone is still firm but lesions are identifiable by radiology, surgical treatment is possible. The skull is opened over the area of the cyst and the cyst removed by first bluntly dissecting it from underlying brain tissue and then lifting it out either by application of suction or by a forceps.

Prevention and Control

Control methods involve efforts to reduce stray dog populations, restriction of feeding small ruminant carcasses and offal to dogs, and strict, regular treatment of stray dogs with anthelmintics for the control of mature tapeworms.

9.9 Echinococcosis (Hydatid Disease)

Hydatid disease occurs in goat, sheep, cattle, swine, horses and humans. Echinococcosis is a disease which occurs when the larval stage of *Echinococcus granulosus* and *Echinococcus multilocularis* are ingested by an intermediate host (goat, sheep, cattle). These larvae then develop into hydatid cysts in various tissues.

Etiology and Pathogenesis

The adult tape worms are found in dogs, cats and other carnivores. They may ingest the hydatid cysts by eating infected organs of the intermediate hosts. The scolex attaches to the intestinal wall. Adult tapeworms develop in approximately seven weeks and eggs are shed in the faeces and are ingested by goat, sheep and cattle. The ova hatch to liberate the *onchospheres* which penetrate the intestinal wall and through the portal venous supply to the liver where they become arrested. In older animals the larvae may reach the lungs and various other organs through the systemic circulation. The most common sites of cysts are the liver and lungs. The cysts are different sizes and shapes and they contain a clear fluid. Due to the growth of the cyst, pressure atrophy is noted in the surrounding tissue. Daughter cysts are found outside the mother cyst and are formed due to trauma or external pressure on the mother cyst. They may or may not be attached to the mother cyst. Daughter-cyst formation may have neoplastic characteristics when there is penetration to the blood and lymph vessels and metastases to various distant organs. Humans get infected with hydatid disease via the ingestion of ova from *Echinococcus* tapeworm in the dog. This usually occurs by touching dog hair that has been contaminated by ova from faeces. It also may occur by the dog transferring ova from the anus to its mouth and then by licking humans.

Epidemiology

Echinococcus granulosus occurs worldwide, but is endemic in portions of Africa, Latin America, the Mediterranean region, India, the former USSR, eastern Australia, and in sheep-raising regions of the western United States. All ages, breeds, and sexes of goats can be infected. The prevalence of disease will be exacerbated when carnivore populations are high and no efforts are made to segregate domestic and wild carnivores from livestock or their grazing areas. The feeding of offal to dogs also enhances completion of the life cycle. Increases in prevalence of hydatid disease in livestock occur after rainy periods, presumably because the rain widely disseminates eggs in feces on grazing land or in water supplies. Prevalence data of hydatid disease in goats is available from abattoir and post mortem surveys around the world. The prevalence range in endemic regions is from 0.26% to 6.5%. In some places, lung cysts are more common than liver cysts in goats.

Clinical Signs

Hydatid cysts are most common in the liver, lung, and spleen, but may be found in other organs at slaughter or necropsy. Cysts may average 5 to 10 cm in diameter and contain a yellowish, serum-like fluid and may have a granular inner wall containing multiple brood capsules. Hydatid "sand," which is an accumulation of detached brood capsules, may be seen in the cyst fluid.

Post Mortem Findings

Multiple Echinococcus granulosus cysts in the liver, lungs and other organs. The animal carcass affected with echinococcosis is approved if edema and emaciation are not found. Otherwise the carcass is condemned. The affected organs are also condemned and must be destroyed. The lungs are most commonly affected and these should be carefully checked because lesions are often missed on routine inspection.

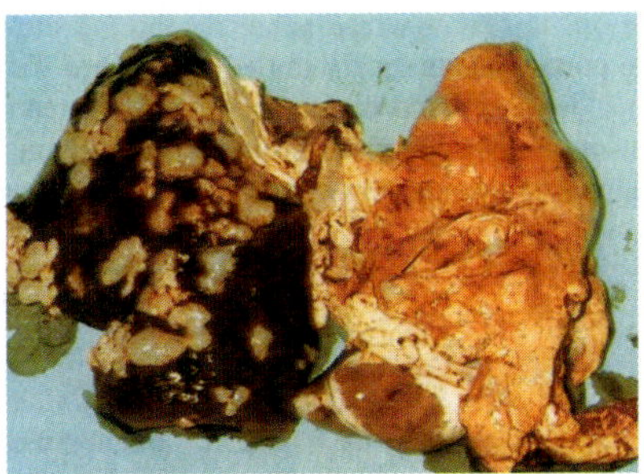

Fig. 35 *E.granulosus* cysts in liver.

Treatment

There is no treatment for hydatid cysts. Control efforts involve disruption of the carnivore/ruminant life cycle. International efforts to reduce hydatidosis, especially with regard to its zoonotic potential, have been reviewed.

9.10 Cysticercosis

Adult *Taenia hydatigena* tapeworms are found in the intestine of dogs, coyotes, wolves and other carnivores. Cysticercosis is a disease which occurs when the bladder worm called *Cysticercus tenuicollis* are found attached to the mesentery, omentum, and serosal surface of abdominal organs of the host.

Etiology

Adult *Taenia hydatigena* tapeworms are found in the intestine of dogs, coyotes, wolves and other carnivores. The eggs are passed in the feces and ingested by goats, sheep and other domestic and wild ruminants. Eggs hatch in the small intestine of the intermediate hosts and enter the blood. Upon reaching the liver, the embryos leave the portal vasculature and migrate through the hepatic parenchyma to the peritoneal cavity, causing distinct hemorrhagic tracts in the liver. The metacestode, in this case is a bladder worm called *Cysticercus tenuicollis*. It matures over a period of 5 to 8 weeks, and is found attached to the mesentery, omentum, and serosal surface of abdominal organs. Occasionally, migration out of the liver is not achieved and the cysticercus will be found in the liver itself. Aberrant migrations sometimes occur with cysticercus found in the lungs or other organs.

Epidemiology

Cysticercosis occurs worldwide. Reported prevalence in slaughtered goats around the world ranges from 0.2% to 23.3% in various countries.

Pathogenesis

The migration of embryos of *T. hydatigena* in the liver produces parenchymal damage and the creation of blood-filled tracts. When many embryos are migrating, acute, fatal hepatic insufficiency may occur in goats. Less severe migrations of embryos may still lead to liver condemnations in otherwise healthy animals, as the blood-filled tracts become fibrotic over time. Occasionally, peritonitis is associated with migration out of the liver and into the peritoneal cavity. The mature cysts in the mesentery, omentum, or liver usually do not produce any clinical disease.

Clinical Signs

Acute cysticercosis may produce signs of depression, weakness, anorexia, and possibly collapse and death caused by hepatic insufficiency. Signs of pneumonia, peritonitis, and anemia can occur as a result of aberrant migration in the lungs and abdominal viscera with resulting hemorrhage. Fever may be recorded if peritonitis occurs. Signs usually occur between 7 and 20 days after exposure. Chronic cysticercosis is usually asymptomatic.

Clinical Pathology

Experimentally, goats infected with eggs of *T. hydatigena* showed marked elevations of ornithine carbamyltransferase (OCT) and AST 10 days after infection, but this would not be specific for cysticercosis.

Indirect hemagglutination and complement absorption tests have been used in the diagnosis of caprine cysticercosis. Radiography or ultrasonography might be useful in identifying hepatic or peritoneal cysts.

Post Mortem Findings

At necropsy, C. *tenuicollis* cysts may be present in the liver, but are common in the mesentery, omentum, and serosal surface of peritoneal viscera. Increased volumes of serofibrinous fluid may be present in the peritoneal cavity. Mature cysticerci have a smooth inner surface and contain only a single invaginated scolex, in contrast to hydatid cysts.Acute cysticercosis is recognized by the presence of tubular, red, blood filled tracts 2 to 4 mm in diameter in the liver. More chronic lesions may appear white as leukocytes fill the tracts.

Treatment

Praziquantel, at a large dose of 60 mg/kg, was reported as 93% to 100% effective in removing C. *tenuicollis* from sheep and goats.

Prevention and Control

Control measures involve disruption of the carnivore/ruminant life cycle as much as possible. The principal efforts should be directed at elimination of the feeding of livestock offal to dogs, burial of livestock carcasses or viscera, and routine anthelmintic therapy to dogs to reduce adult tapeworm populations.

9.11 Setariasis (Cerebrospinal Nematodiasis)

The filarial worm *Setaria digitata* is nonpathogenic in definitive ungulate hosts, but can produce a cerebrospinal nematodiasis in goats, sheep in Asia and India. The disease is commonly referred to as lumbar paralysis or kumri.

Etiology and Pathogenesis

Setaria digitata is a non-pathogenic parasite of the peritoneal cavity of cattle, buffalo, and zebu. The life cycle of these filarids is not completely understood, but is known to be indirect. Mosquitoes of various genera including *Anopheles, Aedes,* and *Armigeres* spp. serve as the intermediate hosts. In the definitive cattle host, adult worms live in the peritoneal cavity and produce microfilaria that circulate in the blood stream. Mosquitoes ingest the microfilaria along with blood during feeding. Within 2 weeks these microfilaria mature into infective larvae and enter the salivary glands of the mosquito. During subsequent feedings they are introduced into the blood stream of new ungulate hosts. In the definitive host these larvae migrate to and mature to adults in the peritoneal cavity to complete the life cycle. In aberrant hosts such as the goat, sheep, and horse, the microfilaria migrates to the central nervous system, sometimes the eye, and occasionally the fetus of pregnant animals, where they cause mechanical damage and an inflammatory reaction. Clinical signs usually develop 1 month after infection. The life cycle is not completed in aberrant hosts, and microfilarias are not detectable in the blood stream.

Epidemiology

The disease was first suspected in imported goats in Japan in 1927, but the parasitic nature of the disease was not confirmed until much later. It has since been reported in goats in Japan, India, Korea, Sri Lanka, the former USSR, and China. The disease occurs as a seasonal epizootic, being associated with increased numbers of mosquitoes. In India, most cases are seen in September through December, after the rains and before winter. Morbidity rates can be high, with, as many as 30% to 40% of susceptible animals affected. The proximity of large cattle populations infected with *S. digitata* predisposes to infection of aberrant hosts.

Clinical Signs

All ages, sex, and breeds of goats may be affected. The appearance of clinical signs may be acute or subacute, and mild to severe. Variable outcomes occur including acute death, progressive worsening of neurologic signs, stabilization of neurologic deficits, and occasionally spontaneous recovery. Most typically, affected animals show evidence of spinal cord involvement with ataxia, lack of proprioception, paresis or paralysis. Any or all limbs may be affected, but most often hind limbs are involved. Sensory deficits are minimal. Additional signs suggesting brain involvement include cranial nerve deficits, nystagmus, and circling. In most cases, animals remain mentally alert. Fever is not observed.

Clinical Pathology

As with other parasitic infections of the central nervous system, eosinophilia of the CSF can occur in caprine setariasis.

Post Mortem Findings

Gross necropsy lesions are rarely identified and diagnosis depends on thorough microscopic examination of the brain and spinal cord, concentrating especially on those areas that correlate with the clinical neurologic examination. In most cases only a single larvae is involved, and rarely more than three. Microscopically, lesions are characterized by a focal malacia that can be identified in serial sections as the migratory tract or channel of the offending parasite. In adjacent areas there may be evidence of axonal degeneration, demyelination, astrocytosis, and perivascular cuffing with mononuclear cells and eosinophils. Hemorrhage is variable. Locating the actual offending larvae in histologic sections is uncommonly achieved.

Diagnosis

Antemortem diagnosis is presumptive based on the seasonal observation of epizootic neurologic disease characterized principally by locomotor dysfunction in countries where setariasis occurs. Definitive diagnosis depends on identification of the parasite in the central nervous system of affected animals at time of necropsy.

Treatment

The filaricide diethylcarbamazine can be an effective treatment in goats and sheep when animals are treated early in the course of disease. Early signs may represent meningeal irritation or radiculitis caused by the parasite before its entry into the brain or spinal cord and the parasite may still be susceptible to anthelmintic therapy at this stage. Diethylcarbamazine is given orally, suspended in water, once at a dose of 100 mg/kg b.w., or at a dosed 40-60 mg/kg daily for 1 to 6 days. Once the parasite has crossed the meninges, response to therapy is poor.

Prevention and Control

To reduce the incidence of setariasis in goats, the parasite should be controlled in the normal host, cattle, particularly at the beginning of the mosquito season. Because the infection is subclinical in cattle, routine treatment of cattle by farmers may be difficult to enforce. When possible, goats should be housed separately from cattle. The use of diethylcarbamazine in sheep and goats at an oral dose of 40 mg/kg given one time every 3 weeks before and during the mosquito season appeared to reduce the number of clinical cases observed in the field compared with untreated controls. Reduction of mosquito populations could also reduce the transmission of the parasite from cattle to goats.

9.12 Intestinal Cestodiasis

Tapeworms occur in goats throughout the world. Relative to gastrointestinal nematodiasis, cestodiasis in goats usually has little clinical or economic significance. However, goat owners are often keenly aware of tapeworms because the proglottids, or egg packets, shed by the adult worm in the goat feces are visible to the naked eye.

Etiology and Pathogenesis

The major intestinal tapeworm of goats worldwide is *Moniezia expansa,* though other *Moniezia* spp. may also infect goats. *Avitenella spp*. occurs in goats in Europe, Africa, and Asia, often in conjunction with *Moniezia* spp., and is sometimes the predominant cestode. *Stilesia globipunctata* occurs in goats and other ruminants mainly in tropical regions of Africa and Asia. It is potentially the most pathogenic of the ruminant tapeworms, producing inflammatory nodules at the sites of attachment in the duodenal and jejunal mucosa that can lead to enteritis and diarrhea. Another *Stilesia* tapeworm, S. *hepatica* occurs in the bile ducts of ruminants in Africa and Asia. Adult tapeworms developing in the goat intestine can be up to several meters long. They consist of a head or scolex and short neck followed by a long, segmented body composed of proglottids. The most posterior proglottids are packed with eggs. These gravid segments separate from the worm and are passed in the feces. These egg packets are white and 1.0-1.5 cm in length. They are visible in the fecal pellets and have the appearance of rice. The proglottid casing gradually disintegrates and oribatid or psocid mites, common in soils and on herbages, ingest the eggs. Infective cysticercoids develop within these arthropod intermediate hosts over a period of 4 months. Infective mites are consumed by ruminants while they feed. Cysticercoids are released from the ingested mites in the intestine and mature tapeworms begin to develop. The prepatent period in the goat is around 40 days for *Moniezia* spp. The goat also serves as an intermediate host for a number of cestode parasites that use canids as definitive hosts. Light infestations of goats by the common tapeworms are generally nonpathogenic. This is largely because tapeworms do not feed destructively with active mouthparts, but rather absorb nutrients from the intestinal lumen through their integument. In goats, at least 50 worms are required to produce deleterious effects. Several hundred may be present in an individual goat. They may compete significantly for nutrients with the host leading to ill thrift. They may produce luminal distension in the intestine resulting in a distended, potbellied appearance. Their presence may prolong transit time of ingesta in the gut. In feedlot lambs on grain rations, this is believed to promote the development of clostridial enterotoxemia. A similar interaction has not been described in goats. Massive tapeworm infections may sometimes even occlude the intestinal lumen. This results in signs of colic and may lead to spontaneous rupture of the intestine in young lambs and kids with dire consequences.

Epidemiology

Infections are most often observed in young kids at pasture during their first summer grazing season. Pasturing is not a prerequisite for infection, however, as the infective mites may be present in the barnyard or on carried forage. The pattern of infection may vary with climate and geography. A natural resistance to cestode infection develops with age and in populations of goats with constant exposure to tapeworms, worm burdens are always less severe in older goats. Because tapeworms are often derived at pasture, heavy tapeworm burdens may also be a sentinel for heavy nematode infections. Therefore, when owners notice tapeworm proglottids in goat feces, deworming for tapeworms may be used as an opportunity to treat more serious nematode infections when broad spectrum anthelmintics are used. Multiple parasitic infections are the rule, not the exception, in goats.

Clinical Signs

When clinical disease is associated with tapeworms, it usually involves young goats younger than 6 months of age. Affected animals may show poor growth rates and a pot-bellied appearance. Proglottids are present in voided feces that are usually normal in appearance, but may be soft or unpelleted. Constipation may also occur. In complete luminal obstruction caused by tapeworms, kids will have symptoms of colic and decreased fecal output. Animals that have ruptured the intestine may be profoundly depressed, or found moribund or dead.

Clinical Pathology

In addition to proglottids in the feces, characteristic tapeworm eggs may be detected by fecal flotation methods as proglottids sometimes degenerate before being shed. *Moniezia expansa* ova are distinctly triangular. There is no other specific clinical chemistry or hematological abnormalities associated with tapeworms.

Post Mortem Findings

At necropsy, the long, white-segmented tapeworms are strikingly obvious in the lumen of the small intestine. In obstructions, they will pack the lumen, and may be free in the abdominal cavity when rupture of the intestine has occurred.

Diagnosis

Tapeworm infections are definitively diagnosed by the presence of proglottids in the feces or worms in the intestine. Attributing disease to tapeworm infection is more problematic. It is essential that nematode parasitism, paramphistomiasis, and liver fluke disease all be thoroughly ruled out before clinical disease is ascribed to tapeworms.

Treatment

A wide range of anthelmintics are currently available for eliminating adult tapeworms in goats. Oral niclosamide (50 mg/kg b.w.) is highly effective and has a wide margin of safety, being nontoxic at 5 times the recommended dose. Praziquantel (5 mg/kg b.w.) is also effective, but the injectable form can be highly irritating to goats at the injection site. Oral febantel (5 mg/kg) is effective against *Moniezia expansa* and a range of nematodes. The newer benzimidazoles are also effective against tapeworms and nematodes. These include mebendazole (15 mg/kg), fenbendazole (15 mg/kg), cambendazole (20 mg/kg), and oxfendazole 10 mg/kg). Older treatments for tapeworms in goats include copper sulfate, nicotine sulfate, and phenothiazine with lead arsenate. Though reasonably effective, these compounds all are potentially toxic if misused or improperly dosed and their use is discouraged.

Prevention and Control

It is difficult to justify a preventive program aimed at adult tapeworm control on economic grounds because the clinical impact on goats is usually minimal. When the appropriate broad spectrum anthelmintics are used for nematode parasite control however, tapeworm control can be included in the program as a bonus. The group of major concem in both cases is the young goat on pasture for the first time. Control of the intermediate host mites is difficult because these arthropods are ubiquitous and often present in enormous concentrations. Plowing and reseeding are considered to reduce mite populations at pasture, but the overall effect on tapeworm populations in goats is questionable. Pasture rotation with at least a 1 year layoff may be helpful as mite populations will decrease over winter.

9.13 Lungworm Infestations (Verminous Pneumonia)

Dictyocaulus filaria is the common goat and sheep lung worm which cause verminous pneumonia or bronchitis.

Etiology

Although *Dictyocaulus filaria* is cosmopolitan, in many parts of the world this lungworm is not very common in goats.

Epidemiology

This may be caused in part by the tendency to keep young goats off pasture under intensive management systems. In India, sheep and goats graze the same infested pastures, and goats are believed to be more susceptible to the parasite.

Life Cycle

Adult worms live in the bronchi where they lay eggs which are coughed up to pharynx and swallowed by the host. The eggs are hatched in the digestive tract and the larvae are then expelled in the faeces. In a moist environment and moderate temperature, the larvae will become infective in 3 - 7 days. Larvae are resistant to cold, although it will cause their maturation to be delayed. Upon digestion by goat (primary host), larvae penetrate the intestinal wall and reach the mesenteric lymph nodes. From the mesenteric lymph nodes via the blood stream, larvae migrate to the lung alveoli and further to the bronchi. They mature in the bronchi and lay eggs. The cycle is then repeated. *Muellerius capillaris* parasitizes in the alveoli and pulmonary parenchyma. Intermediate hosts are snails and slugs which goats ingest during grazing. Larvae reach the lungs and produce small grayish nodules on the back of the lungs.

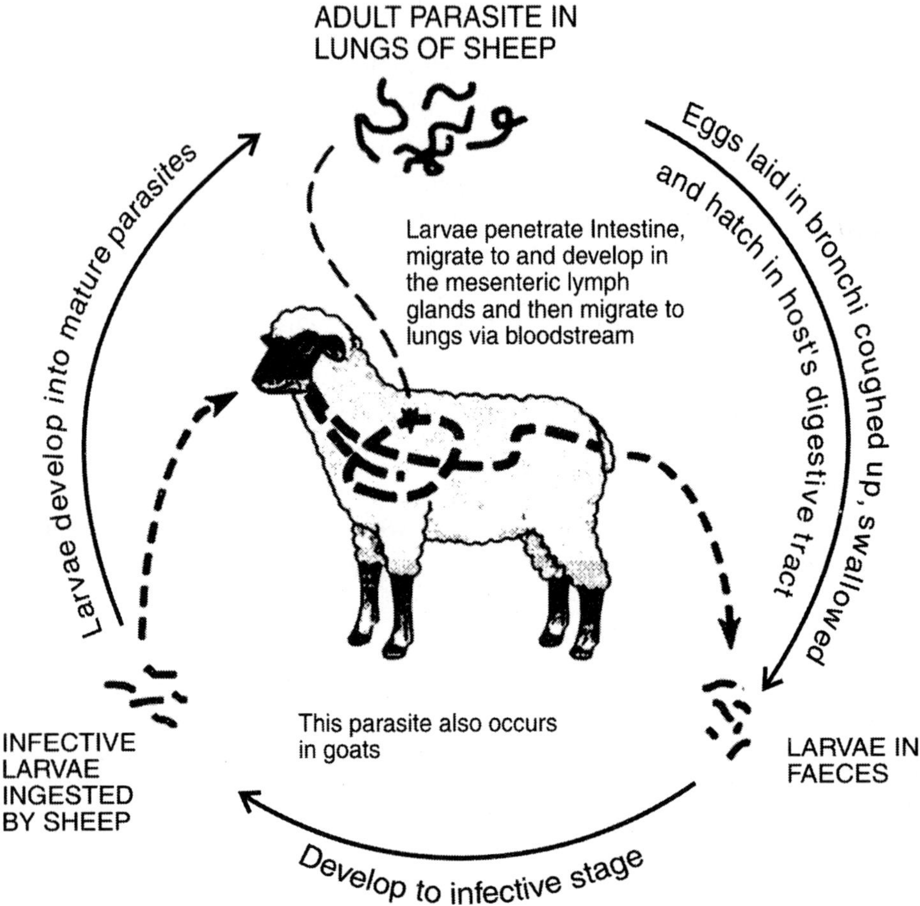

Fig. 36 Life history of lungworm of sheep

Pathogenesis

Dictyocaulus filaria lungworms have a direct life cycle. Adults live in bronchi and are approximately 30 to 80 mm long. First-stage larvae are coughed up and eliminated in the feces. Development to an infective stage takes 1 to 2 weeks. These third-stage larvae are long lived in damp, cool surroundings. Signs can occur as early as 3 weeks after the goats are put on pasture in the spring, but usually occur in autumn. Kids are most severely affected during their first season on pasture; age resistance is via prior exposure.

Clinical Signs

The main sign during the prepatent period is polypnea; dyspnea develops later in the course of infection. The worms cause bronchial irritation, resulting in cough, moderate dyspnea, and loss of condition. Pathologic findings include bronchitis, bronchiolitis, atelectasis, and emphysema. Auscultable crackles and wheezes, fever, and toxemia occur if there is a secondary bacterial infection. Eosinophilia is marked in re-exposed animals. An acute shock syndrome is seen when a significant reinfestation occurs after immunity has weakened.

Diagnosis

The diagnosis of verminous pneumonia is best done by the Baerman technique. Feces are collected fresh from the rectum to avoid confusion with *Strongyloides* larvae. Fecal pellets are put in a tea strainer or wrapped in gauze and half submerged in a funnel of warm 77.7° F water; larvae fall to the neck of the funnel. The larvae are harvested after 4 to 8 hours. The posterior is blunt in *Dictyocaulus,* slender and pointed in *Protostrongylus,* and has a definite dorsal spur in *Muellerius.*

Treatment

Treatments include tetramisole (15 mg/kg) or levamisole (7.5 mg/kg) orally or subcutaneously. The drug is eliminated by the respiratory tract. Also reported to be effective are mebendazole (15 to 20 mg/kg), fenbendazole (5 to 10 mg/kg), febantel (5 mg/kg) and ivermectin (0.2 mg/kg).The withdrawal period for ivermectin should probably last 1 month. Treatment does not allow establishment of immunity, so goats should not be returned to infected pastures.

Prevention and Control

Frequent pasture rotation is an impractical means of control because moving the goats every 4 days would be required. Putting the goats out on pasture early in the season allows progressively increasing exposure and immunity. New goats should be separated from yearlings, and kids should not be placed on pastures inhabited by yearlings within the last year. There is a need to avoid overpopulation, drain wet pastures, and replace ponds with watering devices on concrete slabs. An irradiated *Dictyocaulus* larvae vaccine has protected goats against experimental challenge.

9.14 Protostrongylids

Etiology

Protostrongylid lungworms include at least five genera. *Muellerius capillaris* is diagnosed more frequently than *Protostrongylus rufescens.* Two more *(Cystocaulus ocreatus* and *Neostrongylus linearis)* are recognized, while *Varestrongylus pneumonicus (Bicaulus schulzi)* is found in Eastern Europe and India.

Epidemiology and Pathogenesis

All types have an indirect life cycle and use a variety of snails and slugs as intermediate hosts. The mollusks remain infective for at least 1 year. Goats become infected by eating mollusks or third-stage larvae liberated on foliage by death of the intermediate host. *Protostrongylus rufenscens* adults live in bronchi and are approximately 16 to 35 mm long. Massive infections are unlikely, and only kids show any clinical involvement. *Muellerius* lungworms usually cause clinically insignificant focal lesions in the lung parenchyma. Adults are 12 to 23 mm long and located in subpleural alveoli. Lesions are most common in the caudal dorsal lung lobes and are less well localized in goats than in sheep. Larvae are passed if both sexes are present in the same nodule; the infestation is carried over from one year to the next in the goat. Goats do not develop a marked age resistance.At least forty species of snails and slugs serve as intermediate hosts for *Muellerius.*

Clinical Signs

Muellerius infestation are usually subclinical. However, there are scattered reports of serious, even fatal pneumonia resulting from heavy *Muellerius* infestations. Dyspnea and a persistent cough were noticed in goats with higher than average larval counts, suggesting that the parasite may be pathogenic in heavily infested goats.

Post Mortem Findings

There is a diffuse thickening of alveolar septa with mononuclear cell infiltrates to *Muellerius.* Diffuse (non-nodular) lesions occurred in cases of reduced host resistance, as eosinophils were more prominent in focal nodular lesions. As larvae, eggs, and adult worms in "brood nodules" are readily identified in histologic preparations of the lung, it must be remembered that other conditions (such as progressive retroviral pneumonia) must be excluded before ascribing a clinically important pneumonia to these parasites. Also worth considering is the possibility that *Muellerius* lungworms contribute to clinical retroviral infection by attracting many macrophages to the lungs.

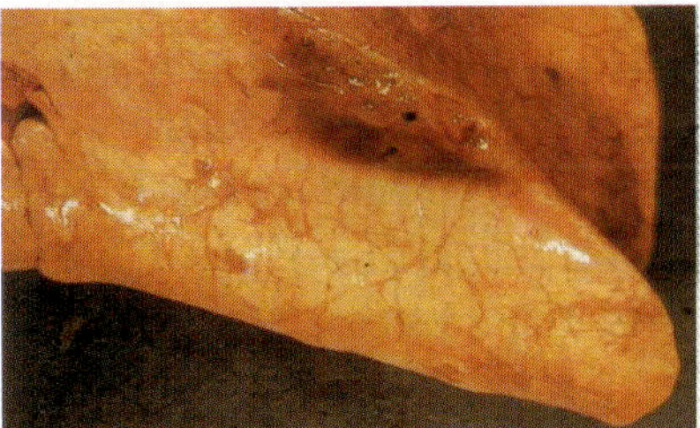

Fig. 37 Numerous greyish nodules of *M. capillaris* in the lung parenchyma.

Treatment

Levamisole and fenbendazole are effective at the same dose rates as used for *Dictyocaulus* infection. Levamisole is not effective against *Muellerius.* Two treatments with fenbendazole at 15 mg/kg 3 weeks apart, completely eliminated shedding of larvae 2 weeks later. Extended studies revealed that larval shedding resumed in most goats by 7 weeks after 30 mg/kg fenbendazole. Another study found that shedding of larvae resumed many weeks after treatment with either fenbendazole or ivermectin and immature *Muellerius* in the lungs resumed development after death of adult worms. Treatment at approximately 35 day intervals has been suggested.

Prevention and Control

During the period when the goats are stabled and not lactating, fenbendazole is fed at 2 mg/kg/day for 14 days. Prolonged low-level treatment reduces larval shedding for 5 to 7 months. This treatment is believed to decrease the prevalence of coughing and to increase subsequent milk production and body weight gains. The possible teratogenic effects of anthelmintics have not been well studied in goats, and it is prudent to avoid treatment during the first 35 days of pregnancy. Wet, undrained pastures should be avoided. It may be helpful to prevent grazing in early morning or evening when the herbage is damp. It is best to treat before pasture season in the spring, to prevent contamination of mollusks. A second treatment in early fall is used in heavily contaminated environments.

9.15 Nematode Gastroenteritis (Haemonchosis, Trichostrongylosis, Ostertagia, Nematodirus, Cooperia and Bunostomiasis)

Nematode infection of the gastrointestinal tract is one of the major causes of wastage and decreased productivity in goats worldwide, particularly under grazing conditions. The natural history of caprine nematodiasis is in many ways similar to that seen in cattle and sheep. Studies in the goat, however, suggest some important species differences in terms of parasite susceptibility, age-related immunity, and the pharmacokinetics of anthelmintic therapy.

Etiology

Many nematode parasites are found in the gastrointestinal tract of goat such as *Haemonchus, Ostertagia, Trichostrongylus, Nematodirus* and *Cooperia* spp. All have a direct life cycle. Adult nematodes in the alimentary tract of the host produce ova that are passed in the feces. First-stage larvae develop in ova within 1 day of passage to the external environment. For most species, first-stage larvae then break out of ova and molt to second-stage larvae. These molt one more time and become infective third-stage larvae. Development to the infective stage usually takes 7 to 10 days but can vary according to environmental conditions; mainly temperature and moisture. *Nematodirus* spp. is distinguished from the other genera in that development to infective third-stage larvae occurs within the ova during the 3 week period after passage in the feces. This adaptation enhances survivability over periods of adverse weather. In *Marshallagia* spp., larvae develop to the second stage before hatching from the ova. Infective larvae of all genera are ingested by the host during consumption of contaminated feed or water. When goats are at pasture, infective larvae will migrate to the tops of grasses during mornings and evenings, enhancing ingestion by grazing hosts. After ingestion, the larvae travels to its predilection site in the alimentary tract. The larvae then burrows into mucosal folds or digestive glands and molts within 1 to 2 days to a fourth-stage larvae. This larvae may remain in place for as long as 10 days and then returns to the mucosal surface to molt finally into an adult capable of producing ova to complete the life cycle. The average pre-patent period is between 3 and 4 weeks.

Trichuris spp. is from trichuridae family (whipworms) and has a direct life cycle. Development to the third larval stage occurs within the egg and release of the larvae does not occur until after ingestion by the host. Embryonation within the egg occurs after passage in the feces, taking 3 weeks or longer depending on temperature and moisture conditions. Ingested eggs release larvae that penetrate the small intestine. Larvae develop for 2 to 10 days before moving to the cecum for maturity to the adult stage. The pre-patent period is 7 to 9 weeks.

Skrjabinema spp. is from oxyuridae family (pinworms) affecting goats. The life cycle is direct. Eggs are fully embryonated when deposited by the adult female on the peri-anal skin of the host. Eggs ingested by the host hatch in the small intestine and larvae migrate to the large intestine to become adults within 25 days of infection.

Oesophagostomum spp. (nodular worms) is from the family of Trichonematidae have a direct life cycle and infective larvae penetrate deep to the submucosa of the alimentary tract reaching the lamina propria for development to fourth-stage larvae. This can occur anywhere from the pylorus to the rectum. After the molt, fourth stage larvae return to the mucosal surface and migrate to the colon to develop into adults. The pre-patent period is approximately 6 weeks. The characteristic nodules that occur in the intestine of affected animals are associated with a host reaction to the deeply burrowed infective larvae, not the surface feeding adults. The third-stage larvae of *Chabertia* spp. develop deep in the intestinal wall, but nodule formation is not a characteristic of chabertiosis. The pre-patent period is approximately 7 weeks.

Bunostomum and *Gaigeria* hookworms of goats are from the *Ancylostomatidae* family. These nematodes have a direct life cycle, but a different route of transmission. Infective larvae penetrate the skin or oral mucosa of the host. Such larvae enter the bloodstream via capillaries and end up in the lungs. They disrupt pulmonary capillaries and enter the alveoli where they molt to fourth-stage larvae. These larvae are coughed up and swallowed. Migration to the small intestine and development of adults then occurs. The pre-patent period is approximately 9 to 10 weeks, but adult worms may be present and feeding in the goat colon by 4 weeks after infection.

Strongyloididae papillosus is the sole caprine pathogen in family of Strongyloididae. The life cycle is unique among the gastrointestinal nematodes. *S. papillosus* is parthenogenic: either free-living or parasitic development can occur. Under adverse environmental conditions, ova produced by parthenogenic females are passed in the host feces and develop only to infective third-stage larvae. Under more favorable environmental conditions, the free-living cycle is more common. Ova are passed in the host feces again, but within 48 hours these develop into sexually mature free-living males and females. After copulation, the females produce a single generation of infective larvae. In both cases, the resulting infective larvae must enter a suitable host to complete a subsequent life cycle. Infection of the host may occur by penetration of the skin, or the oral, or esophageal mucosa. Penetrating larvae enter the bloodstream and localize in the lungs where they enter the alveoli, migrate up the airways, and are swallowed. They mature to adults in the intestine. The prepatent period is 6 to 7 days. In addition, transmission of infective larvae to neonates via the milk of infected dams has been documented to occur in goats, though transplacental transmission was not demonstrated. Very young kids infected via milk or colostrum can pass ova in the feces, and *S. papillosus* may be the only gastrointestinal nematode found in pre-weaned kids.

Epidemiology

The completion of the parasitic life cycle of gastrointestinal nematodes in goats depends on a variety of environmental, parasitic and host-related factors and interactions. The feeding behavior of ruminant species is a major factor in the development of parasitism. Lush, dense pasture in turn provides a protective umbrella for developing larvae, screening out direct sun to reduce desiccation. Direct sunlight has been shown to reduce the survival time of nematode larvae contained in goat fecal pellets. Overstocking of pastures generally promotes increased parasitism. Wild ruminants using same pastures may also transmit nematodes to the goat. Management systems also play a role in the type and intensity of caprine nematodiasis, and a well-recognized advantage of confined housing systems is the marked reduction in nematode parasite loads. Indoor housing was associated with a low incidence of clinical parasitism, attributable mainly to *Chabertia* and *Oesophagostomum*. In goats with access to yards, ostertagiasis was more common and in pastured goats, haemonchosis was the predominant problem. The most dramatic adaptation to hostile environments is hypobiosis or arrested development. In hypobiosis, infective larvae consumed by the host during periods of environmental adversity remain voluntarily dormant and progress to adulthood only when environmental conditions favor survival of ova or larvae outside the host. In temperate regions, increasing temperatures may signal the larva commitment to hypobiosis, while in tropical regions with distinct seasons, hypobiosis is triggered by the onset of hot, arid conditions. Even in tropical regions where temperature and humidity conditions can support free-living larvae year-round, some degree of hypobiosis may occur in response to increasing moisture content of the soil. Environmental temperature is an important factor in the survival of nematode ova. Various nematode ova in the feces of goats died within 6 days at a temperature of 40°C. Ova survived and hatched optimally within 8 to 9 days at temperatures between 30° and 35°C. Hatching was delayed for 14 days at temperatures between 20°C and 25°C. The ova are highly susceptible to desiccation and will not survive in regions where hot, dry summer follows winter rainfall, or where winters are intensely cold. Infective larvae, when developed, are more resistant to weather and can survive repeated periods of desiccation. Warm, humid regions of the world with summer rainfall, and temperate regions with mild winters are conducive to development of infective larvae on pasture. In *H. contortus,* hypobiosis largely replaces over wintering of free larvae as a means of survival for the species. The pattern of development and geographic distribution of *Oesophagostomum columbianum* is similar to that of *H. contortus. Bunostomum trigonocephalum* and *Gaigeria pachysalis* are best suited to humid subtropical and temperate regions. They thrive particularly well under management conditions where housing and bedding are allowed to remain continuously damp because early larval stages are particularly susceptible to desiccation.

Several phenomena in the host-parasite relationship favor success for the parasite. These include the development of anthelmintic resistance in parasites, hypobiosis, and the periparturient egg rise phenomenon.The goat is not totally defenseless, however, and some host resistance is also recognized as well. Parasite resistance to anthelmintics is a growing problem in the control of gastrointestinal nematodiasis in small ruminants. Based on post drenching larval cultures, resistant parasites included *Ostertagia, Haemonchus,* and *Trichostrongylus* spp. On some premises, the nematodes were resistant to anthelmintics in both the benzimidazole and cholinergic agonist classes. The existence of true parasite resistance to anthelmintics in goats, however, is not easy to establish with certainty because the pharmacokinetics of many common anthelmintics used in goats are not completely known. Presumed cases of anthelmintic resistance may be in fact treatment failures that occur when anthelmintics are under dosed, or when doses prescribed for sheep are assumed to be effective in the goat. Goats treated with benzimidazoles continued to pass ova of *Haemonchus, Trichostrongylus,* and *Ostertagia,* while identically treated, commingled sheep had a total suppression of egg output. Thiabendazole administered at the same dose to commingled sheep and goats was moderately effective against *Haemonchus* in sheep but totally ineffective in goats. It is suggested that the anthelmintics used do not achieve therapeutic levels in goats at commonly used dosages, or that resistance patterns of nematodes may change in response to host transfer. In fact, one study has definitively demonstrated that *Trichostrongylus* presumed to be resistant to levamisole in goats were indeed sensitive to levamisole *in vitro*. The presumed resistance was in essence a treatment failure because levamisole given to goats at the sheep dose of 7.5 mg/kg did not maintain adequate plasma levels for sufficient duration to be effective against this particular parasite. The periparturient egg rise commonly observed in sheep results from maturation and ova production by previously arrested larvae, particularly of *H. contortus* and *O. circumcincta.* The intensity of immunity with advancing age and parasite experience varies among ruminant species. Goats show the weakest degree of immunity, sheep somewhat more, and cattle the most resistance to infection as adults. Significant worm burdens have been observed much earlier in kids than lambs, with worm burdens of more than 17,000 noted in kids between 3 and 4 weeks of age. Goats with paratuberculosis are predisposed in increased worm burdens to *H. contortus* infection. The phenomenon of "self cure" is known in sheep that ingest large numbers of infective larvae of *H. contortus* and subsequently expel their existing adult worm populations. Because these animals usually become immediately re-infected, the adaptive significance in terms of parasite control is unclear. Small grazing goats breeds such as the Black Bengal were far less affected than large browsing goats like the Beetal and Jamunapari during a natural outbreak of haemonchosis after unexpected heavy rains.

The development of gastrointestinal parasitism in goats depends on a complex interrelationship of parasite, host, and environmental factors, many known and some unknown. In general, the types and degree of parasitism that develop in goat populations may be predicted on the basis of geographic and climatic location, management system, and prevailing weather conditions. Nematodes, by evolving diverse strategies and rates of development, different mechanisms for feeding, and different predilection sites in the host alimentary tract, appear to have maximized their exploitation of the host, while minimizing competition among each other. This diversity encourages development of multiple nematode infections, which are the rule in goats. The outcome of infection depends on host resistance, level of infection, variety and types of parasites involved the development of parasite resistance to anthelmintics, and the extent of appropriate therapeutic intervention. Acute death, clinical disease, or sub-clinical infection with adverse growth and lowered productivity are all possible outcomes. Worldwide, numerous post mortem studies on causes of mortality in goats, particularly young goats, have confirmed that gastrointestinal nematodiasis is an important cause of morbidity and mortality. However, specific studies concerning the impact of sub-clinical gastrointestinal nematodiasis on production parameters in goats are scarce.

Pathogenesis

Various pathogenic mechanisms are involved in gastrointestinal nematodiasis depending on the genera involved. The principal effect of blood feeding worms on the host is a progressive debilitating anemia. Blood feeders include *H. contortus* and *M. digitatus* in the abomasum, and *B. trigonocephalum*, and *G. pachyscelis* in the intestine, and *Trichuris* spp. in the cecum. Each *H. contortus* adult in the abomasum may be responsible for the loss of 0.05 ml of blood per day, either by active feeding, or by moving to new feeding sites and leaving old sites to continue hemorrhaging. When infection rates are high (more than 10,000 adults per host), and development of large adult populations is synchronous, as can happen after arrested development, then death caused by acute blood loss is possible. In less severe infections, three stages of anemia may be recognized in infected hosts. In the initial stage of blood loss, packed cell volume may decrease markedly because intraluminal blood loss in the alimentary tract is not a strong trigger for hematopoiesis. In experimental haemonchosis of goats the packed cell volume decreased from a mean of 29% to 16% within 19 days of infection with 9,000 to 12,000 larvae. In the second stage, regenerative erythropoiesis begins and the packed cell volume stabilizes, albeit at less than the normal level, for as long as 6 to 14 weeks. During this time, however, iron stores in the host are being reduced by loss in the feces. In the final stage the packed cell volume begins to decrease again as erythropoeisis is impaired by the progressive iron deficiency.

Concurrently, a steady loss of serum proteins occurs as a result of parasite feeding. Serum albumin levels may be maintained initially due to replacement by tissue catabolism, but hypoalbuminemia eventually develops, accompanied by cachexia and clinical evidence of hypoproteinemia such as intermandibular edema. Infective larvae of *Trichostrongylus axei* develop to adults within the abomasal mucosa and the adults feed there and lead to erosion of the mucosal epithelium, catarrhal inflammation, hyperemia, edema, and diarrhea. Plasma loss from the mucosal damage contributes to the hypoproteinemia seen with this infection. Infective larvae of *O. circumcincta* enter the gastric glands of the abomasum to undergo the third and fourth molts to adulthood. When synchronous maturation of large numbers of arrested larvae occurs, a severe gastritis results. Gastric glands become hyperplastic, intercellular tight junctions are weakened, hydrochloric acid secretion diminishes, and the pH of stomach content increases. Hypoproteinemia, and diarrhea are cardinal signs of ostertagiasis. The *Trichostrongylus* spp. that infect the intestine, tunnel under the mucosal epithelium to feed, leading to a protein-losing enteropathy accompanied by diarrhea. Larval stages can be as destructive as adult worms. Over time, marked villous atrophy occurs. *Oesophagostomum columbianum* presents a unique situation in small ruminants. Infective larvae burrow into the submucosa of the small intestine, encyst for the third molt, and then return to the intestinal lumen to migrate to the colon for the final molt. In first-time infections, this process occurs unremarkably. In previously exposed, sensitized hosts, however, encysted third-stage larvae produce a dramatic local inflammatory response around the cyst leading to the formation of caseating nodules. The larvae inside may die or resume migration at a much later time. Nodules may occasionally rupture serosally causing peritonitis, adhesions, and partial or complete obstructions of the intestine. Even without rupture, widespread nodular formation may impair digestion, absorption, and passage of excreta. In addition, adult nodular worms can produce a severe catarrhal colitis with much mucus production. Diarrhea with mucus, weight loss, and hypoproteinemia occur in severe infections. Anorexia and reduced feed intake are consistent findings in all forms of intestinal nematodiasis. Poor growth, decreased productivity, and weight loss also are caused by nematode infections. Failure of muscle growth in young animals results from a decrease in skeletal muscle synthesis, with a shift to increased albumin production by the liver to counteract ongoing protein loss. Muscle wasting in severely affected animals may be associated with muscle catabolism.

Mixed parasitic infections are common and often it is not possible to attribute clinical signs to a single parasite. In general, infections with *Trichostrongylus, Ostertagia, Cooperia* and *Nematodirus* spp. produce a similar clinical picture. Young, grazing animals are most likely to be affected, particularly after weaning. A gradual, progressive loss of condition, poor growth, a dull attitude, and a decrease in feed intake are the most consistent findings. In more severe infections, a dark green to black diarrhea is evident, with staining of the hair and skin of the tail and perineal region. When the course is prolonged, intermandibular edema may develop secondary to hypoproteinemia. Chronically infected animals also develop a pot-bellied appearance, a rough, dry hair coat and flaky skin. Evidence of anemia is usually not pronounced. Deaths may be reported, and are usually spread out over a period of days or weeks. *O. columbianum* infection may result in signs of abdominal pain such as hunched back and reluctance to move when nodule formation occurs and localized peritonitis results. Affected animals may be febrile. Occasionally, nodules may abscess and rupture. Pus may be passed per rectum if the rupture is intraluminal. Diffuse peritonitis may result if the rupture is intraabdominal. Excessively mucoid feces with occasional blood are also associated with *Chabertia ovina* infection. When blood feeding parasites such as *H. contortus* infect goats, clinical evidence of anemia will predominate. When infections are massive, peracute haemonchosis can occur, with animals dying of gastric hemorrhage. Acute and chronic forms are more common. Affected animals will exhibit marked pallor of mucous membranes and conjunctivae, and respiratory and heart rates may be increased. Intermandibular edema, or "bottle jaw" is common. Weakness, reluctance to move, and exercise intolerance will be observed. Constipation is more common than diarrhea in uncomplicated haemonchosis, In prolonged disease, weight loss is also a common finding. Diarrhea may be seen after constipation in hookworm infections. Restlessness and pruritus, particularly on the legs may accompany skin penetration and migration of hookworm larvae in the host.

Clinical Pathology

In goats with clinical gastrointestinal nematodiasis, serum albumin is consistently low. Total serum or plasma protein is also usually low, but in chronic cases, may be normal due to concurrent hypergammaglobulinemia. Anemia is a variable, but important finding.

Packed cell volumes below 9% can occur in severe haemonchosis. In less severe or chronic infections, packed cell volumes in the range of 15 to 25% are likely and red cells may by hypochromic due to iron deficiency. A mild to moderate anemia may occur during severe or prolonged infections with non-blood feeding nematodes. Microscopic examination of feces for parasite ova by direct smear, flotation techniques or quantitative methods can aid in diagnosis. *Nematodirus* and *Marshallagia* spp. ova are distinctly larger than the rest, with average length of 160 to 180 mm. *Trichuris* ova are barrel shaped, with obvious bipolar caps or plugs. Ova may be counted by methods such as the McMaster technique, but direct correlations between egg counts and severity of infection do not always exist. There also may be wide variation in the number of eggs produced by different species, and some prolific egg producers are not always the most serious pathogens. Guidelines for ascribing significance to ova counts from goats are not widely reported, but in lambs, ova counts of 2,000 or more per gram of feces are considered indicative of severe infection requiring immediate therapy.

Post Mortem Findings

At necropsy, nematodiasis in general is suggested by emaciation with reduced fat reserves or serous atrophy of fat around the heart and kidneys. Subcutaneous edema may also be observed, especially in the intermandibular space, when hypoproteinemia is marked. In haemonchosis, the characteristic red and white striped females can be seen on the abomasal mucosa with careful inspection. Multiple sites of hemorrhage, ulceration, or both may be present. In *Ostertagia* infection, the wall of the abomasum is edematous, and the mucosal surface has a grainy. For most of the intestinal worms, gross findings at necropsy are nonspecific, consisting only of catarrhal inflammation. The worms themselves are difficult to see. A catarrhal enteritis and soft, unformed dark feces in the colon is consistent with most forms of nematodiasis. Transmural nodular lesions anywhere along the intestinal tract support the diagnosis of *O. columbianum* infection. Blood staining of intestinal content, particularly in the proximal small intestine is suggestive of hookworm infection. Petechiation in the colon with edema and thickening of the colon wall are suggestive of *Chabertia ovina* infection. Extensive mucus covering the colonic mucosa is associated with adult *Oesophagostomum* spp. infection.

Diagnosis

Any combination of clinical signs of anemia, edema, poor body condition, and diarrhea should suggest gastrointestinal nematodiasis in goats. Evidence of increased fecal egg counts, heavy worm burdens in digestive tract and increased plasma pepsinogen levels supports the diagnosis. When anemia is a predominant sign, various hemoparasites, hepatic fascioliasis, and cobalt or copper deficiency must also be considered in the differential diagnosis. In young goats, coccidiosis is the most important cause of diarrhea that must be differentiated from nematodiasis. Subclinical parasitism will present often as poor growth in young animals or prolonged weight loss in adults with few additional clinical findings.

Treatment

Clinically affected individuals require supportive care to reverse the process of parasitic debilitation and anthelmintic therapy to eliminate existing infections. In severely debilitated goats, anthelmintics of low toxicity such as thiabendazole, fenbendazole, or ivermectin should be used because animals may be more susceptible to the adverse effects of drugs with a narrower margin of safety such as levamisole, or organophosphate compounds. Parenteral administration of iron as iron dextran may promote erythropoiesis because iron deficiency often results from prolonged parasitism. A broad range of anthelmintics have been used for treatment and prevention of gastrointestinal nematodiasis in goats. The benzimidazoles are a useful group of broad spectrum anthelmintics, with the parent compound thiabendazole. The pro-benzimidazoles, febantel, and thiophanate are broken down metabolically to benzimidazoles by the host. Both fenbendazole and the pro-benzimidazole, febantel are metabolized in the goat to oxfendazole. In general, these drugs are highly effective against adult and active immature stages of *Haemonchus, Ostertagia, Trichostrongylus, Cooperia* and *Chabertia* spp., moderately effective against *Oesophagostomum, Nematodirus, Bunostomum, Gaigeria,* and *Strongyloides* spp, and poorly effective against *Trichuris* spp. Importantly, the newer compounds oxfendazole, febantel, fenbendazole, and albendazole are highly effective against arrested larvae of *Ostertagia.* The benzimidazoles are only available for oral use, but a number of oral formulations are available including boluses, pastes, drenches, and feed or salt supplements. The spectrum of activity of these drugs can be a function of dose. Fenbendazole is effective against tapeworms in goats at 15 mg/kg but not at the usual nematode dose of 5 mg/kg. Parbendazole is effective against *Oesophagostomum* and *Trichostrongylus* spp. at 10 mg/kg, *Ostertagia* and *Strongyloides* spp. at 20 mg/kg, and *Nematodirus* spp. at 30 mg/kg. Whipworms that are not commonly killed at recommended doses of benzimidazoles often may be removed by doubling the recommended doses. Albendazole administered either in a single dose of 7.6 mg/kg or two daily doses of 3.8 mg/kg was highly effective against *Ostertagia* and *Trichostrongylus* spp., but the latter dose regimen was less effective against *Oesophagostomum* spp. and neither regimen was effective against *Nematodirus* spp. As a group these drugs are quite safe. Thiabendazole has been used safely in goats at doses up to 100 mg/kg. Ivermectin is effective against adult and immature forms and arrested larvae. Since its introduction, it has become widely popular because of its broad spectrum of activity against gastrointestinal and pulmonary nematodes and some ectoparasites including mange mites and sucking lice. It has no efficacy against cestodes and trematodes. The drug is available for oral or subcutaneous use. The recommended dose is the same as for sheep (200 mg/kg) but doses as low as 50 mg/kg are highly effective in goats. The drug has a wide margin of safety. Ivermectin, however, may be highly irritating to some individual goats when given subcutaneously. Levamisole, an imidazothiazole, is the most widely used drug in this group, which also includes the pyrimidines, pyrantel, and morantel. The spectrum of activity of these drugs against gastrointestinal nematodes is similar to the benzimidazoles and may be better than some of the benzimidazoles against *Nematodirus* and *Bunostomum* spp. However, there is minimal effect against arrested larvae and they are not ovicidal. They also have no activity against trematodes or cestodes. Currently, levamisole is marketed in the pure levo (L) form at a recommended dose of 8.0 mg/kg. The drug is available for either oral or subcutaneous use. The margin of safety is narrow for levamisole so it must be administered carefully. Even at recommended doses, some goats may demonstrate transient symptoms of depression, muscle fasciculation, salivation, or frothing at the mouth. Pyrantel tartrate at a dose of 25 mg/kg in goats is 98% to 100% effective against *Trichostrongylus, Ostertagia, Nematodirus, Bunostomum* and *Strongyloides* spp., 97% effective against *Cooperia,* 91% against *Haemonchus,* and 70% against *Oesophagostomum.* Morantel tartarate given orally at 10 mg/kg to goats was highly effective against *Haemonchus, Bunostomum,* and *Oesophagostomum* spp., but showed little effect against *Strongyloides papillosus* and *Trichuris* spp.

Haloxon, coumaphos, and napthalophos are the organophosphate anthelmintics that have received the most attention in goats. These drugs are most effective against *Haemonchus, Ostertagia* and *Trichostrongylus* spp., moderately effective against *Nematodirus* spp., and have little or no efficacy against other gastrointestinal nematodes. They have been available in drench, paste, bolus, and feed-additive formulations for oral use only. Coumaphos is also available as a topical pour-on, which should not be used in lactating does.

Anthelmintic Resistance

The mechanisms by which parasite resistance develops are not completely understood. It is reasonable to assume that use of any anthelmintic selects for populations of nematodes that can survive its repeated use presumably by counteracting or avoiding the mechanism of action of the particular anthelmintic. The problem is intensified by the fact that when a population of parasites has evolved that shows resistance to a given anthelmintic, it is very likely to show resistance to all drugs in that class because they share a similar or identical mechanism of action. It is therefore necessary to switch to drugs from a different class to overcome a resistance problem, although there is no guarantee that populations resistant to other classes of drugs have not also evolved. It is believed that resistant parasites are most often introduced into a herd or flock via the purchase of new stock. Secondly, when dosing anthelmintics, it is preferable to err on the side of over dosing rather than under dosing to reduce the number of parasites exposed to non lethal doses. Third, the number of treatments administered during the course of a year should be kept to a minimum level because it has been demonstrated that the development of resistance correlates directly with the frequency of dosing. This underscores the importance of identifying and using strategic and tactical treatments based on the ecology of parasites rather than worming simply on a calendar basis. Finally, despite a widespread conviction that rotation of anthelmintics will reduce the development of resistance, experimental data suggests that resistance occurs as rapidly with use of multiple anthelmintics as with a single anthelmintic. Therefore it is suggested that only a single anthelmintic be used during the entire course of a seasonal exposure to infective larvae. Where larvae are present throughout the year, the anthelmintic may be used until evidence of resistance appears. When subsequent changes are made, the change should be to an anthelmintic of a different chemical class.

Anthelmintic Failure

The persistence of infection after anthelmintic therapy is not always caused by the development of anthelmintic resistance. Inadequate dosing is a major cause of treatment failure and may be due to several causes. Ideally, anthelmintics should be administered on a per weight basis to each animal. Practically, this is not commonly done, as weights are guessed or average doses administered when large numbers of animals require treatment. To minimize this problem, animals should be grouped according to size or age and the weight of the largest animal in the group accurately determined. All animals in the group should be given the dose calculated for the largest individual in the group. The site of deposition of the drug may affect anthelmintic efficacy, particularly with the benzimidazoles.

Prevention and Control

In recent years, much emphasis has been placed on the strategy of modifying management of livestock, including goats, to minimize exposure to parasites based on knowledge of parasite life cycles and ecological behavior. Current recommendations for control are considered according to the management system followed for rearing of goats.

Intensive Management System

Eliminating the use of pasture is an obvious control tool when intensive management is preferred. In tropical and subtropical regions, where infective larvae may be continuously present on herbage, goats are often housed in pens with slatted floors, raised off the ground. Many of these animals, however, are fed with fresh herbage gathered daily by caretakers. These feedstuffs may be contaminated with infective larvae from free-ranging goats, sheep, cattle, and it can not be safely assumed that confined goats are in fact parasite-free. Sun drying of fresh cut grasses before feeding may reduce viable larvae on herbage and improve its dry, matter content. In intensive management systems in temperate regions, goats may be continuously housed in barns, fed stored feeds and allowed exercise in well-drained dry lots. Under these conditions, goats, in most cases, can be maintained relatively free of gastrointestinal nematode parasites, although certain species, such as hookworms and *Nematodirus* spp. are adapted to successful transmission away from moist herbage. To be totally effective, confinement should be absolute for all age groups in the herd. However, a drawback of total confinement systems is that, if not carefully managed, the advantages gained by controlling nematodiasis may be lost to an increase in coccidiosis and respiratory problems.

Semi-Intensive Management System

Many goat flock owners, practice only semi-confinement rearing, turning goats out on pasture for 6 hours and give supplementary feeding some fodder and concentrate after, these goats risk exposure to over wintered larvae if the pastures were previously grazed by goats, or other livestock. If the goats are fed fresh cut herbage, however, they may still be exposed to infective larvae. Pasture can be heavily infected when weaned kids begin grazing in earnest. Under conditions of seasonal pasturing, all goats should be strategically treated with anthelmintics before turn out in the spring. Drugs employed should be effective against both adult and arrested larval stages. When possible, spring turnout should be onto pastures not used the past year for grazing by goats or other livestock. Monitoring of infection levels by fecal egg counts from composite fecal samples should be done periodically during the grazing period to determine the need for tactical anthelmintic therapy. Another strategic anthelmintic treatment should be administered to goats when they are finally brought in from pasture at the end of the season to eliminate existing worm burdens and to minimize the persistence of arrested larvae resulting from conditioned larvae ingested toward the end of the pasture season.

Extensive System

Where continuous or prolonged grazing is the fundamental management scheme, the extent and severity of infection depends on multiple factors including stocking rates, herbage quantity and quality, season, and weather conditions. Careful observation and monitoring of body condition, fecal consistency, and fecal ova counts are advisable in grazing animals to identify the effects of management and weather changes on the incidence of parasitic infection. In the case of haemonchosis, if 5% to 10% of the herds have fecal egg counts of 500 eggs per gram of feces or more, then tactical treatment is justified before the onset of clinical signs. When the predominant parasites and their behavior are well known, problems can be anticipated without monitoring. Tactical anthelmintic therapy is automatically given to grazing goats after heavy rains on the reasonable presumption that acute haemonchosis is likely to ensue because of rapid, synchronous larval development. For best effect, tactical treatments should be accompanied by movement of stock to clean pastures to avoid immediate re-infection on contaminated herbage. If strip grazing is practiced, care must be taken that restricted grazing areas are of sufficient size for the animal population. Otherwise the rate and intensity of larval buildup may be severe enough that the presumed beneficial effect of strip grazing is negated. The smaller the strip, the more frequently animals should be moved. Management changes made in response to drought, such as congregation of animals for supplemental feeding and watering, may precipitate an increased incidence of parasitism. Supplemental feed should be fed in feeders, not from the ground, and should be provided at multiple sites, spread out to avoid the intensification of parasite ova in places of high animal density. This is even more important with point source water supplies. If run off or seepage from tanks or troughs is not properly avoided or drained away, heavy contamination of the soil with infective larvae can occur at points where goats congregate to drink. In tropical regions, extensive grazing is often practiced and the lack of seasonal changes in temperature and humidity allows infective larvae to be continuously present in the environment. In this situation, the frequent, regular use of anthelmintics may be the only effective means of limiting resident worm populations and minimizing the effects of parastism in goats. Three week treatment intervals are suggested so that larvae ingested since the last treatment will not have yet matured to adulthood and begun producing ova. Rotation of anthelmintics among the different classes of available drugs should be kept to a minimum, as too frequent rotation of drugs within an active season or cycle of parasite development can itself promote resistance. Where grazing of goats is practiced, mixed grazing of different livestock species is being recommended as means of parasite control on the presumption that even though cross-species transmission can occur, the parasites infecting one species are less suited to an alternative host and will therefore result in a less severe infection. Currently, no vaccines are available for controlling gastrointestinal nematodiasis in goats although the area is one of intense research investigation in other species.

b. Ectoparsites

9.16 Ticks

Ticks are important ectoparasites throughout the world.

Etiology

Argasidae: soft ticks

Otobius megnini

Ornithodorus sp.

Ixodidae: hard ticks

Boophilus microplus

Rhipicephalus haemaphysaloides

Soft ticks belong to the family Argasidae and feed repeatedly on their hosts. Hard ticks, family Ixodidae, have a shield on the dorsal surface and feed only once during each stage. Larval ticks have 6 legs, while nymphal and adult ticks have 8 legs. Ticks are referred to as one-host, two-host, or three-host, according to whether they drop off the host to molt between larval and nymph stages and between nymphal and adult stages. Ticks are not especially host-specific.

Clinical Signs and Diagnosis

The attachment site varies with the tick species and may show papules, pustules, or wheals initially, with crusts and ulcers forming secondarily. Demonstration of the tick will make the diagnosis, but skin biopsy to demonstrate embedded tick mouth parts may be necessary in persistent, nodular lesions. Secondary bacterial infections and myiasis are possible. Other important consequences are blood loss and the transmission of very serious diseases such as anaplasmosis, babesiosis, heartwater, theileriasis, and tick-borne fever. Tick paralysis may also occur.

Treatment

When ticks and goats are few, simple extraction of the tick is recommended. A large variety of sprays, dips, and pour-ons will reduce the population of ticks and provide temporary protection. The use of insecticides can be reduced by local spraying or handdressing of the body area (ear canal, perineum) where ticks are located. Residues in meat, milk, and the environment are a concern, as is the possible development of resistance of ticks to insecticides.

Prevention and Control

Total eradication of a tick species is usually very difficult. A reduction in tick numbers is achieved by several applications of insecticide at 2 to 3 week intervals. Two and three-host ticks require treatments throughout the tick season, while one-host ticks are more likely to be on the host and therefore killed with one or two treatments. Amitraz has been used as a pasture spray to decrease tick burdens. In certain circumstances, burning the pasture or cultivating the land may aid in tick control.

9.17 Tick Paralysis

Ascending paralysis and death can result from the feeding activity of certain species of ticks on domestic animals including goats. The disease has been reported sporadically in goats in parts of Europe, the Middle East, and South America. It is considered a major constraint on small ruminant production in parts of South Africa where the disease is known as Karoo tick paralysis.

Etiology

At least 31 hard ticks in the Ixodidae family and 6 soft ticks of the Argasidae family can produce tick paralysis in mammals and birds. In goats, the condition has long been associated primarily with *Ixodes rubicundus* in South Africa, although more recently an unspeciated tick of the Rhipicephalus genus has also been implicated. In many countries, *Ixodes ricinus* causes the disease in goats. In 1983, tick paralysis was reported for the first time in Brazil, associated with *Amblyomma cajennense*. Most reports of tick paralysis involve adult ticks rather than larvae or nymphs and when such ticks have a 2 or 3 year life cycle, disease outbreaks may be seen only every second or third year.

Epidemiology

The incidence of tick paralysis is tied to the life cycle and ecology of the offending ticks and conditioned by livestock management and grazing practices. For example, in Israel, the disease is seen only every third year between December and February when adult ticks are active, and only in the more temperate, inland regions of the country. It usually occurs only in flocks or herds grazing on the more densely vegetation. Morbidity and mortality can be high in epizootics of tick paralysis.

Pathogenesis

Though not proven in all cases, it appears that ticks responsible for paralysis produce a salivary neurotoxin that is introduced into the host animal during active feeding, particularly when the tick reaches a certain stage of engorgement. At least in tick paralysis of dogs infested with *Dermacentor* sp., the toxin is known to interfere with acetylcholine liberation at the neuromuscular junction. In all cases there is a resulting flaccid paralysis that first involves the limbs and may progress to involve the diaphragm with animals dying of respiratory compromise.

Clinical Signs

The disease is seen most often in grazing animals infested with ticks. Ticks are most common around the head and neck, especially in the ears, and around the tail head, but inspection of the entire body should be carried out. Affected goats show evidence of lower motor neuron involvement including muscular weakness of the limbs, unsteady gait, knuckling, ataxia, and hypoesthesia.There is no fever, and animals usually remain bright and alert and continue to eat. Frequently the hind limbs are affected first, but all four limbs may be affected simultaneously, especially in Karoo tick paralysis. As the disease progresses the animal becomes laterally recumbent due to flaccid paralysis.

9.18 Lice Infestations

All species of goat lice complete their life cycle on the host and are quite host-specific. Eggs (nits) are attached to hairs and hatch in 5 to 18 days. The nymphal stages look like tiny adults; young lice mature 14 to 21 days after hatching. *Damalinia ovis,* the sheep louse, can become established on goats, and thus goats may be a source for reinfesting sheep during louse control programs.

Clinical Signs

Blood sucking lice have relatively narrow heads with piercing mouth parts. Two species have been generally reported from goats i.e. *Linognathus stenopsis* and *Linognathus africanus.* They are bluish-gray and infest most breeds of goats. Blood loss and secondary bacterial skin infections may occur in addition to pruritus. Heavy infestations may kill kids. Biting lice have broader chewing mouth parts. They are pale and difficult to see. They provoke rubbing and scratching and can also bite through the hairs. The consequence is moderate to severe damage to fleeces, depending on the louse numbers. *Damalinia (Bovicola) (Holakartikos) crassipes* and *Damalinia (Bovicola) limbatus* are common on hairy goats, while *Damalinia (Bovicola) caprae* is the biting louse usually found on milch goats. Certain goats appear to have a natural resistance to biting lice, as documented during efforts to cause experimental infestations.

Diagnosis

Diagnosis is by physical examination and demonstration of lice or nits directly or on material collected by plucking or combing. In one study, biting lice were concentrated in the withers area (where the goat had more difficulty grooming itself) while sucking lice were more prevalent over the brisket and shoulders.

Treatment

Numerous insecticides are effective against goat lice, but nits are not killed. Resistance also may develop. Nits are not killed by the initial therapy, treatment should be repeated at 10 to 14 day intervals if possible, to remove lice before they mature. For lactating goats, crotoxyphos 1 % in water spray or 3% dust, coumaphos 0.25% in water spray or 0.5% dust and dichlorvos are commonly used. In cold weather, dusts are preferred to dips. Coumaphos pour-ons may cause milk contamination. Newer permethrins appears to be both safe and effective, with no milk or meat withdrawal requirements. Ivermectin, at 20 mg per 100 kg subcutaneously, is efficacious against sucking but not biting lice; it should not be used in lactating dairy goats. For young kids, rotenone, flea powders labeled for cats, and even flea collars may be more convenient that commercial livestock preparations. Insect hormones have been used experimentally to control lice. Three spray treatments of 0.1 % synthetic juvenile hormone at 2-week intervals controlled lice for 4 months.

Some insecticides used for control of external parasites of goats are as follows:

Amitraz (L)	0.025%-0.05% spray
Coumaphos (L)	0.25% spray, 0.5% dust
Crotoxyphos (L)	0.25%-1.0% spray, 2% dust
Dichlorvos (L)	0.5%-1.0% spray
Fenvalerate	0.05% dip
Ivermectin	20 mg/100 kg subcutaneous injection

Lime sulfur (L)	2.0%-5.0% dip
Lindane	0.06% spray-0.03% dip
Malathion	0.5% spray, 5% dust
Methoxychlor	0.5%-1 % spray/dip,5% dust
Permethrin (L)	0.055% spray
Phosmet	0.15%-0.25% spray/dip
Trichlorfon	0.2% spray/dip

9.19 Fleas

Dog and cat fleas *(Ctenocephalides* spp.) sometimes infest goats in tropical regions. The wingless, laterally compressed, 2 to 4 mm long adult insect suck blood and cause local irritation. Fleas frequently leave their hosts. Eggs are laid on the goat or on the ground. Clinical signs include restlessness, rubbing and chewing, excoriations, alopecia, nonfollicular papules, and crusts. Anemia and weight loss may occur in young or debilitated animals. Treatments suggested for lice also appropriate for fleas, but the environment and all mammalian hosts must be treated.

9.20 Keds

Melophagus ovinus is a wingless blood-sucking fly (6 legs) that infests both sheep and goats. The parasite is 6 to 7 mm long and easily seen, resembling a tick. The entire life cycle is completed on the host in 5 weeks or longer. The parasite causes skin irritation and blood loss but damage to hides may have greater economic significance. Shearing will remove many keds. The insecticides used for louse control are also effective against keds.

9.21 Cutaneous Myiasis

Screw-worms (*Wohlfahrtia magnifica*) are the maggots of flies that must deposit their eggs in wounds or at body orifices rather than in dead carcasses. The parasite is reportable in the most of the tropical countries including India. Infestation of goat increases with age, and the body regions most commonly struck are the genitalia and the extremities. Clinical signs include inflammation, pruritus, apathy, and weight loss.Shearing, dehorning, castration, and ear marking wounds are common targets, as are the mouth and navel of newborns. As a complication to the presence of maggots in foul-smelling, pruritic lesions, toxemia or septicemia may kill the goat. Wounds should be debrided and treated with a topical insecticide such as malathion or coumaphos in ointment form. Fly repellents prevent repeat attacks, and antibiotics are indicated if the animal is systemically ill.

A variety of blowfly larvae (calliphorine myiasis) can be found in the same wounds that attract screw-worms and are differentiated by close examination of larval anatomy. Bacterial activity (fleece rot, fecal contamination of skin, urine scalds) also can make intact skin attractive to these flies. Foul-smelling ulcers are filled with maggots. Cleansing of the wound with a mild solution of pine oil is helpful, as most of the maggots will come out of their holes and drop to the ground. Insecticides, fly repellents and antibiotics are applied as for screw-worms.

9.22 Oestrus Ovis Infestation (Nasal Bots)

The adult fly is a stout, dark gray fly about 1 cm long. Its mouth parts are rudimentary and it can not feed. The flies emerge in the late spring, mate and larviposition. Mature third instars leave the sheep in July or August, pupate and form another generation larvipositing in the autumn. These larvae remain dormant as first instars over the winter to emerge the following spring . In temperate areas there may be one or two generations per year but several generations may be completed in hot areas. The fly deposits larvae in the nostrils of sheep and occasionally in goat, larvae migrate to the dorsal turbinates and sinuses where they remain for some weeks to months before they migrate to the nostrils sneezed out to pupate on the ground. *O. Ovis* are adapted to the various climates prevailing wherever sheep are kept. When winters are cold, the larvae can over winter by remaining dormant in the first instar, but in warmer climates development may continue through winter. The larvae are thick, yellow-white in and when mature there is a dark dorsal band on each segment. The ventral surface has rows of small spines on each segment. The adult fly may emerge from puparium in 4- 5 weeks in summer but this may be led in the winter. O. *ovis* may deposit larvae in the ore or on the lips of man, and in some countries ocular myiasis or infection of the upper respiratory tract is an important zoonosis. Adult flies attempting to deposit larvae at the nares annoy the sheep and cause them to seek shelter. Stamping of the feet and shaking of the head are common. Sheep may bunch together and press their heads into the fleece of others. In bad seasons the sheep may lose a good deal of grazing time. The spiny surface of the larvae causes irritation of the nasal mucosa resulting in catarrhal rhinitis with sneezing, a mucopurulent discharge and difficult snoring respiration. Secondary bacterial infections are uncommon but may cause occasional deaths. The infections are not usually severe enough to necessitate treatment but production losses and mortality have been recorded. Control can be obtained by treatment in late summer to prevent the build-up of heavy infestations and in winter to remove the over wintering larvae. Rafoxanide 7.5 mg/kg , closantel 7.5 mg/kg, nitroxynil 15-20 mg/kg and ivermectin 0.2 mg/kg are effective and the use of these compounds for fluke or worm control also controls nasal bots.

9.23 Warbles

Przhevalskiana (Hypoderma) silenus is a warble-fly that naturally infests goats in Mediterranean countries and Asia. *Hypoderma crossi* infests goats in the dry hilly regions of India. *Hypoderma bovis* and *H. lineatum* (warble flies of cattle) have not been documented to infest goats.

Pathogenesis and Clinical Signs

Adult *P. silenus* flies lay eggs on the hairs of the legs and chest in spring. It is believed that the larvae migrate by a direct subcutaneous route to the back rather than passing through esophagus or spinal cord. First-stage larvae appear under the cutaneous trunci muscle, which becomes necrotic and infiltrated by neutrophils. The larvae penetrate the overlying muscle and skin and molt into the second instar. Larval debris and cellular infiltrates accumulate and a wall of granulation tissue forms around each larva, which is typically 10 to 12 mm long. Third-stage larvae drop to the ground to pupate. As many as 150 larvae have been found in a single goat.

Troatment

Ivermectin, at dosages of 5 to 20 mg/100 kg, has been effective in killing the warbles. In countries where this parasite abounds, modern systemic insecticides are often not available. Shepherds have been taught to remove warbles during the winter months.

9.24 Mite Infestations

The effects of the ear mites infestation vary greatly and generally, only behavioural changes, head shaking and ear scratching are observed in infested herds. However, the infestation can cause exudation, scabs and debris in the ear canal and lesions may spread to the head, neck, body and legs. There is marked pruritis, exudation, scab and scale formation and the effects of infestation appear more severe in older animals. The infestation appears to be transmitted to kids within the first few weeks of life. Removal of crusts and debris from the ear and repeated treatment of affected areas with acaricides is effective. *Sarcoptes scabei* infestation may be common in goats in some countries. Lesions are mostly seen around the eyes and on the pinna of the ears. The infestation may then spread to the head, along the neck and withers, and to the axilla, inguinal region and udder. There is marked hyperkeratosis and extreme pruritis. Some animals seem to be more susceptible than others to severe infestation although they show no defect in cell mediated or humoral immunity responses. Mineral and vitamin A deficiencies have been considered to predispose to the infestation. Sarcoptic mange may be more severe in goats than in other species of animals and may be accompanied by enlargement of the lymph nodes and emaciation. Diagnosis may be difficult since often multiple, deep skin scrapings are required to identify the parasite and on many occasions only eggs and parts of the parasite will be found. Negative skin scrapings are not uncommon even on severely affected animals and the diagnosis of infestation may be complicated by the frequent presence of non-parasitic mites on goats. The extensive hyperkeratotic lesions in some animals make treatment difficult. However, repeated applications of organophosphate or chlorinated hydrocarbon insecticides plus the administration of corticosteroids, topically or parenterally, to reduce the irritation and self-excoriation appears effective. Keratolytic agents or emollients may be useful.

9.24.1 Sarcoptic Mange

The mite that causes sarcoptic mange (scabies) in goats has been referred to as a goat-specific strain of *Sarcoptes scabiei*. It tunnels through the epidermis and feeds on tissue fluids; Experimental infection of sheep with goats of *Sarcoptes scabiei* has been documented and probably occurs naturally when the two species are raised together.

Clinical Signs

The condition is reported to begin with the appearance of small pruritic nodules, especially on the head, several weeks after contact with another infested goat. The skin disease seems to be self-limiting in some goat while others develop extensive severe dermatitis around the eyes and ears and on the neck and thorax, inner thighs, and udder. Hyperkeratosis and alopecia are accompanied by self-excoriation and restlessness. The thickened skin is wrinkled and fissured and may harbor a secondary bacterial infection. Severely affected goats lose weight and have prominent peripheral lymph nodes. Some deaths occur. The value of the hide is markedly decreased.

Diagnosis

Confirmation of the diagnosis usually requires deep scrapings at the margin of lesions; the mites are identified by the presence of long unjointed pedicles on the pretarsi. Even if no mites are seen in multiple scrapings, a biopsy report of eosinophilic dermatitis and tunnels should indicate a presumptive diagnosis.

Treatment

Treatment of lactating animals requires repeated (perhaps 5 to 10) applications of lime sulfur solution every 5 to 7 days. Two treatments of 0.05% amitraz 7 to 10 days apart are recommended for controlling scabies in goats. Sprays often fail to adequately penetrate the long and thick hair coat of unshorn animals. Some animals may need antibiotics for secondary bacterial infections. More recently, ivermectin (subcutaneously) has offered great promise, especially because the inconvenience and stress of bathing a disgruntled goat in cold weather and the dangers of toxicity to the goat are avoided. A cleansing shampoo is recommended to remove crusts even if ivermectin is given. However, excellent response including disappearance of the crusts was achieved in goats were treated with ivermectin (1.0 ml of a 1 % solution per adult of unspecified weight) twice, 1 week apart. Live mites were recovered from treated animals for at least 2 weeks, but all goats yielded negative skin scrapings 3 weeks after the second treatment.

9.24.2 Chorioptic Mange

It is believed that the chorioptic mange mite of goats belongs to the species *Chorioptes bovis* however it is host-specific and designate it C. *caprae. Chorioptes caprae* may cause lesions on the legs, udder, scrotum, flanks, belly, neck, back and base of the tail. There is pruritis, hyperaemia and exudation with crust and scab formation and alopecia.

Epidemiology

The mite lives on the surface of the skin and feeds on epidermal debris. It may exist on nonclinical carrier goats and can also survive in the environment for as long as 10 weeks. This accounts for occurrence of mange lesions in closed herds. Mite populations are affected by the environment, with highest mite numbers and most severe clinical signs in cold weather.

Clinical Signs and Diagnosis

The lesions (nonfollicular papules, crusts, alopecia, erythema, and ulceration) are most commonly or initially found on the lower limbs, udder, scrotum, and perineal region. Pruritus, as demonstrated by chewing at the limb lesions, is usually obvious. Diagnostic samples can be obtained with a flea comb or by skin scrapings; adding rotenone to the mineral oil used for collecting scrapings will prevent the escape of the mites on the way to the microscope. The mites have short pedicles on their pretarsi. Zinc deficiency and bacterial skin infections are important differential diagnoses that may be present concurrently with mange.

Treatment

Treatment of chorioptic mange is sometimes very easy, but hypersensitivity of certain goats to the mites can lead to treatment failure. All in-contact goats should be treated simultaneously to eliminate the carrier state and the premises disinfected. Lime sulfur (four, weekly, total body sprays or dips in 2% solution) is safe for lactating goats. Crotoxyphos (0.25%), Coumaphos (0.25%), Trichlorfon (0.2%), Amitraz (0.05%), and Lindane (0.03%) need to be applied at least twice at 10 to 14 day intervals. In one study, a single dip with fenvalerate (0.05%) killed all mites on goats. Injectable ivermectin can be expected to kill many but not all of the mites, because of their superficial location. Topical application of ivermectin (50 mg/100 kg applied once to healthy skin) has been shown to be effective against C. *bovis* on cattle. Systemic antibiotics for secondary bacterial infections and judicious use of glucocorticosteroids in apparently allergic goats are indicated in select cases.

9.24.3 Psoroptic Mange

Etiology and Epidemiology

The most commonly reported ear mite of goats is the cosmopolitan *Psoroptes cuniculi*. The mite can be found in the external ear canal of kids as young as 10 days old, with most kids infested by the third week. When it causes body mange (in parts of the world where sheep are affected with psoroptic mange), the parasite is sometimes referred to as *P. caprae*. The mite has long, segmented pedicles and is often visible to the naked eye. It does not burrow but feeds on tissue fluids. Prolonged survival times (as long as 12 weeks) in the environment have been reported.

Clinical Signs and Diagnosis

Clinical signs of ear mite infestation include head shaking and scratching, sometimes with alopecia of the part of the ear repeatedly brushed by a hind foot. Occasional goats have flaky or scabby lesions or laminated crusts on the external ear or even on the poll, back and pasterns but most have no externally visible lesions. A large plug of yellowish-white wax is commonly found in the ear and extracting some of this material with a cotton swab for examination is an alternative means of diagnosis. Body mange caused by *Psoroptes* resembles sarcoptic mange but is accompanied by less scab formation. Skin scrapings should be collected from the margin of lesions.

Treatment

Treatment of ear mange is often foregone or limited to removal of any bells from goats showing clinical evidence of pruritus (to avoid continual disturbance of the owner). External lesions may be swabbed with an acaricide. Canine ear mite preparations may be used on goats with mites confined to the ear canal, but removal of crusts and debris is necessary and recurrence is likely. The life cycle probably takes about 3 weeks, and several weekly treatments are advised. In nonlactating goats, a single injection of ivermectin (20 mg per 100 kg) would be a rational therapy.Body mange can be treated with routine acaricides (including amitraz as dips or sprays, Ivermectin (2 injections 1 week apart) also has been found effective in treating psoroptic mange in goats.

9.24.4 Demodectic Mange

Demodex caprae is a cigar-shaped mite that commonly inhabits the hair follicles and sebaceous glands of goats. It has a worldwide distribution.

Epidemiology

Confinement housing and crowding seem to favor development of demodectic mange. The mites appear to survive only a few hours away from the goats, but infection via contact with contaminated feeders might occur. In fact, natural spread from one adult goat to another has not been documented. The appearance of nodules in certain goats in a herd but not in others may reflect the immune competency of the animal.

Clinical Signs and Diagnosis

Demodex caprae causes nodular lesions from which greyish or yellowish caseous material containing the organisms is seen. These are common in the axilla and occasionally may be found on the neck, shoulder and thorax. Lesions may be seen in goats of a year or more in age. Portions of the body exposed to friction, such as face, neck, shoulders, and sides, are commonly involved whereas the ventral abdomen and udder remain free of nodules. Initially lesions are very small and only located by careful palpation. However, defects may be readily detectable in hides at this stage or even earlier. By 18 to 20 months of age the nodules have enlarged to lentil or pea size, but because they remain painless they may not noticed unless the goat is clipped. Occasional animals demonstrate mild pruritus but this is a relatively rare finding.

Each nodule corresponds to a distended pilo-sebaceous follicle. Firm digital pressure will force out a ribbon of yellowish white paste from a central pore. Microscopic examination of this caseous material revel numerous eggs, larvae, and mature mites. The diagnosis is thereby confirmed. The disease process peaks at about 3 years of age. Nodules may reach a diameter of 0.5 to 1.25 cm. At this time, the severely affected goat may apathetic and unthrifty, and may lose weight or milk poorly. It is not clear if these signs are because of mange infestation or underlying (nutritional) problems.

Treatment

Various treatments have been applied. When only few nodules are present, squeezing or incising each one to permit removal of its contents will result in a cure of the hair follicle thus treated. Iodine or other disinfectants is applied when the nodule is expressed to kill residual mites. Some goats carry several hundred discrete lesions. Others have a generalized dermatitis rather than the typical nodule form of the disease. For them, topical or systemic organophosphates, amitraz (0.025%) or ivermectin may be effective. Because the lesions resolve over weeks or months, documentation of a treatment effect is difficult. Nodules may remain long after the mites have died. Because secondary bacterial infections are generally absent in undisturbed lesions, antibiotic therapy is not indicated.

□□□

10

Emerging and Exotic Diseases

Under the fast changing ecology, the epidemiology of infectious diseases of ovine and caprine has also undergone remarkable transition. Introduction of exotic pathogens through exotic germplasm and biologicals have hampered animal production by exposing the native stock to casual agents to which they had no exposure before in the past few decades. The country has experienced heavy economic losses due to entry of one or the other pathogens, thus indicating the emergence of many such diseases and their occurrence and establishment in the country.

A few diseases entered into the country in recent past and these brought heavy losses which include certain strains of blue tongue, PPR, FMD, PI-3, Maedi-visna. Other infections responsible for various types of sickness are adeno, corona, entero and reoviral infections. Caprine arthritis encephalitis has been reported as one of the latest infection of goats in India. Examination of the epidemiological pattern of certain devastating diseases of sheep and goats which entered into the country during the last few decades, revealed the emergence of blue tongue, A22 subtype of FMD and maedi-visna. These diseases have resulted in considerable economic losses due to morbidity, mortality and expenditure on disease research and control programme implementation. In the past, a few more disease has joined the list which includes Aujeszky's disease and Scrapie. Besides these, occurrence of some more new pathogens has also been recorded in India. These include viruses responsible for pneumonia due to PI-3, enteritis due to rota, corona, entero, picorna and pneumoenteritis due to Adeno viruses. Of these, a few may appear as major threat to our small ruminant population if ignored for longer. Many diseases have been described in other sections, the remaining few important diseases of small ruminants are as follows:

10.1 Rift Valley Fever

This arthropod-borne virus disease of ruminants and man is limited to Africa. The disease is characterized by acute, severe hepatic necrosis. Goats are less susceptible than cattle or sheep.

Etiology

The causative agent is an enveloped, single stranded RNA Phlebovirus in the Bunyaviridae family. The virus is very stable near neutral pH in blood or serum and also in aerosol. It is killed by 2% sodium orthophenylphenate and 4% sodium carbonate disinfectants.

Epidemiology

Rift Valley fever was historically limited to East and Central sub-Saharan Africa. However, more recent outbreaks in Egypt, Senegal, and Mauritania indicate its potential for further spread. As many as 300 people died in the Senegal outbreak. The virus is transmitted by mosquitoes in the genera *Aedes, Culex, Mansonia, Eretmapodites,* and *Anopheles* and possibly by other insects. Mosquitoes are capable of Trans-ovarian transmission of the virus, thus sustaining infection. Sub clinically infected, viremic ruminants and man also contribute to the reservoir of infection. Direct transmission by aerosol to people can occur. Epizootics often follow periods of heavy rainfall when mosquito populations increase. Heavy winds that blow infected insects into previously uninfected areas and transport of infected livestock may also play a role in initiation of epizootics. Areas with extensive irrigation systems, warm climate, suitable arthropod populations, and susceptible ruminant and human populations are most susceptible.

Pathogenesis

After infection, there is a viremia with fever and leukopenia. Despite the destruction of liver cells by rapidly multiplying virus, localization of virus in tissues does not occur to any great extent, and blood always contains the highest levels of virus. A carrier state does not develop in ruminants. Lesions are limited to the liver and are characterized by focal hepatic necrosis. Experimental infections in goats have been described.

Clinical Signs

The incubation period is as short as 12 hours in kids and as long as 48 hours in adults. Adult animals are always less severely affected than young stock. In epizootics involving goats, the predominant effects will be abortion in pregnant does and death in kids younger than 1 week of age.Peracute death with no prodromal signs is most likely in neonatal kids. However, some young kids may show fever up to 108.0°F (42.2°C), accompanied by listlessness and inappetance, with death occurring in the next several days.

The acute form of the disease is more likely in older kids and adults, though seen less frequently than in sheep. In addition to fever and depression, variable signs may include vomiting, hemorrhagic diarrhea, unsteady gait, catarrhal stomatitis, and skin necrosis on the udder or scrotum with death in 1 to 4 days. Except for abortion in pregnant does, adult goats are likely to show only subacute disease, with fever in the range of 40.0 to 41.2°C, dullness, and inappetance for 1 to 3 days. Mortality rates are very low in adult goats.

Clinical Pathology

Leukopenia is common in the early stages of disease. Virus isolation is possible from blood of live, febrile goats, or liver and fetal tissues from fatal cases and abortions. When immediate submission to the laboratory is not possible, samples should be frozen at -70°C. After outbreaks, seroconversion can be documented using enzyme-linked immunoabsorbent assay or other serologic methods in serum samples taken 3 weeks apart.

Post Mortem Findings

At necropsy, the liver may appear pale with focal areas of sub capsular hemorrhage. On cut section, 1 to 2 mm gray-yellow necrotic foci are distributed throughout the parenchyma. Other possible findings include visceral and serosal hemorrhage, icterus, hemorrhagic gastritis and enteritis. Similar lesions may be seen in aborted fetuses.Histologically, the liver lesion is one of focal centrilobular or midzonal necrosis of small groups of hepatocytes. As lesions progress, they may become confluent and involve a large portion of the lobule. Eosinophilic, intranuclear inclusion bodies surrounded by marginated chromatin are common in degenerating hepatocytes in advanced lesions.

Diagnosis

Definitive diagnosis is by virus isolation or confirmation of seroconversion in affected goats. The disease is limited to Africa and the most important differential diagnosis is Wesselsbron disease in regions where both occur. In sheep, bluetongue is considered in the differential diagnosis but bluetongue is almost always asymptomatic in goats. When people handling livestock complain of influenza-like illness during epizootics of animal disease, the diagnoses of Rift Valley fever should be pursued vigorously.

Treatment

Antiviral drugs, such as ribivarin, have been used in human cases, but no recommendations for use in animals are made. The intravenous or intraperitoneal administration of 10 to 30 ml of serum from convalescent sheep to 1 to 3 day-old lambs reduced mortality even when lambs were already showing signs of illness.

Prevention and Control

In endemic areas, an attenuated live vaccine, the Smithburn strain, is available but should not be used in pregnant animals, as viremia, abortion, fetal death, and fetal anomalies occur in sheep. Use of the vaccine in goats has not been specifically reported. Immunity in cattle and sheep lasts at least 2 years. Formalin-inactivated vaccines are available for use in nonendemic areas. Work is ongoing in the development of newer live vaccines with less objectionable side effects. Mosquito control is difficult. However, movement of animals out of known mosquito breeding areas after heavy rainfall may reduce the incidence of disease. During epizootic outbreaks, movement of non-vaccinated animals should be restricted. Most human cases can be related to direct contact with blood or tissues of infected animals, or inhalation of aerosols associated with such tissues. Veterinarians should take suitable precautions against infection when performing examinations or necropsies and advise against slaughter and consumption of meat from suspected cases.

10.2 Caprine-Arthritis-Encephalitis and Other Lenti Viruses

Caprine-arthritis-encephalitis (CAE) has emerged as a significant viral disease of goats in several countries which have dairy goat industry. This disease is featured clinically in the form of arthritis, progressive paresis, or other neurologic dysfunction, udder hardness with hypogalactia, chronic interstitial pneumonia, or progressive weight loss and arthritis. A few diseases entered into the country in the recent past and they brought heavy economic losses to goat industry of the country. Caprine arthritis encephalitis (CAE) has been reported as one of the latest infection of goats in India.

Etiology

The CAE virus is a lentivirus belonging to the family of Retroviridae, categorized as slow viruses. Other viruses in this group include the Maedi-Visna (MV) or progressive pneumonia virus of sheep, the equine infectious anemia virus, and the human immunodeficiency virus associated with acquired immunodeficiency syndrome. Different strains of the CAE virus have been shown to exist on the basis of variable response to neutralizing antibodies but this strain variation is not correlated with differences in tissue tropism. The CAE virus is antigenically related to and serologically cross-reactive with the ovine MV virus. Infected goats or sheep develop antibodies that specifically bind to the CAEV. Individual animals may be infected for several months before antibodies can be detected. Antibodies to CAEV are transferred to kids/lambs via colostrum, but these antibodies do not protect the kids/lambs from infection. Rather, they are an indication that the kids/lambs have been exposed to and are likely to be infected with these viruses.

Epidemiology

CAE syndrome was first described in the USA in 1974. Canada, France, Norway, Switzerland, and the United States were most affected countries. The seroprevalence exceeded 65% in all of these countries. In countries actively importing dairy goats, such as Kenya, Mexico, New Zealand, and Peru, the overall prevalence of CAE infection was often less than 10%. However, it is highest in imported dairy goats, and then in local goats in contact with imported goats. The CAE syndrome is caused by a virus which was isolated in synovial membrane from an arthritic goat in 1980. Infection and inflammatory diseases caused by CAE virus in goats are almost world-wide in distribution. However, in India CAE was reported in late eighties. CAE virus belongs to *Lentivirinae* subfamily of the *Retroviridae* family.

Although CAEV is distributed world-wide and is associated with inflammatory disease in goats, it is probable that it could propagate through sheep species concomitantly with ovine lentiviruses (Visna virus), since CAEV like strains have been isolated from sheep with progressive pneumonia, in the USA and Western Europe. These small ruminants viruses do not appear to have any host other then sheep and goats. Infected monocytes and macrophages are the cellular vectors which spread the virus into the specific target tissue within the same animal and probably from animal to animal. In the case of horizontal transmission to kids in the field, colostrum is the major source of infectious virus.

Serological surveys through the group specific antigens of ovine and caprine lentivirus show that the infection is common in goats, with a level of approximately 80% in adult goats. The higher incidence of goats infected with CAEV is probably due to the common practice in the goat industry of feeding kids with pooled colostrum and milk from dairy mothers. In infected animals replication of CAEV is productive only in tissue macrophages of certain target tissues or target organs such as the central nervous system, joints, mammary glands and lungs. The infected macrophages present in the colostrum, milk and lungs of goats housed under conditions are the major source of horizontal infection through the caprine species. Intrauterine transmission (from the infected doe to her fetuses) of CAEV occurs but the frequency is unknown. There is good evidence that CAEV also is transmitted by direct contact with infected goats. This mode of transmission is suspected to occur through ingestion of virus in saliva and feces-contaminated feed or water, or by inhalation of aerosolized virus. CAEV also is present in blood, and can be transferred through blood contamination of needles or syringes. Tattooing equipment also is a potential means by which virus can be transmitted. The major mode of transmission of CAE virus is orally to kids via consumption of virus-infected goat colostrum or milk, through direct suckling of their own infected dam or colostrum from other infected does. Virus may be present in these secretions as free virus or incorporated in somatic cells both subclinically and clinically infected does can shed virus into colostrum and milk. But milking of infected and non-infected do together under normal farm circumstances led to frequent seroconversions in under 10 months time.

Pathogenesis

The disease resulting from CAEV infection is a complicated event involving very complex interactions between the virus and the host. The host environment is also critical and variety of host factors such as immune status, genetic background, age and nutritional status may be determinant factors. As an example of the relevance of the genetic background, it was reported that some class I major histocompatiblity complex (MHC) haplotypes of the Sannen breed of Switzerland may be less sensitive to he CAE syndrome.

Clinical Signs

Although CAEV is associated with several inflammatory diseases affecting the joints, brain, mammary glands and lungs of goats, arthritis and encephalomyelitis are the most typical and most prevalent infections caused by the caprine lentivirus. Arthritic disease is encountered most frequently in adult goats. It progresses over the years with a gradual swelling in the carpal (knee) joints. In advanced cases, carpal swelling may be marked and hard, with extensive periarticular fibrosis and mineralization. Some animals are lame and lean. Encephalitis is the most frequent clinical symptom of infection of goat kids occurring between 2 and 6 months of age. The clinical feature of infection is posterior paresis which sometimes progresses to paralysis of the hind legs and eventually to tetraplegia. When affected kids recover from the neurological disease, they develop severe arthritis at adult age. Slow progressive encephalitis rarely occurs in adult goats.

Post Mortem Findings

At necropsy, affected goats are usually emaciated. Affected joints and adjacent bursae and tendon sheaths are usually enlarged. Foci of mineralization and/ or necrosis as well as fibrous scar tissue may be visible in tendons and other periarticular structures. The synovial membrane appears brown and velvety and is thrown up into fingerlike projections. Cartilage surfaces can be roughened, ulcerated, or eroded depending on the severity of the case. "Rice bodies" are commonly found in the joint space, tendon sheaths, and bursae. Histologically, joint lesions are characterized by synovial hyperplasia with sub synovial mononuclear infiltrates, including lymphocytes, macrophages, and plasma cells. Multinucleated cells are occasionally observed. Fibrin deposition is common on synovial surfaces. There is necrosis and mineralization of synovial, perisynovial, and tendon collagen.

Diagnosis

Diagnosis of CAE arthritis is based on clinical signs and history in the farm, synovial fluid examination and typical histopathology on synovial biopsy but care should be taken to differentiate from any traumatic, metabolic, and infectious causes of arthritis and mycoplasmal infections. Periodic serologic testing, raising CAE free kids administration of heat-treated goat colostrums or pasteurized goat milk or commercial milk replacer are the principle to containment of diseases. The PCR detects the virus genetic material in white blood cells in a sample of whole blood. The AGID and ELISA tests are performed using serum. Some infected goats are antibody negative and PCR positive. These animals may harbor CAEV infected cells in lymph nodes, bone marrow, or neural tissue. The virus in these organs stimulates the immune system to make antibodies. At the time of sampling, however, there may not be any CAEV present in the blood sample, and therefore, the PCR test will be negative. If antibody or PCR positive animals are identified, the herd should be considered to be infected. Uninfected animals will be antibody negative and PCR negative. If negative animals are found in an infected herd, they will need to be re-tested over several months after separation from positive animals. Re-testing is needed because of the sometimes lengthy interval between infection and the appearance of antibodies or virus in blood samples. In countries where it has established, sero-positive reactor rate is about 65-80 per cent. A small project to study existence and incidence of CAE in organized goat flocks has been taken up at HSADL, Bhopal India. There is a need to investigate the presence and prevalence of CAE in India if any.

Prevention and Control

Knowledge of the biology of the CAEV and the availability of reliable serological tests have allowed control management program to be undertaken. These programs consist of separating the kids from adults immediately after birth, so that the kids are deprived of colostrum, and periodically testing the sera of the isolated animals. The controlled kids are then fed with heat-inactivated goat milk. Before the new herd is declared CAEV free, all goats must test serologically negative several times. Such program to control CAEV infection is presently performed in many countries, although the distinction of CAEV from visna virus has not been conclusively investigated. Vaccination is not available for CAEV; unfortunately, a few limited trials to protect goats using, purified inactivated CAE virion on challenge, enhanced clinical symptoms of the disease.

10.3 Pseudorabies

This acute fatal viral disease, also known as Aujesky's disease, or mad itch, is fundamentally a disease of swine and is reported only sporadically in goats. Clinical signs include peracute death or hyperexcitability, intense pruritus, convulsions, and terminal coma. Reported cases of caprine pseudorabies have invariably direct or indirect contact with pigs. The disease is potentially zoonotic.

Etiology

The causative agent is the porcine herpesvirus 1. The virus can survive in saliva or carcasses for several months in winter, and several weeks in summer. Free unprotected virus persists for only 1 to 2 weeks. Sodium hypochlorite, formalin, peracetic acid, and quarternary ammonium compounds are effective disinfectants.

Epidemiology

The virus is extremely well adapted to domestic swine, which serves as the agent reservoir. All other susceptible species, including goats, are essentially dead-end hosts. In reported cases of caprine pseudorabies, the goats have been housed with, or in fence contact with pigs. The pigs may show no evidence of infection, as the latent shedder state is common in swine. Pseudorabies has occurred in sheep that were given parenteral medication using an uncleaned syringe previously used to vaccinate pigs for pseudorabies. Transmission of virus is by inhalation or via skin abrasions. Morbidity can be high, with 80% of exposed goats contracting the disease in an outbreak. The mortality rate is 100%.

Pathogenesis

The incubation period is 1 to 4 days. When the virus is introduced through the skin, it enters peripheral nerves and produces the characteristic localized pruritus that is observed in some cases of ruminant pseudorabies. The virus then moves centripetally and enters the central nervous system, producing a rapidly fatal encephalitis or myeloencephalitis. With inhalation of the virus, signs of encephalitis will occur without a prodrome of pruritus.

Clinical signs

All ages, breeds, and sexes can be affected. Goats may be found dead with no symptoms. Alternatively, they may show signs of intense rubbing or licking and even selfmutilation associated with severe, localized neurogenic pruritus. In cases where pruritus does not occur, initial signs may include agitation or excitation, repeated lying down and rising, hysterical bleating, profuse sweating, and convulsions. Fevers as high as 106.7°F (41.5°C) may be recorded. These excitatory signs give way to recumbency, paralysis with bloat and dyspnea, coma, and finally death. The clinical course may be only several hours.

Clinical Pathology

While serology is useful in the identification of infected swine, it has no application in caprine pseudorabies because goats become clinically ill and die before an antibody response can develop.

Post Mortem Findings

At the time of necropsy, attempts should be made to isolate the virus from brain tissue. There are no obvious gross necropsy findings, but there are histologic lesions of severe, focal, nonsuppurative encephalitis and myelitis. Eosinophilic intranuclear inclusion bodies typical of herpes viruses may be found in degenerating neurons. Immunofluorescence and immunoperoxidase methods can be used to confirm the identity of the porcine herpesvirus 1 in infected brain tissue.

Diagnosis

When pruritus is present, primary dermatologic conditions, particularly ectoparasites, must be ruled out. Scrapie, rabies, can also produce neurogenic pruritus in goats. When encephalitic signs predominate, consider also rabies, heartwater, polioencephalomalacia, and hypomagnesemia. The acute nature of the disease also suggests poisonings, notably cyanide, nitrates, urea, organophosphates, and chlorinated hydrocarbons.

Treatment, Prevention and Control

There is no treatment for pseudorabies. Control depends on recognizing the risks of commingling goats with pigs whose pseudorabies status is unknown. If pigs must be housed with goats, they should be serologically negative for pseudorabies before purchase. Because of the sporadic nature of caprine pseudorabies and the relative ease of reducing the risk of exposure, vaccination of goats with killed vaccines seems economically unjustified. Because of the zoonotic potential of this disease, gloves and a mask should be worn during examination and necropsy.

10.4 Scrapie

Scrapie is a communicable, non-febrile, slowly progressive, degenerative and fatal disease of CNS of sheep and goats. The disease has a long incubation period (18 months of several years). At present it is difficult to demarcate the scrapie free areas in the world as small ruminants of Europe, America, Africa, Asia including India are reported to have the disease. The scrapie agent, which is highly resistant to physical and chemical disinfectants, consists of only a small basic protein without nucleic acid which appears in form of fibrils on electron microscopic examination.

Scrapie is a contagious, degenerative neurologic disease of sheep and goats caused by an infectious agent currently referred to as a prion. The disease is characterized by an extended incubation period of months to years; signs in adulthood are incoordination, progressive debilitation, and pruritus. Goats are far less commonly affected than sheep, but the disease has major significance in the goat industry because of regulatory constraints on the movement of breeding stock from scrapie-infected countries.

Etiology

The prion is a small proteinaceous infectious particle that resists inactivation by procedures known to modify nucleic acids. Prions contain little or no nuclear material. The manner of prion replication is incompletely understood, but it is believed that the prion protein is encoded for by a cellular gene of the host. The prion protein, PrPsc, is not found in uninfected animals, but a similar protein PrPsc is produced by neurons of healthy animals. The PrPsc protein has been demonstrated to be capable of inducing astrocytic gliosis as is seen in scrapie. The agent produces no detectable immune response in the host and the absence of scrapie induced antibody is a major obstacle to development of diagnostic tests and regulatory programs for scrapie control. The scrapie prion is very resistant to physical and chemical destruction and is believed to persist on pastures for months and possibly years when excreted by infected sheep or goats.

Epidemiology

Scrapie is enzootic in Europe, including the U.K. It also has been reported in the United States, Canada, Australia, New Zealand, India, South Africa, Kenya, Colombia, and the Middle East, largely as a result of sheep importations during the last century. Australia and New Zealand are currently free of the disease. Reports of naturally occurring scrapie in goats are quite rare. It is not clear if this is due to a comparatively low incidence, underreporting, or failure to recognize the disease in goats because of possible differences in clinical presentation. In the majority of caprine cases, there is a history of contact between affected goats and infected sheep, though contact with sheep does not appear to be an absolute prerequisite for caprine infection. It is assumed that the principal method of transmission from sheep to goats is via direct contact with infected placentas, or nasal secretions of infected sheep. Indirect contact with pastures or bedding contaminated by placentas may also play a role.

Pathogenesis

Experimentally, the disease has been produced in goats by subcutaneous and intracerebral inoculation of the agent and by feeding of fetal membranes from infected sheep. The incubation period is long, ranging from 30 to 146 weeks, depending on the route of infection. Regardless of incubation time, early replication of prions occurs first in lymphoid tissues, including spleen and intestine, and subsequently in nervous tissue, where progressive neuronal degeneration is responsible for the clinical signs of disease observed.

Clinical Signs

Clinical scrapie has been seen only in adult goats from 2 to 8 years old. The clinical presentation may vary. One documented case showed no signs other than listlessness, progressive weight loss, and premature cessation of milk production. In sheep, intense pruritus manifested by constant rubbing or scratching with resultant alopecia and dermal abrasion is commonly observed. Pruritis is less intense and less frequent in affected goats. Hyperesthesia and incoordination are the more predominant signs.

Typically, the clinical course is slowly progressive over 1 to 6 months. Early on, affected goats often show a characteristic posture of carrying the rear limbs forward so that the rump is elevated and the withers are held low. In addition the tail is held up, cocked forward over the rump, and the ears are frequently pricked forward suggesting heightened alertness. Restlessness increases and goats become hypersensitive to handling. Affected goats may attempt to flee, bleat uncontrollably, or stiffen with muscular rigidity when handled, and lactating does may strongly resist being milked. Left alone, they may hold the head down and stomp as if bothered by imaginary flies. Fine tremors may be observed. Some goats do show localized pruritus, particularly over the withers or tail head. Other signs may include, teeth grinding, salivation or slobbering, regurgitation of rumen contents, and impaired vision. Weight loss may become marked and anorexia develops in the terminal stages.

Clinical Pathology

Currently, there are no laboratory tests or clinicopathologic findings that support a diagnosis of scrapie. Definitive diagnosis is by histologic examination of the central nervous system.

Post Mortem Findings

At the time of necropsy, gross findings are limited to evidence of cachexia, and dermabrasion, if pruritus is a component of the clinical picture. Sections from the entire brain and spinal cord should be examined histologically. The principal lesion is a progressive degeneration of the gray matter. The most consistently affected part of the brain is the tegmentum and its connections. Lesions are most constant in the thalamus, midbrain, cerebellar cortex, and medulla oblongata and are uncommon in the cerebral cortex and spinal cord . They are characterized by neuronal degeneration and shrinkage with vacuolation, and astrocyte hypertrophy.

Diagnosis

The disease is diagnosed on the basis of clinical symptoms and pathological findings on cerebral areas characterized by vacuolization and degeneration of neurons. Inoculation in susceptible goats and also in mice is the most common diagnostic methods. Serological tests are of no use hence diagnosis can not be established prior to appearance of clinical symptoms. Differential diagnosis should be established for Aujeszky's diseas, Maedi, Visna, Louping ill and Rabies. Control programme is based on identification and slaughter of clinical cases.

Diagnosis is presumptive on the basis of slowly progressive neurologic disease in an adult goat with characteristic lesions demonstrated histologically. In many countries, scrapie is a disease reportable to regulatory authorities. In some countries, mouse inoculation studies may be carried out using tissue from suspect cases to strengthen the diagnosis. Fresh brain homogenates can also be examined by electron microscopy for scrapie associated fibrils. If pruritus is a component of the clinical picture in goats, differential diagnoses should include lice, mange, *Parelaphostrongylus tenuis* infection, and primary or hepatic photosensitization. If the onset of pruritis is rapid and the course of disease short, then pseudorabies and rabies are more likely. When tremors and progressive incoordination are the predominant signs, CAE, pregnancy toxemia, enzootic ataxia, paraelaphostrongylosis, gid, spinal cord abscesses, Borna disease, and the dumb form of rabies should be considered.

Treatment

There is no treatment for scrapie. Suspect cases should be killed for diagnostic purposes after contacting appropriate authorities because of the regulatory and herd health implications of the disease.

Prevention and Control

In general terms, the risk of scrapie can be reduced in herds by minimizing outside purchases of stock, and reducing opportunities for horizontal transmission. Sheep and goats should not be commingled when the infection status of sheep is unknown. During kidding season, placentas and fetal membranes should be found and removed immediately. Bedding should be changed in kidding pens between kiddings. Any animals showing tremors, incoordination, or pruritus should be isolated and a definitive diagnosis attempted.The potential for genetic control in sheep has been demonstrated in an experimental flock on the basis of selection of breeding stock from progeny of ewes resistant to experimental challenge. Regulatory programs for control differ from country to country. There are currently no vaccines available for control of scrapie.

10.5 Border Disease

Border disease, also known as "hairy shaker" disease in lambs, is an infectious, contagious viral disease of sheep and goats. It causes abortions, infertility, stillbirths, weak lambs and kids, and a characteristic pattern of tremors, or shaking, in newborn lambs and kids. In lambs, it also produces a characteristic hairy appearance to the fleece and histologic changes in the skin.

Etiology

The causative agent, border disease virus (BDV) is an enveloped, single-stranded RNA *Pestivirus* in the family Togaviridae. It is structurally and antigenically related to the bovine viral diarrhea virus (BVDV) and the hog cholera virus.

Epidemiology

The incidence of border disease in goats and its impact on the goat industry are not clear. The disease in sheep occurs in North America, Europe, Australia, and New Zealand. In contrast, naturally occurring, clinical border disease in goat kids has been reported only from Norway. Nevertheless, serologic surveys performed in Canada, United States, Norway and Nigeria indicates the presence of antibodies to BVDV or BDV in 3 to 16% of goats tested.

Pathogenesis

Horizontal and vertical transmission of the virus occurs. Horizontal transmission is believed to be by ingestion or aerosol inhalation of virus. The fetus is infected transplacentally. The outcome of intrauterine and fetal infection depends to a great extent on the time of infection, although virus strain variation and breed factors are believed to play a part. In an experimental challenge of goats, infection after 40 days of gestation resulted in an 86% abortion rate, infection at 60 days in a 39% abortion rate, and infection at 100 days a 6% abortion rate. Intra-uterine infection produces a more severe placentitis in goats than in sheep and this may account for more frequent fetal death and less frequent delivery of full term, shaker kids when compared with sheep.

Clinical Signs

Shaker kids and lambs show muscle tremors right from birth. These rhythmic tremors are most prominent in the hindquarters but may extend up the trunk and neck. The general impression is one of jerkiness. Affected young have difficulty rising and when they do, they have an awkward gait. Nursing is difficult and affected young may become quickly hypothermic and hypoglycemic. They appear weak, listless, and depressed if not attended to. The long bones of the limbs may feel finer than usual, and the head appears narrowed and exaggeratedly convex in the frontal area. No abnormalities of the hair coat or skin occur in affected goats.

Clinical Pathology

Sero-conversion in association with abortion in a herd is suggestive of border disease. In the one reported field outbreak, 90% of does with serum neutralizing antibodies to the BVD virus had titers of 1:500 or more. In affected kids, absence of antibody does not rule out border disease, due to the possible development of *in utero* immunotolerance.

Post Mortem Findings

At the time of necropsy of affected kids, no gross lesions are observed. Histologic lesions in the central nervous system are characterized by pronounced hypomyelinogenesis, hypergliosis, and vasculitis, particularly in the white matter of the cerebrum and cerebellum. There is perivascular gliosis and infiltration of vessel walls with lymphocytes and histiocytes. In aborting does, there is a marked placentitis with necrotizing caruncultitis. Virus may be isolated from uterine contents or vaginal discharges.

Diagnosis

Diagnosis is based on a history of reproductive failure, occurrence of newborn shaker kids, and laboratory confirmation of infection. Hypoglycemia, septicemia, or meningitis may produce convulsions in newborn kids that could be mistaken for border disease shakers. When kids are recumbent, nutritional muscular dystrophy, congenital vertebral malformations, spinal trauma, and swayback should be considered.

Treatment

There is no specific treatment. Supportive care to insure regular feeding will promote kid survival, but the overall herd status needs to be considered with regard to the advisability of keeping potential carrier animals in the herd.

Prevention and Control

All does that abort, regardless of cause, should be isolated from the herd. The border disease virus is present in vaginal discharges after abortion or kidding. In general goats should be maintained separately from cattle and sheep. Elimination of infection from the herd based on serologic testing is problematic as some nonreactors may in fact be persistently infected, immunotolerant shedders of the virus. In sheep, total depopulation with repopulation from virus-free flocks is suggested as the only possible means of disease eradication. The use of BVDV vaccine has been suggested in sheep flocks suffering annual outbreaks of border disease, but this pattern of disease has not been reported in goat herds. If vaccines are used, vaccine failures may occur because different strains of BVDV and BDV virus may not be cross-protective. Two doses of killed vaccine before breeding are recommended in sheep.

10.6 Maedi–Visna

Maedi-visna is an economically important viral disease of sheep that occasionally affects goats. The maedi-visna virus (MVV), a lentivirus; infects its hosts for life. Although most infections are subclinical, a minority of animals develops progressive, untreatable disease syndromes including dyspnea (maedi) or neurologic signs (visna). Both maedi and visna are eventually fatal. Additional economic costs may include marketing and export restrictions, premature culling, and losses from poor milk production due to indurative mastitis. Economic losses can vary considerably between flocks.

Etiology

Maedi-visna results from infection by the maedi-visna virus, a member of the genus *Lentivirus* in the family Retroviridae (subfamily Orthoretrovirinae). This virus becomes integrated into leukocyte DNA; infected animals become chronic carriers. Several genetically distinct isolates circulate in sheep. Phylogenetic analyses have demonstrated that maedi-visna virus is closely related to caprine arthritis encephalitis virus (CAEV), a lentivirus found most often in goats. These two viruses share many features, and are often considered together as the small ruminant lentiviruses (SRLV). Early phylogenetic studies suggested that SRLV can be divided into six sequence clades, I to VI. Clade I contains the prototype Icelandic visna virus and related MVV strains. Clade II consists of North American lentivirus strains isolated from sheep. Clade III consists of Norwegian SRLV, and clade IV of French SRLV. Clade V contains French and Swiss CAEV strains, North American prototype strains, and North American ovine lentivirus strains. Clade VI contains French SRLV. In this analysis, clades III to VI contain related SLRV from both sheep and goats, while clades I and II are more species-specific. These findings suggested that these viruses might be more closely related to each other, in some cases, than to other CAEV or MVV, but they were based on short sequences of nucleic acids.A new phylogenetic analysis, based on longer genetic sequences, divides these viruses into four principal sequence groups, A to D. Sequence groups A and B are divided further into subtypes. Group A contains at least seven subtypes and group B at least two subtypes. To date, subtypes A5 and A7, and groups C and D have been found only in goats. Subtypes A1 and A2 have been isolated only from sheep. Sub-types A3, A4, A6, B1 and B2 have been found in both species. Recombination between a group A maedi-visna virus and a group B caprine arthritis-encephalitis virus has recently been demonstrated in goats infected with both viruses.

Epidemiology

Maedi-visna affects sheep and, to a lesser extent, goats. Breed susceptibility varies. Serological evidence of SRLV infections has also been reported in wild ruminants including moufflin, ibex and chamois; however, preliminary evidence suggests that these viruses may be distinct from CAEV and MVV. Maedi-visna has been found in most sheep-raising countries other than Australia and New Zealand. MVV has been reported from most of continental Europe, the United Kingdom, Canada, the United States, Peru, Kenya, South Africa, Israel, India, Myanmar and the southern regions of the former U.S.S.R. Most animals become infected early in life, from drinking infected colostrum or milk. The virus can also be spread during close contact, probably by the respiratory route. Co-infection with pulmonary adenomatosis (Jaagsiekte) virus increases MVV titers in the respiratory tract, increasing contact transmission between sheep. Transmission has been reported from water contaminated with faeces, but indirect spread is generally thought to be rare. Intrauterine spread is thought to be negligible or minor. MVV infects sheep or goats for life, but viral burdens vary between individual animals. Both asymptomatic and symptomatic animals can transmit this virus. Sheep can be a source of SRLV transmission to goats, and vice versa. There is little information on the route(s) of transmission between sheep and goats, but the ingestion of contaminated colostrum or milk, or close contact between the two species in crowded barns have been suggested. Under experimental conditions, lambs that have nursed from infected goats can become persistently infected with SRLV. The incubation period for maedi is usually more than two years; clinical signs typically develop when animals are three to four years old. The incubation period for visna is somewhat shorter, and symptoms can appear in sheep as young as two years.

MVV is widespread among sheep in many parts of the world. In the United States, the prevalence varies with the region. Infection rates in sheep are highest in the Western and Midwestern States, and can reach nearly 50% in some areas. Control programs have reduced the incidence of MVV in some countries. Maedi visna is uncommon in goats; clinical cases have mainly been reported as visna in adult dairy goats. Most infections are asymptomatic, but once clinical signs appear, the disease is progressive and usually fatal. When MVV is introduced into a new area, the mortality rate may reach 20-30%. The mortality rate is low in regions where maedi visna is endemic; annual losses rarely exceed 5% in a flock, even when nearly 100% of the flock is infected. Co-infection with Jaagsiekte virus, the retrovirus that causes ovine pulmonary adenocarcinoma, results in more severe symptoms and increased transmission. Management practices can influence the prevalence of infection and, thus, the frequency of disease. Clinical signs are not usually seen in herds with a low prevalence of infection. Genetic factors, including the breed of the sheep, influence the outcome of infection.

Clinical Signs

Most MVV infections are asymptomatic. In animals with clinical signs, the disease can take several forms. Sheep with maedi is the most common form, experience wasting, progressive dyspnea and sometimes a dry cough. Fever, bronchial exudates and depression are not usually seen. Maedi is eventually fatal; death results from anoxia or secondary bacterial pneumonia.Visna occurs less frequently than maedi in sheep, although it is the more common form reported in goats. Visna usually begins insidiously, with subtle neurologic signs such as hind limb weakness, trembling of the lips or a head tilt, accompanied by loss of condition. The symptoms gradually progress to ataxia, in coordination, muscle tremors, paresis and paraplegia. Other neurological signs, including rare instance of blindness, may also be seen. The clinical course can be as long as a year. Unattended animals usually die of inanition. MVV can also cause slowly progressive arthritis with severe lameness, or chronic indurative mastitis with decreased production of normal appearing milk. Weight gain in lambs may be decreased, possibly due to lower milk yields from dams with indurative mastitis.

Post Mortem Findings

In maedi, the lungs are enlarged, abnormally firm and heavy, and fail to collapse when the thoracic cavity is opened. They are typically emphysematous and mottled or uniformly discolored, with pale gray or pale brown areas of consolidation. Mottling may not be obvious in the earliest stages of the disease. Nodules may be found around the smaller airways and blood vessels, and the mediastinal and tracheobronchial lymph nodes are usually enlarged and edematous. Secondary bacterial pneumonia may mask the primary lesions. On histological examination, lung lesions may include chronic, diffuse interstitial pneumonia, perivascular and peribronchial lymphoid hyperplasia, and hypertrophy of the smooth muscle throughout lungs. Apart from wasting of the carcass, the only gross lesions seen in visna occur in the brain and spinal cord. Focal, asymmetric, brownish pink areas may be found in the white matter of the brain and spinal cord, as well as on the ventricular surfaces. The meninges may be cloudy and the spinal cord may be swollen. On microscopic examination, the typical CNS lesion is meningoleukoencephalitis with secondary demyelination. In some cases, the inflammatory cell aggregates may be nodular or granulomatous, and necrotic centers may be noted in severe cases. In ewes with indurative mastitis, the udder is diffusely indurated and the associated lymph nodes may be enlarged. Histologically, the udder is characterized by mononuclear infiltration of the periductular stroma; these cells obliterate the normal mammary tissue.Arthritis may also be seen in some animals, and the kidneys may have microscopic evidence of vasculitis.

Diagnosis

Clinical

Maedi-visna should be suspected in animals that are at least two years old and have a wasting disease with slowly progressive respiratory distress, neurologic signs, indurative mastitis or arthritis.

Differential Diagnosis

The differential diagnosis for maedi includes pulmonary adenomatosis, parasitic lung infections, and caseous lymphadenitis with lung involvement. In cases with neurologic symptoms, scrapie, listeriosis, rabies, louping ill, parasitic central nervous system (CNS) infections and space occupying lesions of the CNS should also be considered. Caprine arthritis and encephalitis can also resemble maedi visna.

Laboratory Tests

Maedi visna may be diagnosed by nucleic acid detection techniques such as polymerase chain reaction (PCR) assays, Southern blotting and *in situ* hybridization. PCR tests are used in some laboratories for rapid diagnosis. Maedi-visna can also be diagnosed using a combination of serology and clinical signs, together with histological examination of tissues when necessary. Agar gel immunodiffusion and enzyme-linked immunosorbent assays (ELISAs) are the most commonly used tests. Immunoblotting (Western blotting) is generally performed only in specialized laboratories, but may be valuable when sera give equivocal results in other tests. Seroconversion generally occurs months after infection. In general, serology is of greater value in screening flocks or other populations than in diagnosing this disease in individual animals. In adult sheep and goats, a positive result indicates that the animal is persistently infected with MVV but, because most infected animals do not become symptomatic, it does not confirm that the symptoms are caused by this virus. Due to these limitations, serology is of greater value in screening flocks than diagnosing this disease in individual animals.In seropositive, symptomatic animals, histology can confirm the diagnosis in biopsy or necropsy samples. Virus isolation can also be useful; however, viral titers are variable and may fluctuate over time. MVV is isolated by co-culturing peripheral blood or milk leukocytes from live animals with sheep choroid plexus cells, sheep skin fibroblasts or other appropriate cell lines. MVV can also be isolated from tissues at necropsy. In co-cultures that display cytopathic effects, the presence of the virus is confirmed with immunolabelling methods and electron microscopy.Adherent macrophage cultures established from post-mortem bronchoalveolar lavage can be tested for virus production by electron microscopy or a reverse transcriptase assay. They can also be co-cultured with indicator cells for virus isolation.

Samples to be Collected

Serum should be collected for serology. Milk can also be tested for antibodies. Virus isolation can be conducted on peripheral blood or milk from live animals. At necropsy, MVV can be isolated from the lung, mediastinal lymph nodes and spleen of animals with maedi. In animals with suspected visna, a sample of the brain should be sent. Alveolar macrophages can also be collected from the lung at necropsy, by post-mortem bronchoalveolar lavage. Samples for virus isolation and alveolar macrophages should be as fresh as possible.Samples for histology should include all affected tissues such as lung, mediastinal lymph nodes, brain, spinal cord, kidney and/or udder.

Treatment

There is no specific treatment for maedi visna. Supportive therapy may be helpful in individual animals, but it cannot stop the progression of disease.

Prevention and Control

Maedi-visna is a contagious disease. Management practices can influence the prevalence of infection and, thus, the frequency of disease. Clinical signs are not usually seen in herds with a low prevalence of infection.MVV is often introduced into a herd in live animals. Additions to uninfected herds should come from MVV negative herds. Other animals should be quarantined and tested before adding them to the herd. Uninfected herds should also be kept from contact with untested or sero-positive herds, as horizontal transfer of the virus contributes to transmission. Goats may also be able to transmit SRLV to sheep. No vaccines are currently available. MVV can be eradicated from a flock, or reduced in prevalence, by isolating lambs permanently from sero-positive dams immediately at birth. The lambs are raised on uninfected colostrum and pasteurized milk, milk replacer or milk from sero-negative ewes. The flock should also be tested frequently for MVV, and sero-negative and sero-positive sheep should be maintained separately. Sero-positive sheep should eventually be culled. In nationwide eradication programs, quarantines of infected herds aid the final stages of the program. Maedi-visna virus cannot survive for more than a few days in the environment, particularly under hot, dry conditions. Lentiviruses can be destroyed with most common disinfectants including lipid solvents, periodate, phenolic disinfectants, formaldehyde and low pH (pH < 4.2). Phenolic or quaternary ammonium compounds have been recommended for the disinfection of equipment shared between sero-positive and sero-negative sheep.

10.7 Pulmonary Adenomatosis

Pulmonary adenomatosis is a contagious viral disease that results in pulmonary neoplasia in sheep and occasionally goats. The economic impact can be significant: up to 60% of the flock can be lost upon first exposure to the virus. There is no treatment and no vaccine.

Etiology

Pulmonary adenomatosis results from infection by the jaagsiekte sheep retrovirus (also known as the pulmonary adenomatosis virus). The virus is a member of the family Retroviridae and appears to most closely resemble the type D oncoviruses. It has not been grown *in vitro,* but has been isolated in immune complexes from the lungs of infected animals. An infectious molecular clone is able to reproduce the disease in sheep.

Epidemiology

Pulmonary adenomatosis affects sheep and, to a lesser extent, goats. All breeds of sheep are susceptible. Pulmonary adenomatosis can be found in continental Europe, the United Kingdom, Africa, India, China, the Middle East, North America, and some parts of South America.Pulmonary adenomatosis is spread by the respiratory route, probably by aerosol or droplet infection. Infectious virus can be found in the respiratory exudates of infected sheep. There is no evidence of transmission *in utero*; however, neonates seem to be particularly susceptible to infection. The incubation period varies from nine months to three years. Most cases of pulmonary adenomatosis occur in sheep over 2 years old; the peak incidence is in 3 to 4 year old animals. Upon first exposure, the mortality rate is usually high: up to 60% of the animals in the herd may die during the first three years. Eventually, the annual rate of loss drops to approximately 1–3%, although the disease usually remains endemic in the herd for a long time.

Clinical Signs

The clinical signs of pulmonary adenomatosis include progressive emaciation, weight loss, and severe dyspnea. There is usually a thin mucus discharge from the nostrils and, if the head is lowered, copious frothy mucus may pour from the nares. Affected sheep often lag behind the flock. Moist rales may be heard on auscultation, but coughing is not prominent. Moist crackles are audible over the affected areas and may be heard at a distance so that the affected group of animals are said to produce a sound like slowly boiling porridge. The symptoms are slowly progressive. Disease usually affects mature sheep between 1-4 years of age and disease onset is insidious. A diagnostic test for the disease is to hold the sheep by the hind legs while lowering the head. In affected animals, a quantity of copious, thin mucoid fluid (up to 200 ml), presumably produced by the neoplastic cells in the lung, pours from the nostrils. Appetite and rectal temperature are usually normal. Sheep die within a few weeks to six months after first exhibiting clinical signs from complicating *Pasteurella hemolytica* pneumonia. The only reported laboratory abnormality in pulmonary adenomatosis is hypergammaglobulinemia.

Post Mortem Findings

The lungs are usually enlarged and, in advanced cases, do not collapse upon opening the chest cavity. Frothy fluid may be seen in the trachea and bronchi, and nodular tumors are found in the lungs. These tumors vary from small nodules to solid masses. They are sharply demarcated, firm, and gray or pinkish-gray. On cut surface, the tumors are glistening and granular; a frothy fluid may be squeezed from them. Most tumors occur in the apical, cardiac, and ventral portions of the diaphragmatic lobes. Metastasis sometimes occurs to the nearby lymph nodes, but the tumors do not usually become more widespread. Secondary pneumonia and fibrinous pleuritis may be found.

Diagnosis

Clinical

Pulmonary adenomatosis should be suspected in sheep with chronic respiratory symptoms, particularly in 3 to 4 year old animals with a frothy mucoid discharge from the nostrils.

Differential Diagnosis

The differential diagnosis includes maedi–visna, bacterial pneumonia, and infestation by lungworms.

Laboratory Tests

Diagnosis is usually based on clinical signs, postmortem lesions, & histopathology. Occasionally, electron microscopy, immunoblotting (Western blotting), or inter–species competitive radioimmunoassay is used to demonstrate type D retrovirus particles in the lungs.Recently, a polymerase chain reaction (PCR) method was able to detect the jaagsiekte sheep retrovirus in the blood before symptoms appeared. This technique could also find the virus in the tumors and lymphoid organs of affected sheep.

Samples to be Collected

Regional lymph nodes and samples of affected lung should be collected aseptically. These samples should be transported cold with wet ice or frozen gel packs. Duplicate samples should also be taken (in neutral buffered formalin) for histopathology.

Prevention and Control

The respiratory secretions of sheep with pulmonary adenomatosis are contagious. Retroviruses are generally fragile in the environment and are susceptible to most common disinfectants. No treatment or vaccine is available.

10.8 Other Exotic Diseases

Other exotic diseases of viral origin infecting both sheep and goats include Nairobi sheep disease, Akabane disease and Ainovirus infection (All from Bunya-viridae family). Some of these viral infections are still limited to African countries except a few such as Louping ill (USSR, Europe and Australia), Akabane disease (Japan & Australia). Details of prevalence will be changing as the disease status of various foreign countries changes. It would therefore be essential to ascertain the disease status from time to time before fixing the priority of disease diagnostic facilities at HSADL Bhopal IVRI centre in coming years. Beside routine diagnostic tests, the diagnosis of exotic and emerging diseases will be based on advanced diagnostic techniques including FAT, ELISA, IPT, electron microscopy and diagnoses at molecular level through advanced techniques such as recombinant DNA technology, radio isotope technology, PCR technique and use of maps in detection and quantification of viral antigen and antibodies.

❑❑❑

11

Production and Metabolic Diseases

Production diseases are of special concern in these animals, particularly those being demonstrated at farm condition, presumably due to instability of the internal environment. Intensification of sheep and goat industry is likely to result in an increase in the number of high producing animals. This intensive method of management and feeding may result in an environmental stress on the metabolic health of the animal resulting the more prone to metabolic disease. It is not surprising that production diseases/metabolic diseases are becoming increasingly serious problems with intensive husbandry/practices. Amongst domestic farm animal, the metabolic diseases achieve the greatest importance in pregnant and lactating sheep and goat. The effect of pregnancy is important in ewes and does and particularly those carrying more than one lamb/kids. The essential metabolite which reduced below the critical level, determines the clinical syndrome. The bulk of economic losses are due to the loss of production and failure to return to full production after recovery in production disease in farm animals including sheep and goats.

The production disease or production related disease comprised of diseases associated with an imbalances between the rate of input of essential dietary nutrients and the output of the production such as Parturient paresis, Ketosis, Hypomagnesaemia and other disease having same pathogenesis and out come. However, still these production diseases are commonly known as metabolic diseases in most of literature. The incidence of production disease has increases with the rise of productivity, increased prolificacy by superovulation technique and low quality of roughages available. All of which are attributable to an imbalances between the input of nutrient and output of the animal production. In recent years, scarcity of animal feed and forage in some part of the country has drastically influenced the production through production disease in domestic animal including sheep and goats. When the imbalances is maintained, it may lead to a change in the amount of body reserve of certain metabolites such as carbohydrate, protein and minerals, which give rise to signs of production diseases. The Kinetics of the mammary glands challenges to the metabolic activities of the body is still explainable even after the advancement of so many theories viz nutritional, endocrine and physiological, because some of the animals accustom themselves to such as metabolic disturbances and do not exhibit clinical disease. In small ruminants, most of these production related disorder are mainly caused by dietary irregularities viz. inadequate feed, irregular feeding, sudden change in feed, spoiled food, change in rumen microbial population, indigestible substances and inadequate nutrients etc. Because of the difference in the soil type, climate condition, livestock husbandry practices and difference in the feed and fodder among countries, the incidence and gravity of disease is also likely to vary country to country. Peculiar problem of sheep and goat industry in India, like hot climate, course roughages, irregular availability of fodder, imbalance of nutrient causing sub-clinical starvation lead to the metabolic diseases, particularly in pregnant and lactating animals. Hence in most cases, the disease picture is so complicated that the diagnosis becomes difficult.

Considerable advance have been made in the laboratory investigations on various aspect of the metabolic and digestive physiology and medicine of ruminant and much still remains to be done in our country particularly in sheep and goat maintained under intensive/ semi intensive system of management. However, little work had been done in our country on metabolic diseases in sheep and goat.

However more work has been done in Ketosis in cow. Pregnancy toxemia, lactational ketosis, hypocalcaemia and hypomagnesaemia are only important conditions in small ruminant that need a detail discussion.

11.1 Pregnancy Toxemia

Also known as pregnancy disease or twin lamb disease or ketosis or pregnancy toxemia is a important metabolic disease of goats and sheep, characterized by increase level of ketone bodies (Acetic acid, Beta hydroxybutyric acid and Acetoacetic acid) in blood, milk and urine and is associated with elevated blood plasma nonesterified fatty acid (NEFA) which are the precursors of ketone bodies. In simple words, ketosis is a metabolic problem caused by the animal's having to live on its own body reserve. Pregnancy toxemia is thought to result from disruption of the dam glucose homeostasis mechanism in response to increased nutrient demand of the rapidly developing fetal placental unit. Commercial production system are comprised of a variety of nutrient metabolic, gastric,physiological,environmental, economic, health and managemental factors that singularly as a group influence the clinical expression of pregnancy toxemia. It is unrealistic to expect a large flock of commercial goats to be totally free of pregnancy toxemia. Late pregnancy ketosis is encountered in sheep and goat carrying multiple foetuses. Goats appear to be more resistant than cow or ewes to ketosis. Lactational ketosis is associated primarily in high producing dairy cow and to a lesser extent in dairy goats. Pregnancy toxemia in sheep and goat is limited to the last 4-6 weeks of gestation. Factors that predispose to develop pregnancy toxemia can broadly be divided in to 3 types: undernutrition, adequate nutrient offering but external or animals factors affecting intake and overnutrition. In starvation ketosis, the animal has not been permitted access to enough nutrients especially energy to meet the demand of itself and developing multiple fetuses. In the condition of adequate nutrient offering to pregnant ewes / does, some external factors like inclement weather, improper feeder space to gravid does, poor quality of drinking water, rapid feed changes, recent transport, preventive diseases such as vaccination in late gestation, may reduce the feed intake and animal become prone to pregnancy toxemia. Besides these, some concomitant diseases such as hypocalcaemia, lameness, gastrointestinal parasitism, Johne's disease, chronic liver disease and dental disease etc reduce fed intake should be considered in management of pregnancy toxemia in small ruminant farms. In over nutrition ketosis, the goats has been overfed until its massive internal fat store plus the full uterus occupy so much of the body cavity space that dry matter intake is severely curtailed at a time,when increased energy consumption is required. The goats kept on excessive grain feeding may also suffer from pregnancy toxemia because of the goats that eats inadequate roughage is apt to go off fed at this critical time, moreover, rumen acidosis also contribute to the development of pregnancy toxemia. These determinant factors had extensively reviewed in goats. In late pregnancy, carrying multiple fetuses in the does/ewes in order to increased demand of energy; the fat mobilization for the availability of glucose is the central point of pathogenesis. This disease is not very common in sheep and goat, usually carrying one foetus but expected to occur in high prolific breed, however, proper nutrient, supply to such animal can remain healthy even carrying twin/triplet/quadruplet. Principal source of energy to the foetus is glucose and utilization by the fetus occurs at the detriment of the mother through maternal hepatic gluconeogenesis. Blood glucose level decreases in pregnancy progress (Hypoglycemia) form the normal 35-45 mg % to 20-25 mg % in late pregnancy. Pregnancy toxemia may develop when level decreases to below 20 mg% in goats. Increased demand of glucose, manufacture either from production of propionic acid precursor in the rumen or from gluconeogenesis of amino acids derived from the diet or body muscle reserve. This prolonged hypoglycemia results in suppressed insulin production, that spare glucose for foetal need, while at same time stimulated lipolysis and gluconeogenesis. This is limited by the availability of glucose precursor oxaloacetate for TCA cycle, result in a build up of acetyl Co enzyme A, which is broken down in the liver to the ketone bodies acetoactetic and β-hydroxybutyric acid.Placental lactose are also greatly increased, particularly with multiple fetuses in goats which have both growth hormone as well as prolactin activity and important to fetus to meet metabolic demand at the expense of dam. Therefore, it had been expressed that pregnant dam is often remain sub clinically ketotic, predisposing factors may trigger the clinical condition at this crucial physiological state. The ketone bodies and free fatty acid do not cross the placenta in any substantial quantity in sheep. As ketosis increase, the bicarbonate level in the blood decreased resulting to the acidotic state. The clinical symptoms of ketosis appeared as the blood bicarbonate level decline sufficiently however, hypoglycemia is a key feature in pathogenesis of pregnancy toxemia but in advance cases when water consumption decrease urine, output affected leading to impairment of kidney function,under the response of the adrenal gland to stress hyperglycemia were observed. A comparative studies on behaviour of serum profile in clinical pregnancy toxemia revealed that glucose level decreased, acetone increased in both the species but difference between two species were significantly differ,as drastic lower level was noticed in sheep as compared to goats. Inorganic phophorus were decreased in both the species might be due to failure to metabolic propionic acid and causes reduction in intake of total digestible nutrient in pregnancy toxemia. The total protein, albumin, globulin value were decreased in pregnancy toxemia in goats. In comparison with healthy twin pregnant ewes, the animals with sub-clinical and clinical pregnancy ketosis showed increased values of bilirubin glutamate dehydrogenase, somatotropin, total estrogen, cortisol and cortisol/ insulin rate, decrease in urea, total protein, increase in creatinine in faster goats.

The course of disease varied but generally develops over 3-10 days. The clinical sign are observed with involvement of the central nervous system originated from decreased utilization of glucose by the doe's brain. Initially decrease food intake, dull, depressed, separate from the herd, grunt and lie off in a corner. Subcutaneous oedema may also appear in goat. Ketotic ewes showed a loss of condition and significantly shortening of the gestation period to 142.0±0 day compared with 150± 0.4 day in normal twin pregnant ewes. Further progression of disease affected animals exhibits neurological signs which include an abnormal gait, apparent blindness, stargrazing posture, teeth grinding, loss of menace response, nystagmus, ataxia, tremors, severe depression followed by succumbing coma, and eventually die. Some of the signs are attributed to ischemic necrosis of the brain cortex secondary to prolonged hypoglycemia. Occasionally animal may show a small period of intermittent period of hypersensitivity having twitching of ear, muzzle or eye lid or spasms of particular muscle. Recumbent animals may have convulsive paddling movement. Temperature and pulse are within normal limit. Respiration is usually normal until the later stage, where animal may breathe more rapidly because of development of metabolic acidosis. Therefore, a careful clinical examination is needed to differentiate with primary pneumonia. In terminal case or not attended animals becomes recumbent, may be death of foetus, which may progress to the development of toxemia. The course of disease in neglected or unattended cases varies from 12 hour to 1 week in goats. Goats with pregnancy toxemia that do not die tend to have dystokia and higher kid mortality. For a clinician, it is also imperative to differentiate the neurological sign of pregnancy toxemia with other clinically related disease i.e. polioencephalomalacia, hypocalcaemia, toxic mastitis, acute rumen acidosis, listeriosis, lead poisoning in sheep and goat. The main necropsy finding in pregnancy toxemia revealed enlarged liver, pink and yellow in color due to infiltration with fat. The carcass dehydrated, uterus may contain two or more foetuses, urine remain in bladder showing strong positive reaction for ketone bodies. The adrenal gland may be enlarged. Besides, clinical sign, clinical diagnosis of pregnancy toxemia, it is not always simple to detect the odour of ketone in the breath of ketotic animals. Others field test must be employed for detection of ketonic bodies in urine, milk and serum. It is also true that urine collection is not easy in goat and sheep; therefore, alternatively plasma or serum can be checked for increased levels of ketone bodies. Serum betahydroxy butyric acid content in blood serum may be used to provide screening test for flock nutritional status and clinical pregnancy toxemia in goat and this test was found superior and reliable over nitroprusside test for detection of ketone bodies in cattle on both milk serum and blood serum. Hypoglycemia below 25 mg% and ketonemia may range from 10 to 100 mg/dl in ketotic goats, However hyperglycaemia generally observed in terminal cases. Increased plasma NEFA levels were a more sensitive indicator of undernutrition in goats then blood ketone or based glucose levels, can also used a health indicator in relation to ketone.

It is important the producer recognized the predisposing factors to pregnancy toxemia and take action to prevent the disease. Does/ewes with reduced appetite and milk depression, without neurological sign may response to propylene glycol (600 mg/ml) at the rate of 60ml *per os* for 3 days, along with improved nutrition, feeding management and management of predisposing conditions. However, some other suggests 175 to 250 ml of propylene glycol twice a day but can be fatal in large doses. Intensive therapy is desired in clinical case showing neurological signs. Intravenous 5-7 gram as 50%glucose may be effective along with B complex 6-8 times a day in sheep and goats. Insulin may be used with these treatment for better utilization of glucose at the dose of 20-40 units a zinc protamine insulin IM for three days. During the late stage of the disease glucose administration may be ineffective or rather detrimental because the blood glucose level may be very high. Therapy requires the correction of fluid, electrolyte and acid base disturbance in addition to glucose therapy. Corticosteroid therapy like dexamethasone and triamcinolone acetonide is also beneficial. For correction of ketoacidosis, interavenous therapy should be continued including isotonic or hypotonic sodium bicarbonate solution or bicarbonate precursor, since hypocalcaemia is often a secondary diseases associated with pregnancy toxemia and advocated therapy accordingly. Other concurrent disease should also be managed to enhance the recovery in sheep and goats. Vasopressin has been used to induce closure of esophageal groove in conjuction with oral administration of glucose with encouraging results. In pregnancy toxemia, therapeutic abortion is a preferred method to stop the energy drain to foetus. A single dose of 10 mg of prostaglandin F2α is sufficient for desired action.

This is a disease that needs to be prevented rather than treated. It is generally a managemental disease and should be initially investigated as a herd problem rather than an individual sick goat. If one doe is clinically ill, may more in the herd are likely at risk. An adequate nutritional level (carbohydarate, protein) throughout the pregnancy will prevent pregnancy toxemia during the last 30-40 days of pregnancy and should meet the does maintenance requirement as well as foetus. Allowing the animal to become excessive fat should be avoided. Appetite problems in pregnant animals and during late pregnancy should be avoided. Avoid sudden feed change, diminish stress, temperature etc. Real time ultrasonograhy may be used to determine number of fetuses. If one ewe/ does suffered from clinical pregnancy toxemia, the diet schedule of rest of herd should be reviewed and corrected accordingly. Prophylactic propylene glycol (60 ml orally) twice a day is only recommendable in does showing diminished appetite and abnormal behavior, otherwise it is itself a stress to a goat. Therefore, management practice, adequate concentrate, best quality roughage and avoidance of predisposing factors are the key points to prevent the occurrence of pregnancy toxemia/ lactation ketosis in ewes and does in farming system. Supplementation of feed with Monensin (20 to 25 mg) or Nicotinic acid (0.8 gram) daily in sheep between 3 to 5 days before lambing and 35 days after words prevented hypoglycaemia and ketonuria. The benefit of Monensin feeding include less mobilization of body fat as evidenced by reduced level of NEFA, ketones and increased glucose levels

11.2 Hypocalcaemia

Also known as milk fever or parturient hypocalcaemia is a metabolic disease resulted due to failure of calcium homeostasis, observed following kidding/lambing characterized by poor milk production, poor appetite,lethargy and low calcium level.This syndrome in ewes is similar to cattle but in goats, it is a rare clinical condition. This may be attributed because goat tends to come into milk production more gradually than do dairy cows. This is a sudden increase in calcium and phosphorus segment at the onset of lactation and the capacity of calcium absorption from intestine increase more slowly than the requirement. Internal absorption requires calcium binding protein and synthesis of this transport protein is controlled by Vitamin D_3 availability. A high level of calcium in the ration during gestation places almost complete reliance on the dietary source of calcium. If the prepartum diet is low in calcium, calcium mobilization from the bone is instituted. If a gastrointestinal dysfunction occurs at parturition, the effect is not very severe since calcium has been mobilized from the bone. The main clinical features consisted of poor appetite, poor milk production, hypocalcaemic tetany, hyper-excitability, sternal recumbency. Death in ewes often occur 6-12 hrs, if not treated. This is more severe in pregnant ewes as compared to lactating ewes; possibly it is complicated with pregnancy toxemia and hypomagnesemia. In high producing goat, clinical milk fever similar to cattle has been reported at the stage of 3-4 week of parturition. Blood calcium level goes down up to 4-5 mg %. Calcium requirement of goat for maintenance, lactation /kg milk and late pregnancy supplement is 0.19, 4.3, 6.0 gram daily. The recommended therapy for hypocalcaemia is slow IV injection of 50-100 ml of 25% calcium borogluconate and for sustains blood calcium level, SC injection may be advocated. Therapy responds dramatically within 30-60 minute of treatment. For prevention of hypocalcaemia oral calcium therapy can be instituted in susceptible animals.

11.3 Hypomagnesemia

Hypomagnesaemia in ewe and does is not a common metabolic disease occurred as a single etiology. Simultaneous hypomagnesaemia and ketosis can occur in ewes after lambing, kept on low quality food availability. Serum calcium may be low in hypomagnesaemia because magnesium is required for release and action of parathyroid hormone. The clinical feature of hypomagnesaemic tetany consisted of tetany, anorexia, hyperexcitability, convulsions, muscular tremor, reduced milk production depending on severity of the disease. Goats are likely to suffer less with hypomagnesaemia because goat compensate to some extent by decrease output through milk. Since disease is commonly associated with calcium deficiency, therefore, it is recommended that calcium-magnesium preparation should be advocated at the rate of 50 ml slow IV to early survival in this disease. However, alone magnesium preparation as 20% solution of magnesium sulphate, achieve immediate magnesium level in blood. For control, magnesium sufficient mineral mixture with good quality green forage for feeding should be ensured. Daily requirement of magnesium for maintenance per kg, lactation per kg milk and late pregnancy supplement is 0.045, 0.7, 0.5 gram daily. In magnesium deficient soil, treatment with magnesium salt may be considered.

In India goat and sheep industry are growing animal sector, new special attention in all aspect of disease syndrome that affecting production in animals. Metabolic diseases are one of them. Considering the fact that small ruminant are thriving on poor quality ration with high prolificacy, such animal are bound to face metabolic stress. It is suggested that:

- Metabolic profile consist all essential nutrients should be prepared in all physiological state in each breed of small ruminant to compare with clinical data for immediate and correct interpretation, to frame preventive measures.
- High yielded animals in late pregnancy and early lactation should ensured feeding ration with sufficient nutrient requirements.
- Predisposing factors associated with metabolic disease should be minimized by modifying the management in farm.
- A constant and close vigil is desired on occurrence of metabolic disease in farming condition.
- Genetic predisposition in metabolic disease should be studied in sheep and goats for superior selection.

11.4 Mineral and Vitamin Deficiency Diseases

Nutritional disorders in goats are common in several parts of world including India though well defined deficiencies are not common in goats. Countries providing informations on animal diseases, 80%

reports on nutritional disease of moderate or high incidence and trace element deficiency or toxicities are involved in more than half of those whose cause were unidentified. The soil- plant- animal interaction in relation to the incidence of trace element deficiencies in livestock is being examined. In case of vitamins, the influence of climatic conditions is considerable. Nutritional requirement should be well understood for the species such as types of production include maintenance, growth, fattening, gestation, milk production for raising kids, maximum milk production and mohair or cashmere production etc. The trace element are involved as component parts of many tissues and one or more enzyme activities and their deficiency leads to a wide variety of pathological consequences and metabolic defects. These are summarized in Table 5.

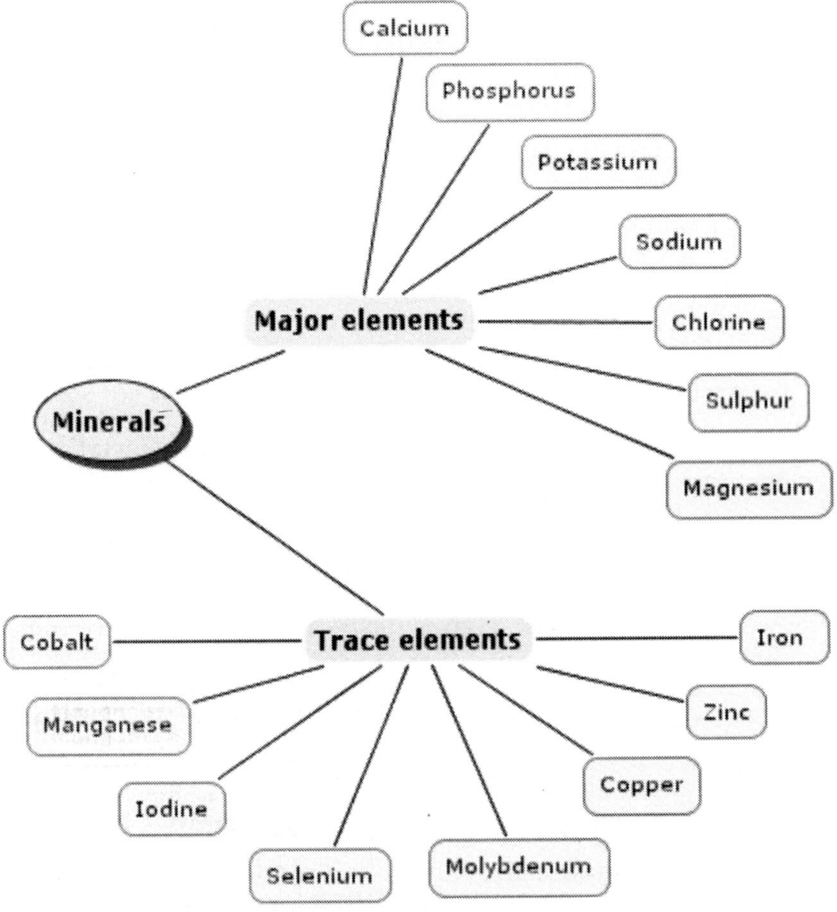

Fig. 38 Major mineral and trace elements

Copper and Cobalt

All these trace element have their main function to synthesize hemoglobin and prevent nutritional anemia. From dietary point of view, copper and cobalt are deficient in feed in India and their requirements are not met unless supplemented. Milk is deficient in iron and copper; therefore, its supplementation is important from neonatal feeding point of view. There are problem areas where feeding are not able to provide adequate quantities of these element to provide adequate quantities of these element to plants and inturn bring sub clinical and clinical syndrome. Excess in molybdenum in certain areas may bring about indirect copper deficiency in lambs and kids and they suffer from sway back, a paralytic syndrome and enzootic ataxia are related to defective myelination. Molybdenum interferes metabolically in the copper intake by liver cells and this may disturb the primary intracellular metabolism of ceruloplasmin and other Cu protein compounds.Low level copper in lush green pasture compared with conserved forage and inhibitory effects on absorption of molybdenum and sulfur antagonism from iron ingested in soil. In adult goats, abortion and still birth, anemia, frequent bone fracture, poor appetite, weight loss, decreased production are the symptoms of copper deficiency related to decreased cytochrome oxidase activity. Deficiency symptoms occur when dietary copper is less than 7 ppm and Mo is normal, therefore a suitable ration formulation is 10 ppm and it is generally recommended that to keep the Cu:Mo ratio above 2:1 and below 10:1. Enzootic anemia due to copper deficiency has been reported in young goat kids. Kids may be affected at birth or the clinical signs may be delayed until the animals are several weeks of age. Cerebellar hyperplasia is a frequent finding in goats. Adult goats are not as susceptible to copper toxicity as sheep. Younger kids are more susceptible to copper toxicity because pre ruminant kids absorb copper more efficiently than adult goats. Cobalt is a component of vitamin B_{12} and in the absence of cobalt rumen microflora can not synthesize the essential vitamin and also required as a coenzyme for transformation of propionic acid to succinic acid. The disease is caused by deficiency of cobalt in the diet and characterized by anorexia, poor production, weight loss, anemia and, diarrhea may be noted and increased susceptibility to strongly, White liver disease, (Hepatic dystrophy) with low Vitamin B_{12} has been reported in young goats in New Zealand and responded to Vitamin B_{12} and cobalt therapy. Similar to sheep, 0.1 ppm of cobalt in the diet is necessary in goats. Primary cobalt deficiency occurs only on soil deficient in cobalt. In such areas, salt should be fortified with cobalt chloride or cobalt sulphate at the rate of 12 g/ 100 kg salt and alternatively to dietary supplementation, top dressing of deficient pasture with cobalt salt may be instituted.

Iron

Unless goats are anemic, iron deficiency in caprine not a problem in foraging goats. Stomach worm (*Hemonchus*), suckling lice and blood loss are the most common cause of anemia in goats. An increased proportion of *Allium* (onion type) plants in the diet can also cause iron deficiency because n-propyl disulphide in these plants induces hemolytic anemia. 30 mg/kg dry matter (ppm) in the ration is the standard ruminant recommendation. Young kids on all milk diet routinely become deficient and could be treated with iron dextran (150 mg/kid at 2-3 week interval) if diet modification is not possible.

Sulfur

Sulfur is the part of sulfur containing amino acid such as methionine and cystine. Dietary supplementation of Cashmere goat with dl-methionine increased hair yield much more than it increased cashmere yield. Deficiency sign are poor growth rate, salivation, lacrimation and alopecia with marked deficiency. Consider offering sulfur in the form of a sulfur-salt block free choice.

Iodine

Iodine is as essential in caprine diet as it is in human. Iodine deficiency has been reported from hill areas in India such as Himalaya region., Jabalpur and even Delhi region etc, where goiter cases have been reported. The major clinical manifestation is neonatal mortality, with alopecia and visible and palpable enlargement of the thyroid glands. The iodine deficiency may be primarily due to deficient intake or secondarily by a high intake of Calcium. The diet high in linseed meal (20% of ration) given to pregnant goats may also result in a high incidence of goitrous in kids. Soybean products are also considered to be goitrogenic. Iodized salt is a good preventive treatment against goiter, its requirement is 0.1ppm and it should be given with caution because high level of iodized salt in diet may also cause goiter, if given in excess. Potassium Iodide and sodium iodide are good iodine supplement for livestock. Certain crops which are high in levels of goiterogens would need supplementation of extra iodine as a preventive measure against goiter.

Sodium, Potassium and Chlorine

All these three elements occur in soft tissues and body fluids to maintain osmotic pressure, acid base equilibrium, controlling the passage of nutrients into cells and water metabolism. Sodium and potassium are not stored in the body; therefore, their daily supply in the feed is essential. Subclinical syndromes to effect growth and milk production and even infertility used to occur due to their deficiency. Both are required 0.2% each of total DM intake. Sodium which is commonly supplemented as NaCl is required at a level of 0.5% of the total DM. Goats out on forage usually get all the potassium they need. Emaciation and muscle weakness are signs of severe potassium deficiency.

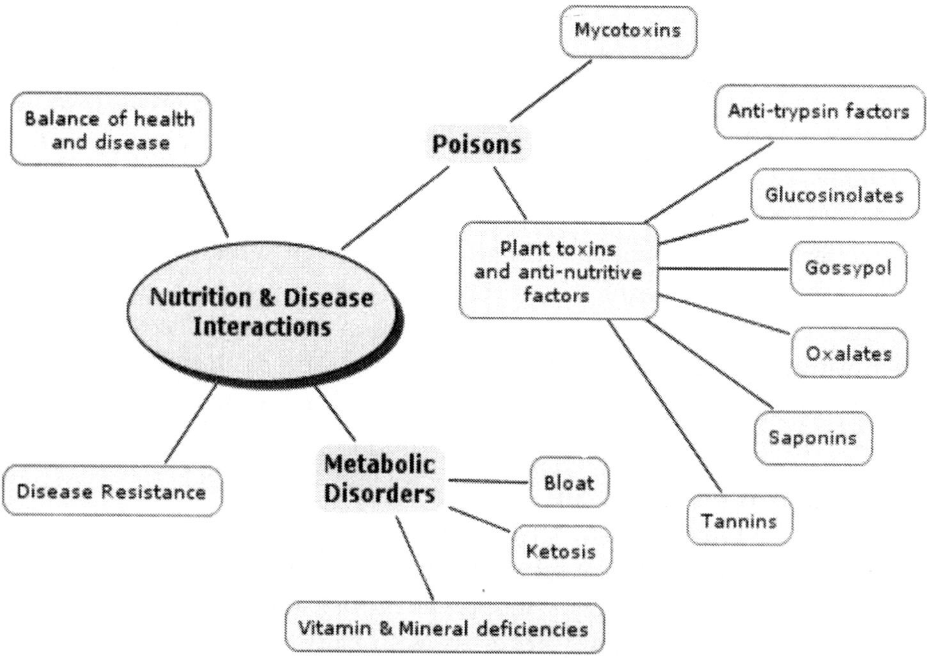

Fig. 39 Nutrition and disease interaction

Magnesium

Magnesium has received far less study than have calcium and phosphorus, although the metabolism of these mineral is interrelated. It is required for many enzyme system, necessary for energy metabolism.

The physiological levels for goats are 2.8-3.6 mg/dl. Deficiency of magnesium may lead to anorexia or hyperexcitibility. Goats deficient in Mg have lowered urine production and milk production, which seems to a compensatory mechanism in deficient state. Magnesium tetany has been recorded in livestock in India. "Grass stagger" problem did not seem to be prevalent. Magnesium requirement is 0.35% of the total DM intake. Grasses growing rapidly in cool, wet weather or after heavy fertilization are often low in magnesium, need special attention in feeding strategy. Whole goat milk also occasionally supplies inadequate magnesium, and kids might develop tetany, if not supplemented.

Zinc

Zinc content of the body is only approximately 3 mg% because it is not stored. Certain soils have been declared deficient in zinc because of intensive cultivation and no attention has been paid towards its supplementation. It causes deficiency disease in plants and animals named "Khera" disease in paddy and "parakeratosis" in livestock. Sub-optimal levels may lead to poor growth, testicular size and spermatogenesis reduced. There is a marked delay in wound healing. Apart from its function as a co- factor in activation of many enzymes, it is helping in mobilization vitamin A from liver, which is the storage organ for vitamin A after absorption. It is also a factor in preventing vitamin A resistant night blindness. Zinc is considered essential for the optimum rate of microbial fermentation and is deficient in milk which makes it all the more important as a dietary essential for neonates.

Normal diets in certain deficient areas including India do not contain more than 25 ppm zinc, whereas the requirement is 70 ppm. Further more, its requirement will very depending upon the Ca level and phytic acid in the diets. The oral administration of zinc at the rate of 250 mg zinc sulfate daily for 4 weeks resulted in a clinical cure of zinc deficiency in goats in 12-14 weeks.

Selenium and /or Vitamin E deficiencies

Earlier selenium was identified more as a toxic factor than its usefulness eg. Blind staggers or Alkali disease in USA in livestock. Similarly to this toxicity "Degnala" disease in India has been recognized as a result of selenium toxicity. In the last about 25 years, selenium research has revealed that it is a dietary essential along with vitamin E as an antioxidant property and the diets for this purpose may contain 0.1 ppm selenium. This level can prevent white muscles disease or stiff lamb disease or encephalomalacia. In selenium deficient regions, selenium by injection is an alternate to oral supplementation. It is common to incorporate a prescription vitamin E / selenium preparations.There are some areas in United States such as South Dakota and some areas in India particularly north western region which are affected with selenium toxicity. Whenever the soils are high in selenium, the plants take up higher quantity and thus prove toxic. On the other hand, quite a many areas in good number of countries have been identified as selenium more than 5 ppm is potentially dangerous and is recognized as a seleniferous area. Several out-breaks of selenium deficiency have been reported with paralysis, reproductive failure and white muscle diseases. In selenium deficient goats, the production of leukotrine B4, a product of neutrophil arachidonic acid lipoxygenation and a potent chemotactic and chemo kinetic stimulus for neutrophils is decreased which results in dysfunction of the neutrophils rendering them more susceptible to infectious diseases.

Fluorine and Molybdenum

Both of these elements are dietary essentials and have physiological function in the animal's metabolism. At higher levels, they are toxic. Up till now molybdenum toxicity has not been recorded in India where as fluorine toxicity is quite widespread in certain states such as Andhra Pradesh, U.P. and Rajasthan involving all species of animals including human beings. Fluoride is one of the potent anions which is considered essential trace element, but with a narrow margin of safety. It is established that where fluorine occurs in abnormal amount in soil or water, a chronic endemic fluorosis, characterized by mottling, excessive wear of teeth and osteoporosis is likely to exist. Some investigations are of opinion that of al pollutant which affects farm animals, fluorine has caused most severe and wide spread damage. Dental lesions in young ruminants which occur on the intake of 10-14 ppm DM of hay contaminated. The osseous changes fluorine has been described as osteosclerosis, exostosis, hyperstoses, osteomalacia and rickets. Fluoride might act as a possible mutagen in the ova of animals taking excess of it with feed and water. Goats are also susceptible to high fluoride levels causing dental lesions and its accumulation in bones and teeth, while the levels in soft tissues and milk remain unaltered and the reproductive performance of adult and the growth of kids were unaffected.

Calcium, Phosphorus and Vitamin D

A dietary deficiency or disturbance in the metabolism of calcium, phosphorus or vitamin D, including imbalance of calcium phosphorus ratio, is the principal cause of the osteodystrophies. A severe imbalance of calcium and phosphorus (relatively too much phosphorus from mostly grain diet) can cause osteodystrophia fibrosa in goats, against the normal Ca: P ratio of 2.2:1, which clinically characterized by soft enlargement of lower jaw, commonly encountered when owners increases the grain in the diet with the thought of speeding the younger's goats growth.The normal level of calcium and phosphorus for goats is approximately 9.0-11.6 mg/dl and 4.2-9.9 mg/dl respectively. Hypocalcaemia in goats is occurring most commonly at the time of parturition. It is characterized by low levels of blood calcium, general muscular weakness, circulatory collapse and depression of consciousness, low energy and low protein feed given to goats leads to decreased retention of calcium and phosphorus. The maternal demand for calcium and phosphorus; and their absorption increases during pregnancy and lactation as a result of additional requirement for the fetus and for milk. Hypocalcaemia is one of the possible cause of non - dilation of cervix in particular in parturient goats. A primary deficiency due to the lack of calcium in the diet seldom occurs, although a secondary deficiency due to marginal calcium intake aggravated by high phosphorus intake is common. Phosphorus deficiency is usually primary under field conditions but may be exacerbated by deficiency of vitamin D and possibly by an excess of calcium. Vitamin D deficiency alone in young growing kids, particularly raised in dark conditions develop rickets, rachitic rosary on the ribs, enlarged joints and bowed limbs etc. Growth rate and body condition are poor. Adult deficient in vitamin D develop osteomalacia or osteoporosis.

Manganese

Deficiency of manganese in goats consisted of deformed forelimb, reluctances to walk and reduced fertility and abortion, and reduced growth rate in growing kids.

Vitamins

Vitamin A

Deficiency of vitamin A may be caused by an insufficient supply of vitamin A in ration or its defective absorption from the alimentary canal. In young animals, the manifestations of the deficiency are mainly those compressions of the brain and spinal cord, especially in young animals a deficiency predisposes to diarrheal disease and respiratory disease and to parasitoses, moreover, coccidiosis affected kids require high vitamin A, which related to impaired absorption. In adult goats, the syndrome produced includes night blindness, opacity of cornea or cornea keratinization, ptyriasis defects in the hooves, loss of weight and infertility, which particularly related to inadequate steroid hormone synthesis. In hypovitaminosis a, desquamation of kidney epithelium, which form a nucleus for further development of calculi. Animal confined indoor and fed on prepared diets may suffer a severe deficiency. Deficiency of vitamin A also occur due to lack of green fodder during prolonged drought, pelleting of feed may also cause a serious loss upto a 32% of the vitamin A in the original feed stuffs.Green fodder contain sufficient vitamin A, where as stored feed substantially decreased content. For kids, colostrums are a very good source of vitamin A. In deficiency of vitamin A in kids was better managed by oral vitamin A palmitate suspension, preferably with the injectable vitamin A because, it rapidly peroxidised at injectable site.The minimum daily requirement in all species of animals is 40IU of vitamin A per kg body weight. During pregnancy, lactation or rapid growth, the allowances are usually increased by 50-70% of the requirements.

Vitamin B

There are two instances where B vitamin deficiencies have primary clinical importance. One with cobalt deficiency and other associated with vitamin B1 (Thiamin). Thiamine is made by the normal bacteria in the rumen. Kids or does on high carbohydrate may have an upset in normal flora. A change in bacterial type may cause a deficiency of thiamine or production of an enzyme which inhibit thiamine activity, resulting polioencephalomalacia (softening and necrosis of grey matter of the brain). Overdosing of amprolium in the treatment of coccidiosis and exposure to high level of sulfur in diet may result same clinical situation. Such cases are responded to thiamine injection.

Table 5: Principal pathological and metabolic defect in essential trace element deficiencies.

Deficiency	Pathological consequences	Associated metabolic defects
Copper	Defective melanin production, defective, Keratinization (hair, wool), Connective tissue defects, Ataxia, Myelin aplasia, growth failure, Anemia, Uricaemia	Tyrosine/DOPA oxidation, SH Oxidation to S-S Lysyl oxidase, Cytochrome oxidase, Urate oxidase
Cobalt	Anorexia, Impaired oxidation of propionate, Anemia	Methyl malonyl CoA mutase, Tetrahydrofolate methyl transferase
Selenium	Myopathy (Cardiac and Skeletal), Liver necrosis, Defective neutrophil function	Peroxidase/ Hydroperoxidase destruction, Glutathione peroxidase, OH; O2 generation
Zinc	Anorexia, Growth failure, Parakeratosis, Prenatal mortality, Thymic involution, Defective cell mediated immunity	Polynucleotide synthesis, Transcription, Translation
Iodine	Thyroid hyperplasia, Reproductive failure, Hair and wool loss	Thyroid hormone synthesis
Manganese	Skeletal/ cartilage defect, Reproductive failure	Chondroitin sulphate synthesis

11.5 Polioencephalomalacia (Goat Polio)

Polioencephalomalacia (also known as Cerebrocortical Necrosis) is thiamine (Vitamin B1) deficiency. Any change in the rumen's environment that suppresses normal flora activity can lead to decreased thiamine production.

Etiology

From improper feeding, particularly feeding too much grain and too little roughage decreases the pH of the rumen, predisposing the animal to Goat Polio. Thiamine must be present in order for glucose to be metabolized. If thiamine is either not present or exists in an altered form (thiaminase), then brain cells die and severe neurological symptoms appear. Causes of thiamine deficiency include feeding moldy hay or grain, overdosing with amprollium when treating for coccodiosis, feeding molasses-based grains, ingesting some species of ferns, sudden changes in diet, the dietary stress of weaning, and reactions to dewormers thiabendazole and levamisole. Each of these can interfere with Vitamin B1 production. Even the usage of antibiotics destroys flora in the rumen and can lead to thiamine deficiency. This is why it is so important to repopulate the gut with live bacteria after using antibiotics or scour medications. Goat Polio is generally seen most often in weanlings and young adults.

Clinical Signs

It is an afebrile condition with excitability and progressive nervous system involvement developing either acutely or over a period of about a week with blindness, opisthotonus, extensor rigidity, nystagmus, recumbency and convulsions. Some times diarrhoea may be present. The other symptoms of Polioencephalomalacia are excitability, "stargazing," uncoordinated staggering and/or weaving (ataxia), circling, muscle tremors, and apparent blindness. Initial symptoms can look like Enterotoxemia (overeating disease). There is a component of "overeating" involved in that the rumen flora has been compromised. As the disease progresses, convulsions and high fever occur, and if untreated, the goat generally dies within 24-72 hours.

Diagnosis

Diagnosis is available via laboratory tests, but the producer does not have the luxury of the time that such tests take. Thiamine is the only effective therapy, and treatment can result in improvement within two hours, if the disease is caught early enough.

Treatment

A dramatic response to intravenous or intramuscular, thiamine (250 to 500 mg which may be repeated at six-hourly intervals) is often seen. Thiamine at the dose rate of 10 mg/kg should be given every six hours for at least 24 hours. Initially, IV dosage is best, but S/C or IM can be used. Treatment should be continued for 5 days and then weekly intervals as required. Thiamine can be given orally after the initial treatment. If thiamine is unavailable then multiple B vitamins can be given. The dosage is based upon the amount of thiamine in the multiple B vitamins. The key to overcoming Goat Polio is early diagnosis and treatment. Complete recovery is possible under such circumstances.

Prevention

To try to avoid this disease, decrease grain, increase roughage, avoid moldy hay and grain, and don't feed molasses-based (textured) feeds.

11.6 Obstructive Urolithiasis

In goats, clinical obstructive urolithiasis is most frequently seen in young, castrated males and the calculi are usually comprised of phosphate salts, especially calcium phosphate and magnesium ammonium phosphate.

Etiology and Pathogenesis

Obstructive urolithiasis is the inability to void urine normally because of obstruction of the urinary outflow tract by calculi. The formation of urinary calculi results from the interaction of numerous physiological, nutritional, and management-related factors. The tendency for urinary calculi to become lodged in the urethra derives from anatomic factors and castration practices in male ruminants.

The urine is a highly saturated solution of mineral solutes. Under normal circumstances, these solutes usually remain in solution. However, numerous factors can predispose to precipitation of minerals from urine leading to calculi formation. These include increased urine concentration from decreased water intake or increased insensible water loss, urine stasis, increases in urine pH that favor precipitation of phosphate calculi, increased mineral excretion in the urine related to diet composition, and decreases in the concentration of protective colloids in the urine that ordinarily inhibit precipitation by transforming the urine into a stable gel.

Calculus composition frequently reflects diet. Silicate stones are common where grass and cereal hays are the main dietary constituents, particularly in arid regions. Forages high in oxalates promote oxalate calculi, with certain plants such as *Halogeton* contributing significantly to the problem. Concentrated grain rations are commonly high in phosphorus and relatively low in calcium. This imbalance promotes the formation of phosphate calculi, as frequently seen in feedlot animals. The calculogenic potential of such diets is increased when the concentrate is also high in magnesium.In the goat, several relationships of diet to calculus formation have been observed in the field or confirmed experimentally. Obstructive urolithiasis caused by magnesium ammonium phosphate calculi was observed in goats fed a concentrate ration and goats fed a pelleted feedlot fattening ration with a calcium to phosphorus ratio of 1:15. Goats fed rations containing magnesium in excess of 0.6% developed calculi even when the calcium to phosphorus ratio was properly balanced.

Additional factors come into play in the actual formation of calculi. A nidus may be necessary for calculus formation to occur. Desquamated epithelial cells in the urinary bladder are commonly considered to function as a nidus. Factors leading to increased desquamation of epithelial cells include vitamin A deficiency and infections of the urinary tract. A role for vitamin A deficiency in obstructive urolithiasis has been specifically confirmed in goats. Desquamated epithelium in sufficient quantity may obstruct the urethra even without the formation of mineral calculi. Infections can alter urine pH, which promotes precipitation of mineral salts.

Once calculus formation is initiated, growth occurs by the process of concretion. Increased mucoproteins in the urine serve as a matrix and favor concretion. The feeding of high concentrate rations, particularly in pelleted form, promotes an increase in urine mucoproteins. Obstruction is not a necessary result of calculus formation, but is predisposed by anatomic factors. The short, straight urethra allows calculi to pass easily with the urine. In contrast the long, convoluted course of the urethra through the sigmoid flexure of the penis, ending in the small urethral process provides ample opportunity for even small calculi to become lodged in the male urethra. This tendency is aggravated by early castration of male ruminants that results in penile hypoplasia with attendant reduction in urethral bore size, and failure of the maturing urethral process to separate completely from its distal attachment to the preputial mucosa. The effect of castration on reduced urethral diameter has been documented in goats. Exogenous estrogens either in the diet or from growth promoting implants can also lead to reduced urethral diameter by promoting swelling of the surrounding accessory sex glands in the pelvic region. When complete obstruction of the urethra occurs, forceful attempts to urinate will produce little or no voided urine. Animals so affected soon become azotemic because of decreased renal function, secondary hydroureter, and hydronephrosis. Unless the obstruction is relieved spontaneously or by intervention, the urinary bladder or the urethra will eventually rupture. In the former case, abdominal distension may result as urine fills the peritoneal cavity. In the latter case, urine fills the subcutaneous space in the perineal or preputial region depending on the site of urethral rupture. At this point, the uremic process accelerates and the animal ultimately dies.

Epidemiology

Obstructive urolithiasis is a well recognized, highly prevalent, and costly disease of goats and sheep maintained in feedlot conditions and fed high concentrate rations. Most affected goats are young males, often castrated. There was an increased prevalence in summer and winter, times when dehydration is more likely to occur. Phosphatic calculi were most common and occurred in the form of either sand or sediment. Calcium carbonate calculi were less frequently identified, were larger, and more likely to occur as stones. To promote growth and condition in time for sale before the breeding season, the animals are fed on a pelleted ration comprised of sorghum or corn and alfalfa. The diet is recognized as highly calculogenic and obstructive urolithiasis is avoided by appropriate adjustment of the calcium to phosphorus ratio, and addition of salt and urinary acidifiers.

Clinical Signs

In the first phase of obstruction, male goats will show restlessness and anxiety. Tail twitching is an early sign. There may be excessive vocalizing and animals will strain frequently and forcefully to urinate, often stretching out to full length with a dip to the back, followed by arching. Marked abdominal pressure may produce some degree of rectal prolapse. Palpation of the urethra on the median raphe in the perineal region reveals distension and pulsation during straining. Drops of bloody urine and crystals may be seen attached to preputial hairs. Animals with partial obstruction may be able to void small intermittent streams of urine, but show discomfort.

Careful examination of the urethra should be undertaken in an attempt to locate the site of obstruction. Visualization of the urethral process may reveal calculi at this common site of blockage. Deep palpation of the urethra along the prepuce and perineum may indicate pain or swelling at more proximal sites of obstruction such as the sigmoid flexure.

When the obstruction goes uncorrected, rupture of the bladder or urethra usually results within 24 to 48 hours. Rupture produces alleviation of discomfort and anxiety and animals resume a more normal attitude, at least until the signs of uremia supervene. Subcutaneous filling of the preputial or perineal region becomes noticeable when the urethra is ruptured. Bilateral, ventral abdominal distension secondary to bladder rupture is slower to develop and more difficult to detect. Ensuing uremia causes affected goats to become anorexic, weak, and depressed. Advanced cases often present in a moribund state and the condition is fatal if left untreated. The entire clinical course is usually 2 to 5 days.

Diagnosis

Diagnosis is based on history and clinical signs. During the early stages, when straining is present, gastrointestinal obstructions should be considered. When stranguria is recognized, cystitis and ulcerative posthitis must be ruled out. When the bladder or urethra has ruptured and signs of uremia predominate, other causes of profound depression and weakness must be considered, especially hepatoencephalopathy and enterotoxemia. Confirmation of the presence of urine in the abdomen or subcutaneously quickly narrows the diagnosis.

Obstructed goats usually demonstrate some degree of azotemia. Elevations in BUN and creatinine increase more markedly after rupture of the bladder or urethra. Levels of BUN in affected goats have been reported in the range of 11.7 to 47.5 mmol/l (70-285 mg/ dl). Urine sediment should be examined whenever a urine sample can be obtained. Marked crystalluria supports the diagnosis of obstructive urolithiasis. Aspiration of subcutaneous or peritoneal fluid can be helpful; though confirming that the fluid is urine can be problematic. A urea nitrogen or creatinine level equal to or greater than that of blood suggests urine. An odor of urine may be helpful in field diagnosis.

Plain radiographs mayor may not reveal cystic or urethral calculi, but contrast studies can be effective diagnostic aids. When possible, introduction of contrast material via the penis rather than use of an intravenous pyelogram is preferable for illuminating defects in the distal urinary tract. Currently, ultrasonography offers the most rapid, noninvasive, and safest means of assessing the extent and location of calculi.

Post Mortem Findings

At necropsy, careful dissection of the urinary tract will reveal the presence of calculi, urethral trauma and bladder or urethral rupture. Calculi may be present in the bladder or renal pelvis as an incidental finding in goats dying of causes other than obstructive urolithiasis and suggest the need for reviewing nutritional management in the source herd.

Treatment

The approach to treatment depends on the stage of the disease, the nature and extent of calculi present, the intended long-term use of the animal and frequently, financial constraints. When conservative treatment does not alleviate the condition and surgical intervention becomes necessary, a guarded to poor prognosis should be given for successful long-term outcome. Before rupture of the bladder or urethra, conservative management can be attempted. If the offending calculus is lodged in the urethral process, the process can be removed to re-establish patency. When sandy material is present in the urethral process, it sometimes can be successfully milked out without removing the process itself.

If calculi are lodged proximal to the urethral process, attempts can be made to pass a catheter far enough into the urethra to permit infusions of sterile saline in attempts to distend the urethra and dislodge offending stones. This procedure is often traumatic and frequently unsuccessful. Tranquilizers and antispasmodics may help to promote urethral relaxation and facilitate natural expulsion of the calculus by the pressure of attempted urination. Diazepam, at a dose range of 0.1 to 0.5 mg/kg intravenously, acepromazine (0.1 mg/kg intravenously), and aminopromazine (2.0 mg/kg intramuscularly) each have been used for this purpose, with variable results. When conservative treatment fails, or if urinary tract rupture has occurred, then some sort of surgical intervention becomes necessary. Before any surgery is attempted the patient should be evaluated for uremia and electrolyte imbalances and stabilized accordingly. If the bladder has ruptured, urine should be drained from the abdomen via abdominocentesis to slow the uremic process if surgery is delayed.

Radiographic contrast or ultrasound studies can be helpful in determining the appropriate surgical management of the case. When such studies indicate sand or multiple calculi in the bladder, the likelihood of recurrent obstruction is high. In such cases, cystotomy is recommended as a preliminary surgical procedure. Sand and calculi can be removed from the bladder and the urethra can be flushed from the open bladder, providing an increased rate of success in clearing existing obstructions than presurgical retrograde flushing. By using cystotomy, the need for subsequent penile amputation proximal to the obstruction site and the likelihood of recurring obstructions are both reduced. If patency can not be reestablished by flushing, then an appropriate urethrotomy or urethrostomy procedure or euthanasia can be elected at this time depending on the specifics of the case at hand. Prevention of recurrence is a major goal of post-surgical management. An indwelling catheter should be left in place during the healing stage after urethrostomies to prevent premature stricture or reobstruction caused by blood clots and scabs. Antibiotic therapy is indicated to prevent ascending infection. Severely uremic goats require aggressive medical management before and after surgery.Whenever possible, structural analysis of calculi removed at time of surgery or by conservative treatment should be performed. This permits appropriate dietary and management recommendations to be made for prevention of new calculi on the basis of their composition.

Prevention and Control

Dietary management is key to the control and prevention of obstructive urolithiasis. The calcium to phosphorus ratio of the overall diet should be in the range of 2:1 to 2.5:1. Rations comprised of grass hay and heavily supplemented with grain concentrates are most likely to provide excessive phosphorus and inadequate calcium. When possible, grain should be eliminated from the diet and the animal provided a ration of high quality, free choice, mixed alfalfa/ grass hay with salt and trace mineral available. When alfalfa is not available or owners are unwilling to eliminate grain from the diet, the volume of grain should be reduced relative to grass hay, and calcium supplements, such as dicalcium phosphate or calcium carbonate, should be made available free choice or mixed into the feed at quantities sufficient to properly adjust the calcium to phosphorus ratio of the diet.

Additional dietary management factors include provision of an adequate and continuous supply of clean, fresh water, increasing the concentration of salt in the ration up to 4% to promote water consumption and diuresis, and insuring an adequate level of vitamin A to inhibit epithelial desquamation in the bladder.Because phosphate calculi are more likely to form in the normally alkaline urine of herbivores, the prophylactic use of urinary acidifiers is advocated. Continuous administration of ammonium chloride at a dose of 10 grams per head per day or at a level of 2% in the concentrate ration has been recommended in goats and sheep at risk to decrease urine pH levels. Concentrations as low as 0.5% have been used successfully to control urolithiasis. Ammonium sulfate is sometimes used in place of ammonium chloride at a rate of 0.6% to 0.7% of the total ration. Urinary acidifiers however, are not foolproof. Delaying castration may be another management tool for reducing the prevalence of obstructive urolithiasis. However, the early precocity of goats requires that castration be performed by 3 months of age to avoid unwanted sexual activity.

11.7 Bloat

Ruminal tympany or bloat, is less common in goats than cattle and sheep. When it occurs however, it must be considered a medical emergency. Frothy (primary, or nutritional) bloat is more likely to be encountered than free gas (secondary) bloat.

Etiology and Pathogenesis

Gas production is a normal outcome of rumen fermentive activity. Ordinarily, gas rises to the dorsal sac of the rumen and is discharged by the orderly process of rumination and eructation. Frothy bloat derives from nutritional causes. Under certain dietary conditions, such as the feeding of legumes or finely ground grains, gas bubbling through the rumen content can no longer freely coalesce in the dorsal sac but rather is trapped in a stable foam in the liquid phase of the rumen content. Factors contributing to the formation of stable foam are complex. Stable foam can not be freely eructated and a progressive distension of the rumen ensues as additional gas becomes trapped. This progressive ruminal distension can be fatal as a result of subsequent respiratory and cardiovascular compromise. Experimentally in goats it has been documented that ruminal insufflation leads to decreased cardiac output, increased blood pressure and increased total peripheral resistance due to obstruction of venous return by the distended rumen.Free gas or secondary bloat occurs when gas accumulating normally in the dorsal sac of the rumen can not be expelled because of extraruminal obstructions to outflow. When free gas can not be eructated, progressive ruminal distension ensues.

Epidemiology

Bloat occurs in goats throughout the world. All goats with a mature functional rumen are at risk. Factors that have been identified as predisposing goats to frothy bloat include unaccustomed ingestion of lush legumes such as clover or alfalfa, either as green feeds or as new hay, recent turnout to legume pastures, and even turnout to grass pastures if they are wet. As such, bloat may have a seasonal occurrence. Sudden access to grain either as concentrate or from gleaning of grain fields can also lead to frothy bloat. Turnout to grain fields results in a seasonal increase of bloat.

Free gas bloat occurs only sporadically in goats, usually secondary to esophageal choke. Pieces of apple or carrot are common foreign bodies. A cud can lodge in the esophagus, particularly in debilitated or sick animals, and in animals with inadequate water supply. Internal abscesses in the anterior abdomen or mediastinum associated with caseous lymphadenitis or other causes can also produce secondary or free gas bloat by compression of the cardia or esophagus.

Clinical Signs

Frothy bloat can occur within hours of exposure to the offending feed or feeding situation. Many cases of bloat are first recognized by finding animals dead at pasture. In the initial stages of the disease, affected animals become anxious and uncomfortable and stop eating. The most characteristic sign is a progressive distension of the abdomen, particularly apparent on the left side and high in the left paralumbar fossa. Percussion of this area reveals the tight, tympanic feel of the distended rumen and a drum-like sound. As the distension progresses, animals become more uncomfortable, stamping their feet, vocalizing, salivating, urinating frequently, and moving with a stilted gait. Without intervention, these animals become recumbent; exhibit marked, labored respiration and will die within hour.

In free gas bloat, signs are similar. Salivation is more pronounced if the esophageal obstruction is high and complete and saliva can not be passage to the rumen. If intraluminal obstruction is not complete or if the pressure of accumulating gas in the rumen can override an extraluminal obstruction, gas will be intermittently expelled and the abdominal distension caused by the rumen may be less pronounced with less severe tympany. Such cases often result in a prolonged bloating that is not immediately life-threatening.

Diagnosis

Bloat is easily confirmed by physical examination. It is important to determine whether the condition is primary or secondary as this will affect treatment and subsequent management decisions. The difficulty in the diagnosis of bloat comes when animals are found dead. Because bloat may be a normal post mortem change, it is necessary to rule out all possible causes of sudden death under the circumstances that the animal is found.

Post Mortem Findings

Because of the acute nature of bloat laboratory investigations are not performed. Because the condition often proceeds to the fatal stage without detection, necropsy is commonly employed in the diagnosis of bloat. In a fresh necropsy of a case of frothy bloat, the ruminal distension will be obvious and large amounts of froth will be present within the rumen. In free gas bloat, rumen distension will be present but the rumen will not contain froth. Careful inspection of the alimentary tract and surrounding structures from the mouth through the forestomachs should identify the source of secondary obstruction leading to free gas bloat. When animals are found several hours after death, diagnosis becomes more problematic, particularly in warm weather. Rumen distension with gas caused by the continued fermentive action of rumen flora after death eventually produces bloat in all carcasses. Froth will also break down after death causing frothy bloat to appear as free gas bloat. Examination of the carcass may help to establish the presence of antemortem bloat. In bloat cases, there is congestion of tissues in the anterior portions of the carcass and blanching or paleness of tissues in the caudal portions reflecting peripheral vascular changes induced by the distended rumen.

Treatment

Timely intervention is essential. Simple passage of a stomach tube, though often effective in relieving free gas bloat, will not correct frothy bloat unless the foam is first broken down. Cooking oils or mineral oil given orally in a dose range of 100 to 200 ml is effective. If given as a drench, care must be taken to avoid aspiration pneumonia. Linseed oil should not be used because it is associated with indigestion in goats. Oil of turpentine is a common home remedy. Though effective, it can taint meat and milk for 5 days after use.

There are a number of commercial surface-acting agents that are also effective in degrading foam and results may be seen within minutes. Poloxalene is available as a concentrate to be diluted for use as a drench. It is dosed at a rate of 100 mg of poloxalene per kg b.w. Dioctyl sodium sulfosuccinate (DSS) (15-30 ml), and polymerized methyl silicone are other effective compounds. When large numbers of animals are affected and treatment time becomes a critical factor, these products can be injected directly into the rumen through the distended left paralumbar fossa using an 18-gauge needle rather than administering them orally by stomach tube.Forced exercise after administering oils or surfactants will help break down foam and promote expulsion of gas. If animals are recumbent, rolling the goat or massaging the rumen may help distribute the oil and break down the foam. A stomach tube should be passed to facilitate gas removal as the foam breaks down. Trocharization of the rumen through the left paralumbar fossa can be life saving in goats in the terminal stages of bloat. However, this invasive treatment should be reserved for the most advanced cases due to the potential sequelae of peritonitis and prolonged rumen dysfunction. Goats trocharized for relief of bloat should be placed on a 3 to 5 day course of broad spectrum antibiotic therapy after the procedure.

Prevention and Control

Goats should be adapted gradually to pastures, being turned out initially for short periods on a full stomach of hay to limit pasture consumption. A similar recommendation applies to animals turned out to glean grain fields. Whenever green feeds are fed, dry hay should be available simultaneously to avoid over consumption. Concentrate feeds should not be too finely ground and should not be abruptly introduced into the ration. When animals show a tendency to bloat, concentrates can be top dressed with some oil, such as peanut or corn oil, to reduce the bloating potential of the feed.

11.8 Indigestion and Rumen Acidosis

When a high proportion of concentrates is fed in the ration, there is a marked day to day variation in consumption. This is probably evidence of indigestion. Marked rumen acidosis is life-threatening.

Pathogenesis

In general, roughages have an increased fiber content compared to concentrates and stimulate long periods of rumination. This in turn results in buffering of the rumen fluid by bicarbonate in saliva. The rumen pH remains in a range (6.0 to 6.8) favorable for cellulolytic bacteria. The major end product of roughage digestion is acetate. Grains lack long fiber. Because of a decreased rumination rate, less saliva reaches the rumen and the pH of the rumen contents drops when grain is consumed. This is desirable within limits, as the bacteria for degrading starches and sugars multiply best at a pH of 5.5 to 6.0. Also, the volatile fatty acids propionate and butyrate are produced in large quantities by digestion of concentrates and are absorbed faster at lower pH levels.

If too much grain is eaten at one time, lactic acid producing streptococci overgrow in the rumen and the pH decreases even more to a point where lactobacilli can proliferate. The ensuing lactic acid production lowers the rumen pH below the physiologic limit of 5.5 and kills the normal flora and fauna. Bacteria produce both dextro- and levo-rotatory isomeric forms of lactic acid, but the mammalian liver is only able to efficiently degrade the L form. This allows the D form to build up in the blood, causing a systemic acidosis. Increased production of histamine and other toxins is associated with the occurrence of laminitis in some animals. Deranged synthesis of B vitamins or production *of* thiamine antimetabolites can lead to polioencephalomalacia. It should be noted that liver abscesses in goats are usually caused by *Corynebacterium pseudotuberculosis,* rather than other bacteria passing through a rumen epithelium damaged by toxic indigestion.

Clinical Signs

Mild overeating of grain results in an off feed condition. Rumen motility decreases but does not stop. The goat may grind its teeth. Milk production decreases and diarrhea may develop. Because the goat stops eating for 1 or 2 days, the rumen pH levels increase to a near-neutral range and recovery follows. Severe overloading is accompanied by a systemic and often fatal acidosis. Rumen motility ceases and the contents are initially firm; mild bloat may be present. Constipation followed by diarrhea, muscle tremors, teeth grinding, groaning, increased heart and respiratory rates, and even a low fever may be noted. As the disease progresses, splashy (abnormally liquid) rumen contents and dehydration (evidenced by sunken eyes, decreased skin turgor, and increased packed cell volume) develop. This is because the rumen contents become hyperosmotic and water came out from the systemic circulation into the rumen. The urine is acidic and blood pH and bicarbonate values may be markedly decreased. Blood lactate levels greater than 40 mg/ dl can occur in severe cases, with levels in normal goats of 8 to 10 mg/ dl. Significant decreases in blood calcium sometimes occur. Some cases are fatal within 24 hours, while less severe cases may develop over 24 to 72 hours.

Diagnosis

Rumen acidosis should be suspected when a goat is depressed and off feed and has had chronic or acute access to large quantities of readily fermentable carbohydrates. A sample of rumen fluid obtained with the aid of a stomach tube (up to 1 cm diameter) or even a needle aspirate can be helpful both for establishing a diagnosis and for determining if a rumenotomy is justified. A milky gray fluid with a pH level less than 5.0 is conclusive evidence of rumen acidosis. Slightly increased pH levels may result from contamination of the rumen fluid sample with saliva during collection. A pH level in the normal range (5.5 to 7.0) does not rule out rumen acidosis because absorption of acid, passage of ingesta to the abomasum, and flow of saliva will cause increase of the pH level if the goat survives long enough. A Gram-stained smear of rumen fluid demonstrates an overwhelming preponderance of Gram-positive organisms when rumen acidosis has occurred. Rumen protozoa are dead. The advanced stages of rumen acidosis may mimic toxic mastitis or milk fever. Animals recumbent with milk fever will not be dehydrated or diarrheic and will respond to parenteral administration of calcium salts. Enterotoxemia occurs under similar circumstances of accidental or excessive exposure to carbohydrates and may look clinically similar. Enterotoxemia generally has a more rapid clinical course and the degree of systemic acidosis and dehydration is not as severe.

Post Mortem Findings

At necropsy, there may be evidence of large quantities of grain in the rumen or of rumenitis and localized peritonitis. Dehydration is evidenced by sunken eyes.

Treatment

Free access to drinking water is somewhat protective before signs occur. Early in the course of the indigestion, oral antacids should stop progression of the disease. Various antacids include 10 to 20 g magnesium oxide, 50 g magnesium hydroxide or 20 g sodium bicarbonate. Oral tetracycline (single dose of 0.5 to 1 g) also helps to inhibit additional bacterial proliferation. Concentrate is withheld and palatable forage is offered free choice.

A rumenotomy to remove the rumen contents and rinse the mucosa may be life-saving in acute, severe cases. If the rumen fluid pH is near neutral, and especially if the rumen contents have a splashy consistency, rumenotomy is no longer of great benefit. Instead, as much fluid as possible is drained off with a stomach tube. Then warm water, antacids, and brewer's yeast are administered through the tube.If available, determination of acid-base status is helpful in guiding therapy; the severely affected goat has a metabolic acidosis. Dehydrated goats should be treated intravenously with 3 to 5 liters or more of physiologic saline with added bicarbonate. A goat weighing 50 kg with moderate acidosis would need approximately 15 g of sodium bicarbonate. Thiamine (300 to 500 mg several times a day), other B vitamins, and calcium gluconate administered subcutaneously are commonly given. Glucose solutions are not necessary, and lactated Ringer's solution is contraindicated. Repeated oral administration of a liter of rumen fluid from a healthy cow, sheep, or goat (obtained fresh from a slaughterhouse or from a fistulated animals very beneficial in restoring normal rumen flora and fauna.

Prevention and Control

If the goat has 2 to 4 weeks to adapt to gradually increasing quantities of grain, then many cases of indigestion can be avoided. Adaptation occurs in two ways. First, the rumen microorganisms that digest readily fermentable carbohydrates to volatile fatty acids increase in number. At the same time, certain bacteria (such as streptococci) produce lactic acid from starches and sugars, but still other species of bacteria with the ability to metabolize lactate to propionate have time to proliferate. Second, increased concentrations of propionic and butyric acid stimulate the development of rumen papillae so that there is an increased surface area for absorption of fatty acids. Rapid removal of volatile fatty acids from the rumen helps to keep the pH within physiologic limits.

Another management technique that is very helpful with high concentrate, feeding is to distribute the grain over three or more meals a day. This will result in a less drastic pH level decrease after each meal than if all the grain is fed at once. Grains that are digested very rapidly (e.g., wheat and high moisture corn) can be replaced by slower digesting grains (e.g., sorghum, dry corn). Whole grain will be digested more slowly than finely ground grain, with more surface area for bacterial attack. Roughage should be fed before grain, first thing in the morning. Regular supplementation with buffers (bicarbonate of soda or calcium carbonate at 1.5% to 2% of concentrate) and magnesium (0.2% to 0.3%) also reduces the risk of acidosis. Corn silage should also be distributed three or more times a day; otherwise, the goat may initially select out a large quantity of grain from the silage, risking acidosis.

□□□

12

Toxicities

12.1 Organophosphates and Carbamates

Etiology

Organophosphate and carbamate pesticides toxicities are commonly found in farms animals. They are used to treat livestock for internal and external parasites, and to protect soil and crops from insect pests.

Epidemiology

Toxicities occur when animals are overdosed either by parenteral or topical administration or when livestock gain accidental access to chemicals that are carelessly stored, accidentally spilled, or inadvertently added to feed or water supplies.

Pathogenesis

Organophosphates and carbamates exert their toxic effects by competitive inhibition of acetylcholinesterase, the enzyme ordinarily responsible for the degradation of the neurotransmitter acetylcholine. This inhibition leads to accumulation of acetylcholine at neuromuscular junctions, autonomic ganglia, and effector cells, and causes an increased, prolonged stimulation of skeletal muscle, the entire parasympathetic nervous system, and the postganglionic cholinergic nerves of the sympathetic system. These sustained muscarinic and nicotinic effects are responsible for the clinical signs associated with toxicity.

In goats, a delayed neurotoxicity caused by the organophosphate anthelmintic haloxon is recognized. The pathogenesis is different than in acute toxicity and is believed to involve an inherited esterase deficiency in goats. A distal axonopathy with dying back of neurons, particularly long axons in the spinal cord, develops several weeks after administration of haloxon. The clinical picture is one of progressive, symmetric, spastic paresis of the hind limbs. This syndrome has been suspected in goats, but additional documentation is needed to substantiate the disease.

Clinical Signs

The clinical signs of acute toxicity in goats are similar to those seen in other species, and include restlessness, frothing at the mouth, dyspnea, tremors, frequent urination and defecation, bloat, head pressing, teeth grinding, lacrimation, staggering gait, intermittent convulsions, paresis, and finally, recumbency and death.

Clinical Pathology

Identifying a marked decrease in acetylcholinesterase activity supports the diagnosis of organophosphate and carbamate poisoning. Whole blood is a suitable sample in goats because 89% of acetylcholinesterase activity in blood is in red blood cells. Heparinized blood samples should be submitted refrigerated but not frozen. When carbamates are suspected, blood must be refrigerated immediately and tested as soon as possible because inhibition of acetylcholinesterase by carbamate is reversible. The laboratory should be notified of your suspicion of carbamates so that test methods that require dilution of the sample are not used.Traditional test methods determine the change of pH that occurs when acetylcholinesterase hydrolyses acetylcholine or a substitute ester under controlled conditions. Newer titrimetric methods, which do not require sample dilution, report acetylcholinesterase activity as the p-mol of substrate hydrolyzed/ min/ml of whole blood or plasma. Normal values in goats using acetylcholine iodide as a substrate have been reported in the range of

4.20 to 5.60 p-mol/min/ml of whole blood and 0.60 to 1.00 p-mol/min/ml in plasma. Activity was reduced to levels of 20% to 50% of normal after oral administration of toxic doses of diazinon, phosmet, phosphamidon, or trichlorfon, with reported activity ranging from 0.90 to 2.50 p-mol/min/ml in whole blood and 0.27 to 0.32 p-mol/min/ml in plasma. Dialkyl phosphates are hydrolytic breakdown products of the phosphate moieties of organophosphate insecticides. Detection of dialkyl phosphates in the urine of goats proved to be a reliable indicator of exposure to diazinon over a range of challenge doses. The test was more sensitive than measurement of blood cholinesterase activity in goats not showing clinical signs of toxicity.

Post Mortem Findings

There are no specific necropsy findings in organophosphate or carbamate poisoning. Chemical analysis of tissues is often unrewarding because organophosphate and carbamate insecticides are rapidly metabolized. Analysis of stomach or rumen content, suspected feeds or other formulations for insecticide content is preferred.

Diagnosis

Diagnosis is based on the characteristic signs of acute toxicity in conjunction with documentation of decreased or absent acetylcholinesterase activity. In the absence of laboratory support, a favorable clinical response to atropine or oxime therapy supports the diagnosis. Other acute poisonings, such as cyanide, nitrate and urea toxicity, and anaphylactic reactions, must be considered in the differential diagnosis. The pyrethroid insecticide, fenvalerate, has been shown to produce clinical signs suggestive of organophosphate toxicity when administered orally to goats at doses more than 112.5 mg/kg.

Treatment, Prevention and Control

Treatment can be successful, especially with early intervention. The treatment of choice to counteract signs of toxicity is atropine sulfate given at a dose of 0.6 to 1 mg/kg b.w. One fourth to one third the total dose should be given intravenously and the remainder subcutaneously or intramuscularly. The higher end of the dose range can be used in severe cases. Repeated dosing may be necessary every 4 to 5 hours for as long as 2 days in severe cases, but dosage should be cut back when possible to avoid serious bloat. Oximes, such as trimedoxime bromide, 2-pyridine aldoxime methiodide (2-P AM) and pralidoxime chloride, will free acetylcholinesterase from organophosphate, but not from carbamate complexes. They are particularly useful in combination with atropine for the treatment of coumaphos, ronnel, dimethoate, and crufomate toxicity, where atropine alone is not always effective. The use of these drugs in goats is not reported. In other ruminants, recommended doses are as follows; 2-PAM, 50-100 mg/kg b.w.; trimedoxime bromide 10-20 mg/kg b.w.; and pralidoxime chloride at 20 mg/kg b.w. intravenously. When poisoning is by the oral route, oral administration of 0.5 kg of activated charcoal per adult goat may help to reduce the additional uptake of insecticide. When exposure is dermal, washing animals with soap and water will help to reduce additional absorption. Control of these toxicities is difficult because most outbreaks involve accidental exposure. They should be stored securely where animal contact is impossible.

12.2 Chlorinated Hydrocarbons

Etiology

These insecticide compounds, also known as organochlorines, have been used widely in agriculture for treatment of soils, water supplies, crops, seeds, and livestock. Methoxychlor, lindane, DDT, aldrin, dieldrin, chlordane, and toxaphene are common examples.

Epidemiology

Use of these insecticides directly on livestock is increasingly restricted. Nevertheless, accidental exposures or inappropriate use still lead to toxicities in livestock, including goats. Because goats are used for both meat and milk production, the potential for relay toxicity to man from consumption of goat products is similar to cattle. These compounds are noted for persistence in fatty tissues over long periods. Residue levels in goat milk and tissues for various chlorinated hydrocarbons have been reported.

Pathogenesis

The pathogenesis of toxicity for these compounds is not completely known and may be different for each. At least for chlorophenothane, the compound acts on axonal membranes to prolong the depolarized state by interference with sodium and potassium influx. Poisoning can occur by single large-dose exposure or chronic lower-dose exposure due to the ability of these insecticides to accumulate in tissues.

Absorption of these chemicals can occur through the skin, or by oral ingestion, aspiration, or inhalation. Neurologic signs in all cases are the usual clinical manifestation.

Clinical Signs

Clinical signs are similar in all species. Hypersensitivity, apprehension, and/or aggressiveness are frequent early signs followed by fasciculations of muscles, especially in the head and neck that extend to muscular spasms over the entire body. Snapping of the eyelids and continuous chewing or teeth grinding are common, sometimes with excessive salivation. Moderate bloating may occur. Loss of coordination with staggering gait, aimless wandering or circling may be observed. Severe and prolonged convulsions frequently develop. It is not unusual for animals to have a high fever due to seizure activity. Though animals may die during convulsions, death is usually preceded by terminal coma. The severity of signs is to some extent dose dependent. Experimental poisoning with aldrin at a daily oral dose of 2.5 mg/kg b.w. led to clinical signs of depression, anorexia, teeth grinding, salivation, staggering gait, and hypersensitivity in one goat after 18 days of dosing but no signs of toxicity in three others. Goats are given daily aldrin orally at a dose of 20 mg/kg showed signs of hyper excitability, incoordination of movement, muscle tremors, and convulsions by the ninth day of dosing, and died 1 to 3 days later.

Diagnosis

Presumptive diagnosis depends on a history of exposure to chlorinated hydrocarbons followed by the development of typical clinical signs. Definitive diagnosis depends on identification of the offending agent by laboratory analysis of hair samples, rumen content, fat biopsies, or milk samples antemortem or in tissue samples post mortem, especially liver, brain, and fat. Concentrations of the various chlorinated hydrocarbons considered diagnostic of toxicity in goat tissue are not well documented.

Post Mortem Findings

There are no specific necropsy findings, although pulmonary congestion and ecchymotic hemorrhages of the heart and other serosal surfaces may be observed. When there have been severe prolonged convulsions with fever, the intestines may have a blanched or cooked appearance.

Treatment, Prevention and Control

Therapy is primarily supportive. When topical poisoning is involved, animals should be washed thoroughly with soap and water. In cases of ingestion, gastric lavage may be performed or activated charcoal given at a dose of 0.5 kg per adult goat as soon as possible after exposure. Seizures may need to be controlled by use of long acting barbiturates. Daily oral dosing of recovered animals with small volumes of mineral oil may help to clear chlorinated hydrocarbons from the intestinal tract but will have little or no effect in mobilizing the agents in tissue. Specific guidelines for the safety of meat and milk products from goats recovered from poisoning are not available. Because of the potential harmful effects in humans, the consumption of meat or milk from these recovered animals should be discouraged. Control essentially involves client education concerning the danger, proper usage, and storage of chlorinated hydrocarbons.

12.3 Copper Toxicity

Copper (Cu) toxicity was suspected when several lambs are appeared lethargic and anemic. Sheep are unique in that they accumulate copper in their liver more readily than other farm animals. As a result, they are very susceptible to Cu toxicity (poisoning). Mature ewes appear to be the most vulnerable and have a tendency to accumulate more Cu in their liver than other breeds.

Clinical Signs

Copper is essential for life. It is required for normal iron metabolism, synthesis of elastin and collagen, melatonin production and integrity of the central nervous system. It is essential in keratin (wool) production. More recently, it has been shown that copper is one of the key trace minerals required for an effective immune response. Signs of deficiency include anemia, brittle or fragile bones, loss of hair or wool pigmentation and poor wool growth. In sheep, stringy wool and "swayback" are commonly reported. Affected sheep are lethargic and anemic. They may grind their teeth incessantly and experience extreme thirst. Membranes are very pale and may appear yellow, as jaundice sets in. Urine is a bloody color. Death usually occurs 1 to 2 days after the onset of clinical symptoms.

Pathogenesis

Generally, sheep require about 5 ppm (parts per million or mg/kg) of Cu in their total diet. Toxicity can occur at levels above 25 ppm.

However, dietary molybdenum (Mo) levels also affect copper requirements, as Mo forms an insoluble complex with Cu to prevent copper absorption. If molybdenum levels are low (less than 1 ppm), sheep are more susceptible to Cu toxicity. If Mo intakes exceed 10 ppm, Cu deficiency may occur on diets that would normally be adequate. Sulfur (S) further complicates the Cu: Mo relationship by binding with the Mo. Copper toxicity in sheep usually results from the accumulation of excess Cu in the liver over a period of a few weeks to more than a year with no clinical signs, followed by a sudden release of liver Cu stores to cause toxicity (rapid breakdown of red blood cells). In these situations, chronic Cu poisoning may result from excessive Cu intakes or from low intakes of Mo, S, zinc, calcium or following liver damage.

Epidemiology

Stresses, such as weather, environment, poor nutrition, transportation and handling, can also cause the liver cells to die and release the stored copper into the bloodstream.

Post Mortem Findings

At post-mortem, tissues are pale to dark yellow and the kidneys are a very dark color.

Treatment

The most common treatment is to give a drench daily containing 50 to 100 mg of ammonium molybdate and 0.5 to 1.0g of sodium sulfate.

Prevention and Control

It is recommended that sheep not to be fed poultry litter or other waste products which contain high levels of copper. In addition, there have been instances where high levels of Cu have been traced to the fertilization of pasture with pig manure.

If and when copper toxicity is suspected, the diagnosis needs to be confirmed by a veterinarian or diagnostic laboratory. Sources of copper needed to be promptly identified. It is important to know the Cu and Mo status of all feeds and forages being fed. Grains are lower in copper than forages. Most forages will contain copper at levels equal to or above the NRC requirement for ruminants; however, as plants mature, the bio-availability of the copper decreases. Errors in feed mixing should also be considered as a possible source of excess copper. Water should not be overlooked as a source. Observations on data suggest that goats are more tolerant of copper than sheep. In copper load studies goats consumed more copper and retained 6 to 9 times less in their livers than their trial lamb mates, indicating differences in utilization and resistance to toxicity between the species. Goats not to be fed minerals labeled for sheep or sheep and goats, without some other form of copper supplementation.

12.4 Cyanide Poisoning

Hydrocyanic acid (HCN) or prussic acid is present as a glycoside in certain plants. Hydrolysis in the course of wilting, frosting or digestion within the rumen releases free HCN. A plant enzyme, glycosidase, hydrolyses the terminal glucose on a cyanogenic glycoside such as amygdalin, producing the aglycone compound hydroxynitrile. Another enzyme catalyzes a further breakdown into HCN and benzaldehyde. The cyanide is absorbed directly from the rumen.

Etiology and Epidemiology

A variety of plants are potentially cyanogenic; the level of HCN can be affected by genetic selection. Immature or rapidly growing plants, and plants heavily fertilized with nitrogen tend to have higher concentrations of cyanogenic glycosides. Damage to the plants by drought, wilting, frosting or chewing increases toxicity because glycoside and enzyme are combine more rapidly. Other ingesta in the rumen may react with cyanide and prevent absorption. There are few reports of cyanide poisoning in goats, but toxicity levels should be comparable with those reported in other ruminants. The goats grazing on pasture having cyanogenic shrubs and trees have higher potential hazard of cynide poisoning. A list of some potentially toxic plants follows.

Acacia leucopholea	*Ramja*
Sorghum spp.	*Jowar, Sodan grass*
Cynodon spp.	*Dub grass*
Manihot esculenta	*Cassava*
Trifolium repens	*White clover*
Zea mays	*Corn*

Pathogenesis and Clinical Signs

The cyanide ion binds to the ferric ion in cytochrome oxidase, yielding a stable complex that cannot transport electrons in the process whereby O2 is used in metabolic respiration. Oxygen is not released from hemoglobin in the blood and thus is not available within cells; cellular asphyxiation occurs. The oxygen-laden blood is bright red but the animal rapidly becomes severely dyspneic. Cerebral anoxia leads to the clinical signs; initial excitement and muscle tremors are followed by gasping and convulsions. Pupils are dilated. Signs are generally peracute, with death occurring within 15 minutes to a few hours after plant consumption.

Diagnosis

Venous blood is bright red in peracute cases but cyanosis of mucous membranes supervenes if death is delayed. There may be an odor of "bitter almond" because of benzaldehyde in the rumen contents. Feed, blood, rumen contents, liver, or muscle tissue may be analyzed. Quick freezing or immersion in 1 % to 3% mercuric chloride will prevent additional release and loss of HCN from the sample. Levels in plants of 200 or more are considered potentially toxic, and with cyanide, toxic usually means lethal. A field test using picrate paper has been described, but this detects cyanide at less than the toxic level.

Treatment

Because death occurs so rapidly (often in 2 to 3 minutes), treatment is rarely possible. However, prompt intravenous injection of sodium nitrite (22 mg/kg) will convert some hemoglobin to methemoglobin, which will preferentially pick up cyanide ions from the cytochrome oxidase enzymes. Simultaneous sodium thiosulfate (67 mg/kg) converts cyanide to stable, less toxic thiocyanate. More recently, an increased intravenous dosage of sodium thiosulfate (660 mg/kg) has been used. Oral or intraruminal sodium thiosulfate (6 g to a large goat) at hourly intervals is recommended to fix free HCN in the rumen of affected or exposed animals.

12.5 Nitrate Poisoning

Ruminant digestion converts ingested nitrates to nitrites and then reduces the nitrites to ammonia. Nitrite is considered to be 10 times more toxic than nitrate.

Etiology and Epidemiology

Atmospheric nitrogen is converted to nitrate (NO3) by nitrogen fixing bacteria, including those associated with the roots of certain plants such as legumes. The plants can then reduce nitrates to nitrites and eventually convert the nitrogen into plant protein. Animal wastes (urea and ammonia) can also enter into the nitrogen cycle. Animals most often consume excess nitrates by eating plants containing increased nitrate concentrations. Water contaminated with animal wastes or run-off from fertilized fields and direct consumption of fertilizers are other possible sources. Sources of nitrate are cumulative. Some plants that may accumulate toxic levels of nitrates are listed below.

Amaranthus	*Pigweed*
Avena sativa	*Oats*
Beta vulgaris	*Beet*
Chenopodium spp.	*Bathua, Goosefoot Alfalfa*
Medicago sativa	*Alfaalfa*
Sorghum spp.	*Jowar*

Heavily fertilized soils will make more nitrates available for plant uptake. Uptake is also favored by acid and moist soils, soils deficient in certain minerals (molybdenum, phosphorus or sulfur) and by low temperatures. Rapidly growing plants after drought or hormonal herbicide application accumulate increased nitrates. Finally, decreased light results in accumulation of nitrates within plants because nitrate reductase enzyme requires light for normal activity. Nitrates accumulate in vegetative tissue rather than fruits or grains and nitrate levels in the whole plant decrease rapidly after flowering and setting of fruit. Corn, for instance is not dangerous after teaseling out.

Pathogenesis

Nitrate toxicity is caused by nitrite formation within the rumen or occasionally, in ensiled forages. Starved animals are at greatest risk because they are less selective grazers. Also, active rumen microorganisms supplied with rapidly digestible carbohydrates (the well-fed animal) can convert nitrates to microbial protein and thereby escape toxicity. Nitrite is absorbed from the rumen. The nitrite causes oxidation of ferrous hemoglobin to ferric hemoglobin (methemoglobin) that cannot transport oxygen. The blood turns a dark, chocolate brown color. In addition to cyanosis, clinical signs include weakness, trembling, severe dyspnea, and frothing at the mouth.

Clinical Signs

Clinical signs are visible when 30 to 40% of the hemoglobin is converted to methemoglobin. Death occurs at 80 to 90% conversion, although this is influenced by stress and exertion. Death usually occurs within 12 to 24 hours after ingestion.Plants containing more than 1.0% nitrates (as KNO3) on a dry weight basis and water with more than 1500 parts per million nitrates may cause acute toxicity.

Diagnosis

Unless there is a history of direct nitrate consumption (e.g., with fertilizer), diagnosis usually requires laboratory confirmation. Dark brown blood and cyanotic membranes indicate the need for methemoglobin determination. Use of a phosphate buffer as a preservative is recommended if testing of heparinized blood will be delayed longer than a few hours. Nitrate levels may be determined in feed, water, rumen contents, or body fluids (including aqueous humor). Nitrite is converted to nitrate very rapidly after death and will often be undetectable in field specimens. Samples should be preserved by freezing.

Treatment

If nitrate toxicity is likely and respiratory signs are present, the clinician should not delay therapy while awaiting laboratory confirmation. Affected animals are treated intravenously with 4 to 15 mg/kg of methylene blue as a 1 % solution in distilled water; this is repeated if necessary. The methylene blue is reduced in blood and body tissues to leucomethylene blue, which in turn rapidly reduces methemoglobin to hemoglobin. When methylene blue is given intravenously to goats, it passes readily into the milk. Ruminal levage with cold water or administration of oral antibiotics or a laxative may decrease additional ruminal reduction of nitrate to nitrite.

12.6 Urea Toxicity

As similar to cattle, urea can be fed to the goats as NPN source which converted in to microbial protein in rumen, if adequate energy sources are available. Urea toxicity can occur in over mixing of urea in feed or improper mixing and / or consumed at the level of 30-50 g/

100 kg bwt in a single meal. Excess urea is degraded in to ammonia and absorbed systemically and further converted in to urea, and cause interference in to citrate cycle. The main clinical signs consisted of muscle and skin tremors, salivation, frequent urination and defecation, incoordination, bloat and death. Urea toxicity can be best managed by administration of acetic acid (Vinegar) @ 200-500 ml by stomach tube, which reduces the rumen pH and retard absorption of urea.

12.7 Subabool Feeding

Sole feeding of subabool caused anemia in goats. In addition, varying degree of degenerative changes occur in liver, kidney, brain and ovaries. Changes in adrenal were indicative of stress. Thyroid exhibited goitrous changes. Reproduction problems (Abortion, non-conception) were correlated with hypothyroidisim which was caused by the goitrogenic effect of subabool.

□□□

Appendix

Normal Physiological values and clinical chemistry of goats

Blood volume	90 ml/kg
Erythrocytes	11-15 X 10^6/mm
Diameter	3. 3 μm
Packed cell volume	28-35%
Leukocytes	8000-11000 X 1O3/mm3
Lymphocytes	50-70%
Neutrophils	30-50%
Eosinophils	1-10%
Monocytes	0- 6%
Basophils	< 1%
Plasma proteins	60- 78.5 g/l
Albumin	32- 49 g/I
Globulin	23- 46 g/I
Temperature	39.4°C (38.6- 40.6°C)
Heart rate	70- 90 beats/min
Respiratory rate	15- 30/min. Polypnea may be seen.
Urine: Specific gravity	1.015 - 1.050
pH	7.0 - 8.0

Volume		10- 40 ml/kg body weight/day
Milk:	Fat	3.5% or more
	Protein	3.1- 3.5%
	Casein	2.7%
	Whey	0.8%
	Lactose	4.6%
	Ash	0.8%
Semen:	Volume	0.5 - 2.5 ml
	Sperm concentration	1000 - 5000 X 10 6/ml

Blood values

Glucose	2.2 - 3.3 mmol/l
Calcium	2.4 - 2.6 mmol/I
Magnesium	1.0 - 1.4 mmol/l
Potassium	3.5 - 6.3 mmol/l
Chloride	100 - 120 mmol/l
Phosphate, inorganic	1.2 - 2.5 mmol/I
Urea	4.0 - 9.9 mmol/I
Creatinine	80 - 159μ mol/ l
Bilirubin	8.55μ mol/ l
Aspartate aminotransferase (AST)	32. 3 IU/l
Alkaline phosphatase (ALP)	56.9 IU/l
Lactate dehydrogenase (LDH)	289.9 IU/I
Ketones	10 mg/100 ml
Total lipids	298.6 mg/100 ml
Uric acid	1.0 mg/100 ml

□□□

Bibliography

Ackermann, M. R. and Brogden, K. A. (2000). Responses of the ruminant respiratory tract to *Mannheimia* (*Pasteurella*) *haemolytica*. *Microbes and Infection,* 2(9):1079-1088.

Adlakha, S.C.; Bansal, M.P. and Malik, B.S. (1971). Studies on sheep and goat-pox viruses. I. Adaptation of the viruses in laboratory animals and tissue culture. *Indian Journal of Animal Sciences,* 41: 171–175.

Ameh, J.A. and Tari, I.S. (2000). Observations on the prevalence of caprine mastitis in relation to predisposing factors in Maiduguri. *Small Rumin. Res.,* 35: 1–5.

Arora, N.; Sharma, S.D.; Vihan,V.S. and Kumar, A. (2006). Seasonal changes in gastro intestinal parasitic infection of goats reared at an organized farm in semi-arid zone. *Indian Journal of Small Ruminants,* 12:37-41.

Arya, H.P.S.; Yadav, M.P. and Tiwari, R. (2002). Livestock technologies for small farm systems. In: Technology Options for Sustainable Livestock Production in India. Eds. P.S. Birthal and P.P.Rao ICRISAT, Patancheru and ILRI, Ethiopia.

Bandyopadhyay, S.K. (2002). The Economics appraisal of a PPR control programme in India. In 14th Ann. Conf. Nat. Symp. on Mang. of Viral diseases with emphasis on global trade & WTO regime. *Ind. Vir. Soc.* 18-20, Jan, Bangalore.

Bandyopadhyay, S.K.; Gajendragad, M.R.; Dhal, N.K.; Gupta, A.R. and Yadav, M.P. (1984). Some observations on a goat pox virus isolated from an outbreak in Orissa. *Indian Journal of Animal Sciences*, 54: 961–964.

Barua, S.; Kumar, A.; Vaid, R. K. and Rana, R. (2005). A clinical appraisal of an outbreak of Sheep pox in Mathura, Uttar Pradesh. In :XIIth Annual Conference of IAAVR in College of Veterinary Sciences, Bikaner on 31st Dec.-01st Jan., 2005.

Barua, S.; Vaid, R.K.; Rana, R. and Kumar, A. (2005). Pox in sheep and goats: Pock marks in small ruminant health. In. Proceedings of National Academy of Sciences, India. Infectious diseases of domestic animals and Zoonoses in India. Ed. V. Tandon and B. N. Dhawan. 75(B) (Biological Sciences). Special issue: 41-54.

Bath, G. F.; Van Wyk, J. A. and Pettey, K. P. (2005). Control measures for some important and unusual goat diseases in South Africa. *Small Ruminant Research*, 60: 127-140.

Batta, M. K.; Katoch, R. C.; Mandeep Sharma; Joshi, V. B. and Nagpal, K. B. (1999). Epidemiological observation on goat pox in Himachal Pradesh. *Indian Veterinary Journal*, 76 : 683-684.

Berge, A. C.; Sischo, W. M. and Craigmill, A. L. (2006). Antimicrobial susceptibility patterns of respiratory tract pathogens from sheep and goats. *Journal of the American Veterinary Medical Association*, 229(8): 1279-1281.

Bhambani, B.D. and Krishnamurthy, D. (1963). An immunodiffusion test for the diagnosis of sheep and goat pox. *Journal of Comparative Pathology and Therapeutics*, 73: 349–357.

Bhat, P.N.; Bhat, P.P.; Khan, B.U.; Goswami, O.B and Singh, B. (1981). Animal Genetic Resources in India. IVRI (ICAR) Izatnagar, Publication No.192, pp 1-111.

Bhat, P.P. (1993). Molecular characterization of pox (viral) DNA. *Indian Journal of Animal Sciences*, 63: 857- 862.

Bickhardt, K. (1994). Clincial studies of kidney function in sheep and effect of pregnancy, lactation and feed restriction and metabolic disease on kidney function. *Dtsch Tierarztl Wochenschr*, 101:467- 471.

Blackwell, T. E. and Butler, D. G. (1992). Clinical signs, treatment, and post mortem lesions in dairy goats with enterotoxemia. *Journal of the American Veterinary Medical Association*, 2: 214 -217.

Bostedt H. and Hamadeh, M.E. (1990). Significance of pregnancy induced ketonuria in sheep and goat. *Tierarzti Prax*, 4 (18): 125-129.

Bretzlaff, K.; Haenelein, H. and Huston, E. (1991). Common nutritional problems feeding of sick goats in large animal clinical nutrition. Edited by J.M. Naylor and S.L. Ralston, St Lois, Mosby year book Inc. 1991 pp 351-355.

Brogden, K. A.; Lehmkuhl, H. D. and Cutlip, R. C. (1998). *Pasteurella haemolytica* complicated respiratory infections in sheep and goats. *Veterinary Research,* 29 (3-4), 233 -254.

Buck, W. B. (1986). Copper-molybdenum. In Current Veterinary Therapy Food Animal Practice 2 edited by J. L. Howard, Philadelphia, W B Saunders Co, 437-439.

Buriel, R. (1997). Dynamics of intramammary infection in the sheep caused by coagulase-negative *staphylococci* and its influence on udder tissue and milk composition, *Vet. Rec.,* 140 : 419–423.

Caple, I.W. and Mclean, J.U. (1984). Pregnancy toxemia. In Current Veterinary Therapy Food Animal Practice 2 edited by J.L. Howard, Philadelphia, WB Saunders Co. pp 320-323.

Carn, V.M. (1993). Control of capripoxvirus infection. *Vaccine,* 11: 1275–1279.

Celer, V.; Zanoni, R. G. and Paterhans, E. (1993) Comparison of various antigens in the diagnosis of caprine arthritis-encephalitis virus using the ELISA test. *Vet. Med.,* 38:237-244.

Clavijo, A.; and Thorsen, J. (1995). Serologic diagnosis of caprine arthritis-encephalitis by ELISA with two recombinant proteins in a parallel testing format. *J. Immunoassay,* 16:419-436

Daniel, J. A.; Held, J. E.; Brake, D. G.; Wulf, D. M. and Epperson W. B. (2006). Evaluation of the prevalence and onset of lung lesions and their impact on growth of lambs. *American Journal of Veterinary Research,* 67 (5), 890-894.

Datta, S. and Soman, J.P. (1990). Dot-immunobinding technique in differential diagnosis of pox diseases of sheep and goats. *Indian Journal of Animal Sciences,* 60: 1186–1187.

Dhar, P.; Sreenivasa, B.P. and Barrett, T. (2002). Recent epidemiology of peste des petits ruminants virus (PPRV). *Vet Microbiol.,* 25: 88 (2):153-159.

Dobbelar, P.; Lange, L.L.M.; De Wuuchkhise, L.A. and Schuvke, Y.H. (1998). Detecting hyperdetomia in cows by means of milk yield data and two test for ketone bodies in milk. *Tiajdschrift Voor Dierjance Skunde,* 123: 140-150.

Dubey, S. C.; Mondal, D. and Tripathi, S. (2006). Reproductive tract diseases (RTD's) and gonadal transmitted diseases (GTD's) of small Ruminants. Proc. "National Seminar on Innovations and Recent Advances in Reproduction for Augmenting Small Ruminant Production" 28-30 Dec. 2006. Avikanagar.

Dubey, S. C.; Sonawane, G. G.; Tripathi, S.; Saravanan, P. and Singh, R. K. (2006). Seroprevalence of PPR in small ruminants of semiarid Rajasthan. *Ind. J Anim. Sci.* (Personnel communication).

Duff, G. C. and Galyean, M. L. (2006). Recent advances in management of highly stressed newly received feedlot cattle. *Journal of Animal Science,* 82:123-127.

Dutta, N. K.; Nayak, B. C. and Patnaik, R. K. (1983). Further studies of an outbreak of goat pox in Orissa and its diagnosis. *Indian Journal of Animal Health.* 22: 121-123.

FAO (1994). A manual for the primary animal health care worker. Food and Agriculture Organization of the United Nations, Rome.

FAO (1995). FAO production year book, Vol. I, (48). Food and Agriculture organization, Rome.

FAO (2002). FAO Statistics. Food and Agriculture Organization, Rome, Italy.

Fernandez-Miyakawa, M. E and Uzal, F. A. (2003). The early effects of *Clostridium perfringens* type D epsilon toxin in ligated intestinal loops of goats and sheep. *Veterinary Research Communications*, 3: 231-241.

Ford, E.J.H. and Evan . (1986). The effect of triamcinolone on glucose metabolism in ketotic sheep. *J. Agri. Sci.,* 106 : 337-340.

Garner, M. G.; Sawarkar, S. D.; Brett, E. K.; Edwards, J. R.; Kulkarni, V. B.; Boyle, D. B. and Singh, S. N. (2000). The extent and impact of sheep pox and goat pox in the state of Maharashtra, India. *Tropical Animal Health and Production,* 32 : 4.

Gershon, P.D.; Ansell, D.M. and Black, D.N. (1989b). A comparison of the genome organization of capripoxviruses with that of the orthopoxviruses. *Journal of Virology*, 63: 4703– 4708.

Goranov, K.H.; Gabrzovski, P.; Zhelyazkov and Popgeogieva, M. (1984) Theraputic and prophylactic effect of monensin and nicotinic acid in cow with sub clinical ketosis. *Veterinarnomeditsinski Narki*, 21: 43-50.

Guo, H.; Guo, Z.Y.; Shang, Z.P. and Liu, Q.G. (1986). Goat pox cell culture attenuated vaccine. *Chinese Journal of Veterinary Science and Technology*, 3: 8–10.

Gupta V. K.; Kumari, R.; Vohra, J.; and Vihan, V. S. (2009). Comparison of serum and whole blood PCR for diagnosis of goat brucellosis. *The Indian Veterinary Journal*, 86 (4) 335-338.

Gupta V. K.; Vohra, J.; Kumari, R. and Vihan, V. S. (2007). PCR-ELISA for diagnosis of brucellosis in goats milk. *Indian Journal of Animal Sciences*, 78(8): 797-800.

Gupta V. K.; Vohra, J.; Kumari, R. and Vihan, V. S. (2009). Evaluation of an ELISA using recombinant outer membrane protein 31 for the diagnosis of ovine brucellosis. *Indian Veterinary Journal*, 79 (1) 41-43.

Gupta V.K.; Rana, R. and Vihan V.S. (2002). Dot-ELISA for detection of *Brucella melitensis* antibodies in goats. *Indian Journal of Animal Sciences*, 72(3): 237-239.

Gupta, V. K.; Kumari, R.; Verma, D. K.; Singh, K.; Singh, S.V. and Vihan V. S. (2006). Detection of *Brucella melitensis* from goat tissues employing PCR. *The Indian Journal of Animal Sciences*, 76 (10): 793-795.

Gupta, V. K.; Rout, P. K. and Vihan,V. S. (2006). Induction of Immune Response in Mice with a DNA Vaccine Encoding Outer Membrane Protein (OMP31) of *Brucella melitensis* 16M. *Research in Veterinary Sciences*, 87:305-317.

Gupta, V. K.; Verma, D K.; Kumari, R.; Singh, S.V. and Vihan V. S. (2006). Single-step PCR for detection of *Brucella melitensis* from tissue and blood of goats. *Small Ruminant Research*, 66:169-174.

Gupta, V. K.; Verma, D K.; Rout, P.K., Singh, S.V. and Vihan, V. S. (2006). Polymerase chain reaction (PCR) for detection of *Brucella melitensis* in goat milk. *Small Ruminant Research*, 65:79-84.

Gupta, V. K.; Verma, D. K.; Singh, S.V. and Vihan, V. S. (2007). Serological diagnostic potential of recombinant outer membrane protein (Omp31) from *Brucella melitensis* in goat and sheep brucellosis. *Small Ruminant Research*, 70: 260-266.

Gupta, V.K.; Rana R. and Vihan, V.S. (2002).Enzyme-linked immunosorbent assay (ELISA) for detection of *Brucella melitensis* antibodies in goats. *Indian Journal of Animal Sciences*, 72 (3):259-261.

Gupta, V.K.; Rout, P.K.; Chandrasekhar, T and Vihan, V.S. (2004). Induction of immune response in goats immunized with *Brucella melitensis* 16M DNA. *Indian Journal of Animal Sciences*, 74 (8):813-817.

Gupta,V.K.; Rout, P.K.; Chandrasekhar, T and Vihan,V.S. (2005). Induction of cytotoxic T cell response in goats immunized with *Brucella melitensis* 16 M DNA. *Indian Journal of Animal Sciences*, 75 (2) : 168-170.

Gupta,V.K.; Verma, D. K.; Rout, P.K. and Vihan,V.S. (2005). Detection of *Brucella melitensis* in goats milk by polymerase chain reaction (PCR). *Indian Journal of Animal Sciences*, 75 (10) : 1163-1164.

Halyal, K.G.; Chetani, L.M. and Popat, M.N. (1980). Adoption of improved animal husbandry practices in ICDP of Junagarh district. *Rural India*, 100: 225 – 227.

Hartelius, E. (1985). Management of pregnancy toxemia in the goats, Senior Seminar, New York State College of Veterinary Medicine.

Heckert, R. A.; McNab, W. B.; Richardson, S. M. and Briscoe, M. R. (1992). Evaluation of an enzyme-linked immunosorbent assay for the detection of antibodies to caprine arthritis-encephalitis virus in goat serum. *Can. J. Vet. Res.*, 56:237-241.

Heine, H.G.; Stevens, M.P.; Foord, A.J. and Boyle, D.B. (1999). A Capripoxvirus detection PCR and antibody ELISA based on the major antigen P32, the homolog of the vaccinia virus H3L gene. *Journal of Immunological Methods,* 227: 187–196.

Henze, P.; Bickhardt, F.; Fuhrmann, H. and Sallmann, H.P. (1988). Spontaneous pregnancy toxemia (ketosis) in sheep and role of insulin. *J. Vet. Med. Series* A 45: 255-266.

Henze, P.; Bickhardt, K.B. and Fuhrmann, H. (1994). The influence of insulin cortisol, growth hormone and fetal oestrogen on the pathogenesis of ketosis in sheep. *Deutsche Tierartiliche Wochenschrift,* 101 (2) (Cited Vet. Bull 64: 7854).

Hoffsis, G.F.; Saint-Jean, G. and Rings D.M. (1989). Hypomagnesaemia in ruminant. *Comp. Contin. Educ.Pract. Vet.,* 11: 519-523.

Joshi, R.K.; Chandra, R.; Rao, V.D.P. and Garg, S.K. (1989). Coagglutination test. A simple and rapid diagnostic technique for goat pox. *Tropical Animal Health and Production,* 21:233–235.

Joshi, R.K.; Chandra, R.; Rao, V.D.P. and Garg, S.K. (1991). Passive haemagglutination test with goat pox virus. *Indian Journal of Veterinary Medicine,* 11: 56–57.

Joshi, R.K.; Garg, S.K.; Chandra, R. and Sharma V.D. (1995). Growth and cytopathogenicity of goat pox virus in MDBK cell line. *Indian Journal of Virology,* 11: 31–33.

Kaufmann, A. and Bergman, E. (1978). Renal studies in normal and toxemic pregnant sheep. *Cornell Vet.,* 68: 124 -134.

Khan, B.U.; Arora, A.L. and Sharma, R.C. (2002). Enhancing productivity of sheep: technology dimensions. In: Technology Options for Sustainable Livestock Production in India. Eds. P.S. Birthal and P.P.Rao ICRISAT, Patancheru and ILRI, Ethiopia.

Khan, B.U.; Vihan, V.S. and Sahni, K.L. (1978). A note on mortality in Jamunapari kids. *Indian Veterinary Medical Journal,* Vol. II 225-226.

Kitching, P. (1983). Progress towards sheep and goat pox vaccines. *Vaccine,* 1: 4–9.

Kitching, R. P. and Taylor, W. P. (1985). Clinical and antigenic relationship between isolates of sheep and goat pox viruses. *Tropical Animal Health and Production,* 17 : 64-74

Kumar, A.; Rana, R.; Vihan, V.S.; Barua S.; Vinod Kumar Gupta and Tiwari, H.A. (2004). Epidemiological features observed in contagious pustular dermatitis in kids. *Indian journal of Animal Sciences,* 74 (11) 1113-1115.

Kumar, A.; Arora, N.; Vihan, V.S.; Rana, R. and Vinod K. Gupta (2005). Tolerance level of egg per gram in haemonchus infection in goats under farming system. *Intas Polivet,* 6 (II): 243-245.

Kumar, A.; Barua, S.; Rana, R. and Vihan, V.S. (2004). Epidemiological features of Foot and Mouth disease virus (Type O) in goats. *Veterinary Practitioner*, 5 (2) 134-137.

Kumar, A.; Rana, R.; Vihan V.S. and V.K. Gupta (2004). Incidence of Caprine pneumonia in relation to climatic factor in semi arid zone. *Indian Journal of Small Ruminants*,10: 131-133.

Kumar, A.; Rana, R.; Vihan, V.S. and Arora, N. (2004). Cerebrospinal fluid analysis and serum biochemistry in *Coenurus Cerebralis* in goats for clinical appraisal. *Veterinary Practitioner*, 5 (1) 27-28.

Kumar, A.; Rana, R.; Vihan, V.S., Arora, N. and Tiwari H.A (2004). A clinical note on occurrence of staphylococcal dermatitis in goats. *Veterinary Practitioner*, 5 (1) 57-58.

Kumar, A.; Rana, R.; Vihan, V.S.; Barua, S.; Gupta V.K. and Tiwari, H.A (2004). Epidemiological features observed in contagious pustular dermatitis in kids. *Indian Journal Animal Sci.*, 74 (11):1113-1115.

Kumar, A.; Singh, S.V.; Rana, R.; Vaid, R.K.; Misri, J. and Vihan, V.S. (2001). PPR outbreak in goats: Epidemiological and therapeutic studies. *Indian J. Animal Sci.*, 71 (9): 815-818.

Kumar, A.; Vaid, R. K.; Barua, S.; Rana, R.; Vinod K. Gupta and Vihan, V.S. (2005). An outbreak of pneumonic pasteurellosis in sheep: Clinical behavior, meteorological variables and management. *Vet. Pract.*, 6 (2) 119-122.

Kumar, A.; Vaid, R.K.; Rana, R.; Vihan, V.S. and Gupta, V.K. (2002). Descriptive epidemiology of colibacillosis outbreak in Barbari kids under semi-intensive system. *Indian Journal of Animal Sciences*, 72(3):233-236.

Kumar, A.; Vihan, V.S. ; Chaudhary, U.B. ; Vinod K. Gupta and Tiwari H.A. (2004). Rumen disorders in goats reared under organized farming system-A retrospective study. *Intas Polivet*, 5 (II): 302-308.

Kumar, A.; Vihan, V.S. and Rana, R. (2006). Clinico-biochemical and therapeutic observations on spontaneous *Acacia leucopholea* poisoning in goats. *Indian Vet. J.*, 83 (1) 41-43.

Kumar, A.; Vihan,V.S.; Rana, R. and Vinod K. Gupta (2005). Blood biochemical changes in some important parasitic infestations in goats for clinical appraisal. *Indian Journal of Small Ruminants*, 11 (2) 156-160.

Kumar, S. and Deoghare, P.R. (2002). Goat rearing and rural poor: A case study in southwestern sem-iarid zone of Uttar Pradesh. *Annals Arid Zone*, 41 (1): 79- 84.

Kumar, S. and Pant K.P. (2002). Goats in India: Status and technological possibilities for improvement. In: Technology Options for Sustainable Livestock Production in India. Eds. P.S. Birthal and P.P.Rao ICRISAT, Patancheru and ILRI, Ethiopia.

Kumar, S., Vihan, V.S. and Deoghare, P.R. (2003). Economic implication of diseases in goats in India with reference to implementation of a health plan calendar. *Small Ruminant Research*, 47: 159-164.

Kumar,A.; Mahour, K.; Vinod Kumar Gupta and Vihan,V.S. (2006). Susceptibility and relative resistance in tick population against cypermethrin in organized farm and field animals. *Veterinary Practitioner*, 7:41-43.

Kusiluka, L.J.M. and Kambarage, D.M. (1996). Diseases of Small Ruminants in Sub-Saharan Africa: A Handbook. Animal Health Programme,Overseas Development Administration, and VETAID.

Lamand, M. and Ligoelements, O. (1978). Alimentation des ruminants. Edit. by R. Jarrige. Versailes, France, INRA pub, 143-159.

Leitner, G.; Chaffer, M.; Zamir, S.; Mor, T.; Glickman, A.; Winkler, M.; Weisblit, L. and Saran, A. (2001). Udder disease etiology, milk somatic cell and NAGase activity in Israel Assaf Sheep throughout lactation, *Small Rumin. Res.* 39: 107–112.

Linzell, J.L. (1965). Milk fever in goats. *Vet. Rec.* 77: 767-768.

Lloyd, W. E. (1986). Urea and other non protein nitrogen sources. In. Current Veterinary Therapy Food Animal Practice 2 Edited by J. L. Howard, Philadelphia, W B Saunders Co, 354-356.

Lodh, C.; Chakrabarti, A. and Mukhopadhyay, S. (1993). Factors affecting kid mortality in West Bengal. *Indian Veterinary Journal*, 70 : 48-50.

Marteniuk, J.V. and Herdt, T.H. (1988). Pregnancy toxemia and ketosis in ewes and does. *Vet. Clin. North Am. Food Anim. Pract.*, 4: 307-315.

Maynard, L. A. and Loosli, J. K. (1969). Animal Nutrition, 6th Ed. McGrew Hill Book Co. Inc Newyork.

Menzies, P.I. (2000). Ruminal Health Management Group. Ontario Veterinary College, University of Guelph.

Menzies, P.I. and Ramanoon, S.Z. (2001). Mastitis of sheep and goats, *Vet. Clin. North Am. Food Anim. Pract.*, 17: 333–358.

Merck and Company (2006). Pasteurellosis of Sheep and Goats: Introduction. Merck Manual Whitehouse Station, NJ: Merck & Company.

Mirshamsudin; Joshi, B.P.; Rai, P and Vihan, V.S. (1988b) Dynamics of serum immunoglobulin in Jamunapari goats under semi arid condition. *Acta Veterinaria Brno.*, 14:134-36.

Miserez, R.; Frey, J.; Buogo, C.; Capaul, S.; Tontis, A.; Burnens A. (1998). Detection of alpha and epsilon toxigenic *Clostridium perfringens* type D in sheep and goats using a DNA amplification technique (PCR). *Letters in Applied Microbiology*, 26(5): 382-386.

Misri, J.; Vihan, V.S. and Kumar, A. (2000). Caprine ruminal acidosis- A retrospective studies. *Vet. Practitioner*, 1 (2):127-130.

Misri, J.; Vihan, V.S. and Kumar, A. (2003). Toxicity studies on *Prosopis juliflora* in goats- Haematobiochemical and pathological profile. *Indian Journal of Animal Sciences*, 73 (4):349-352.

Morand-Fehr, P.; Bas, P.; Hervien, J. and Sauvant D. (1984). Observation on ketosis in goat: Etiology and metabolic state. In les maladies de la cherre. (Cited by Smith and Sherman, Goat Medicine, Lea and Febiger, Philadelphia.

Nagpal, A.K.; Vihan, V.S.; Singh, D.; Singh, S.V. and Singh, N. (1990). Epidemiological study of an outbreak of goat pox and its effect on the performance on kids. *Indian Journal of Anim. Sci.*, 60 (5): 544-545.

Nelson, D.R. (1992). Metabolic and nutritional disease, S.B. Guss, Pennsylvania State University Pennsylvania.

Ngichabe, C.K.; Binepal, Y.S.; Njiru, J.W. and Carn, V.M. (1999). Evaluation of an immunocapture enzyme-linked immuno-sorbent assay for detection of capripoxvirus. *Veterinary Record*, 145: 231–232.

OIE. (2002) Office International des Epizooties/World Organization for Animal Health). Peste des petits ruminants. Technical disease card database. Apr. 22.

Oros, J.; Fernandez A.; Rodriguez, J. L.; and Poveda, J. B. (1997). Bacteria associated with enzootic pneumonia in goats. *Journal of Veterinary Medicine*, Series B, 44(2): 99-104.

Overby, I. and Odegaard, S.A. (1980). Hypokalsemi hos geit. *Norsk Veterinaertid*, 91: 21-25.

Ozkul, A.; Akca, Y.; Alkan, F.; Barrett, T.; Taner, K.; Dagalp, S.B.; Anderson, J.; Yesilbag, K.; Cokcaliskan, C.; Gencay, A. and Burgu, I. (2002). Prevalence, distribution and host range of peste des petits ruminants virus in Turkey. *Emerging Infectious Diseases*, 8:708-712.

Pawaiya, R. V. S.; Misra, N.; Bhagwan, P. S. K. and Dubey, S. C. (2004). Pathology and distribution of antigen in goat naturally infected with PPR virus. *Indian Journal Animal Sciences*, 74: 35-40.

Phelps, S. L. and Smith, M. C. (1993). Caprine arthritis-encephalitis virus infection. *J. Am. Vet. Med. Assoc.*, 203:1663-1666.

Prasad, M. C. and Sastry, M. S. (1989). National Symposium on Pathology and Biotechnology in the diagnosis of diseases of livestock and poultry, IVRI, Izatnagar.

Radostits O.M. ; Blood, D.C. and Gay, C.C. (1994) Veterinary Medicine, 8th Edi. ELSB, BT, London.

Radostits, O.M.; Gay, C.C.; Blood, D.C. and Hinchcliff, K.W. (2000). Veterinary Medicine- A text book of Diseases of Cattle, sheep, pigs, goats and horses, 9th Edition, Saunders, London. pp. 1077-1079.

Rajak, K. K; Sreenivas, B. P.; Hosamani, M.; Singh, R. P. and Bandopadhayay, S.K. (2005). Experimental studies on immuno suppressing effect of PPR virus in goats. *Comp. Immun. Mico*, 28(4):287-96.

Ram Kumar, S. (1989). Adoption pattern of selected animal husbandry practices by dairy farmers: An extensive study. *J. Edu. Res. and Ext.*, 25 (4):192–197.

Ramirez-Romero, R. and Brogden, K. A. (2000). The potential role of the Arthus and Schwartzman reactions in the pathogenesis of pneumonic pasteurellosis. *Inflammation Research*, 49 (3): 98-101.

Rana, R.; Gupta, V.K. and Vihan, V.S. (2002). Isolation and characterization of *Brucella melitensis biovar-1* from goats. *Indian Veterinary Medical Journal*, 26:257-259.

Rana, R.; Kumar, A. and Vihan V.S. (2003). Study of goat pox out break and its clinical management. *Indian Journal of Animal Research*, 39 (2): 149-150.

Rana, R.; Kumar, A.; Vihan, V.S. Tiwari, H.A. and Vinod K. Gupta (2002). Serological evidence of occurrence of Bluetongue virus in goats of Uttar Pradesh. *Indian Vet. Med. Journal*, 26: 157-158.

Rana, R.; Mehra, S.; Vijay kumar; Kumar, A. and Vihan, V.S. (2009). A Study on sero-prevalence of Contagious pleuropneumonia in goats using native *M. mycoides* Sub species *capri* Isolated. *Indian Veterinary Journal*, 115-117.

Rana, R.; Vaid, R.K.; Bhatia, A.K.; Kumar, A. and Vihan, V.S. (2001). Bacteriological study and antibiogram profile of infantile diarrhoea in kids. *Indian Vet. Med. Journal*, 25(4):333-335.

Ranveera, A.; Ford, E.J.H. and Smad, A.R. (1972). The effect of Triamcinolone acetonide in plasma glucose and ketone concentration and a total entry of glucose in twin pregnant hypoglycaemic ketotic sheep. *Res. Vet. Sci.*, 26: 12-16.

Rao, T.V.S.; Poonam Malik; Nandi, S. and Negi, B.S. (1997d). Evaluation of immunocapture ELISA for diagnosis of goat pox. *Acta Virologica*, 41: 345–348.

Reid, R.L. (1968). The physiopathology of under nourishment in pregnant sheep with particular reference to pregnancy toxemia. *Adv. Vet. Sci.*, 12: 163-238.

Rekib, A. and Vihan, V.S. (1997). Economic losses in goat production due to diseases. 3rd National Seminar on Small Ruminant Diseases. Pub. CIRG, Makhdoom, pp1-9.

Rimstad, E.; East, N.; DeRock, E.; Higgins, J. and Pedersen, N. C. (1994). Detection of antibodies to caprine arthritis-encephalitis virus using recombinant gag proteins. *Arch. Virol.*, 134: 345-356.

Romero, C.H.; Battett ,T.; Evans, S.A.; Kitching, R.P.; Gershon P.D.; Bostock, C. and Black, D.N. (1993). Single capripoxvirus recombinant vaccine for the protection of cattle against Rinderpest and Lumpy skin disease. *Vaccine,* 11: 737–742.

Rook, J.S. (2000) Pregnancy toxemia of ewes, does and beef cows. *Vet. Clin. North. Am.,* (16): 393-397.

Rosati, S.; Pittan, M.; Tolari, F.; Erre, G. and Kwang, J. (1995). Genetic and antigenic characterization of CAEV (caprine arthritis-encephalitis virus) recombinant trans-membrane protein. *Vet. Microbiol.,* 45:363-370.

Saha, A.; Lodh, C. and Chakraborty, A. (2005) Prevalence of PPR in goats. *Indian Vet. J.,* 82:668-69.

Saha, G.R.; Roy, S.S.; Mukherjee, N. and Mitra, K. (1985). Clinical and epidemiological features of pox in Black Bengal goats of West Bengal (India). *Indian Journal of Animal Health,* 24: 29–34.

Saini, A.L.; Khan, B.U.; Prakash, B. and Vihan, V.S. (1986). Effect of management systems on the growth performance of Barbari kids under semi-intensive conditions. *Indian Journal of Anim. Sci.,* 56: (9) 990- 993.

Saini, A.L.; Prakash, B. and Vihan, V.S. (1987). Management systems in relation to some production traits in Barbari kids. *Indian Vet. Journal,* 64: 596-598.

Sambrook J. E.; Fritsh, F. and Maniatis, T. (1989). Molecular cloning: a laboratory manual, 2nd ed. Cold Spring Harbor Laboratory Press, Cold Spring Harbor, N.Y.

Sanjay, S.; Rao, V. D. P. and Rajesh Chandra (2004). Characterization of capripox virus isolated from field outbreak in goats. *Indian Veterinary Journal.* 81: 241-244.

Sauvant, D.; Chilliard,Y. and Morand-fehr, P. (1991). Etiological aspect of nutrient and metabolic disease in goats. In Goat nutrition edited by P. Morand-fehr, wageningen, Netherland, pp 124-142.

Sauvant, D.; Morand Fehr, P. and Bas, P. (1984). In goat diseases, International Conference, Noist, France, INRA, Pub. pp 369-378.

Shaila, M. S.; David, S.; Forsyth, M. A.; Diallo,A.; Goately, L.; Kitching, R. P. and Barrett, T. (1996). Geographic distribution and epidemiology of peste des petits ruminants viruses. *Virus Research,* 43: 149-153.

Shankar, H.; Gupta, V. K. and Singh, N. (1998). Occurrence of pest des petits ruminants like disease in small ruminants in Uttar Pradesh. *Ind. Journal of Anim. Sci.,* 68(1):38-40.

Sharma, G.; Singh, S.V.; Sevilla I.; Singh, A.V.; Whittington, R.J.; Juste R.A.; Kumar, A.; Gupta,V.K.; Singh, P.K.; Sohal, J.S. and Vihan, V.S. (2008). Evaluation of indigenous milk ELISA with m-culture and m-PCR for the diagnosis of Bovine Johne's disease (BJD) in lactating Indian dairy cattle. *Research in Veterinary Science*, 84:30-37.

Simard, C. and Briscoe, M. (1990). An enzyme-linked immunosorbent assay for detection of antibodies to Maedi-Visna in sheep. II. Comparison to conventional agar gel immunodiffusion test. *Can. J. Vet. Res.*, 54:451-456.

Simard, C.; and Briscoe, M. (1990). An enzyme-linked immunosorbent assay for detection of antibodies to Maedi-Visna virus in sheep. I. A simple technique for production of antigen using sodium dodecyl sulfate treatment. *Can. J. Vet. Res.*, 54:446-450.

Singh, N.; Kala, S.N.; Vihan, V.S. and Singh, S.V. (1990). Genetic study on the susceptibility to Johne's disease in goats. *Indian Journal of Anim.* Sci., 60 (10): 1164-1167.

Singh, N.; Singh, S.V.; Sinha, N.K. and Vihan, V.S. (1989). Ovine wool shedding syndrome in subabool toxicity. *Indian Journal Vet. Med.*, 9: 67-68.

Singh, N.; Vihan, V.S.; Singh,S.V. and Gupta, V.K. (1998). Prevalence of Johne's disease in organized goat herds. *Indian Journal of Anim. Sciences*, 68 (1): 41.

Singh, N.; Vihan,V.S.; Sengupta, U. and Singh, S.V.(1999). Evaluation of enzyme-linked immunosorbent assay using wild type species-specific sonicated antigen for diagnosis of *Mycobacterium paratuberculosis* in goats. *World Review of Animal Production*, 34 (2) 78-81.

Singh, S.V. and Vihan V.S. (2003). *Mycobacterium avium* subsp. *paratuberculosis*: A major threat and possible association with Crohn's disease of human beings. *Intas Polivat*, 4 (1) 15-17.

Singh, S.V. and Vihan, V.S. (2003). Incidence *of Mycobacterium avium* subsp. *Paratuberculosis* in clinically suspected small ruminants. *Indian Journal of Small Ruminants*, 9:56-59.

Singh, S.V. and Vihan, V.S. (2004). Detection of *Mycobacterium avium* subsp. *paratuberculosis* in goat milk. *Small Ruminant Research*, 54:231-235.

Singh, S.V.; Singh P.K.; Singh, A.V.; Sohal, P.K.; J.S.;. Gupta,V.K and Vihan, V.S. (2007). Comparative efficacy of an indigenous 'inactivated Vaccine' using highly pathogenic field stain of *Mycobacterium avium* subspecies *paratuberculosis* ' Bison type' with commercial vaccine for the control of Capri-paratuberculosis in India. *Vaccine*, 25 :7102-7110.

Singh, S.V.; Singh, A.V.; Shukla, N.; Singh, P.K.; Sohal, J.S.; Gupta, V.K. and Vihan, V.S. (2007). Sero-prevalence of Johne's disease in prospective young bulls of Haryana breed by indigenous ELISA kit, using protoplasmic antigen from "Bison type' genotype of *Mycobacterium avium* subspecies *paratuberculosis* of goat origin. *The Indian Journal of Animal Sciences*, 77 (8) : 659-662.

Singh, S.V.; Singh, N.; Gupta, V.K.; Shankar, H.; Vihan, V.S.; Gupta, Vinod K. and Tiwari, H.A. (1998). Sero-prevalence of brucellosis in a few important Indian goat breeds. *Small Ruminant Research*, 30: 93-98.

Smith, M.C. and Sherman, D.M. (1994). Goat Medicine, Lea and Febiger Philadelphia.

Sohal, T.S. (2000) Constraints in the transfer of technology in livestock sector. Paper presented in summer school on "Animal Husbandry and Veterinary Extension Education", May 8-28, IVRI, Izatnagar.

Sotillio, J.; Ceron, J.; Dornal, A.; Benedeto, J. and Gutterrez, P. (1994). Variation in blood chemistry in fasted and normally fed lactating goats, In proceedings of 18th World Buiatrics congress 26th congress Itallian Association Buiatrics, Bdogna Italy, 29th Sept., Vol. 2 1313-1316 (Cited Vet. Bull 66: 1780).

Stevenson, D.E. and Wilson, A.A. (1963). Metabolic disorders of domestic animals. Blackwell scientific publication, Oxford.

Sunil kumar, N. S.; Ravishankar, C.; Jayaprakasan, V.; Mini, M. and Krishnan Nair, G. (2005). Seroprevalence of peste des petits ruminants in Kerala. *Indian Vet. J.*, 82:570-571.

Swarup, D. (1989). Lead Paper. Proceeding National Symposium on emerging diseases of livestock due to environmental pollution. IVRI, Izatnagr.

Taneja, V.K. (2002). Keynote address at V National Seminar of ISSGPU on "Strength, Challenges and Opportunities in Small Ruminant Diseases in New Millennium" at Jaipur, December, 30.

Taylor, W.P.; Diallo, A.; Gopalakrishna, S. (2002). Peste des petits ruminants has been widely present in southern India since, if not before, the late 1980s. *Prev. Vet. Med.*, 52 (3-4):305-12.

Tibbo, M.; Woldemeskel, M. and Gopila, A. (2001). An outbreak of respiratory disease complex in sheep in central Ethiopia. *Trop. Anim. Hlth. Prod.* 33(5):355-65.

Tiwari, A.K.; Rao, T.V.S. and Negi, B.S. (1996). Spot agglutination test for the rapid diagnosis of goat pox. *Tropical Animal Health and Production*, 28: 213–215.

Uzal F. A. and Kelly W. R. (1998). Protection of goats against experimental enterotoxemia by vaccination with *Clostridium perfringens* type D epsilon toxoid. *The Veterinary Record*, 142(26): 722-725.

Uzal, F. A.; Pasini, M. I.; Olaechea, F. V.; Robles, C. A. and Elizondo, A. (1994). An outbreak of enterotoxemia caused by *Clostridium perfringens* type D in goats in Patagonia. Veterinary Record, 135(12): 279-280.

Uzal, F. A.; Plumb, J. J.; Blackall, L. L. and Kelly, W. R. (1997). PCR detection of *Clostridium perfringens* producing different toxins in faeces of goats. *Letters in Applied Microbiology*, 25(5): 339-344.

Vaid, R. K.; Barua, S.; Kumar, A.; Dwivedi, D. S. and Vihan, V. S. (2008). Isolation and molecular identification of multi drug resistant *Pasteurella multocida* isolate from sheep. *Veterinary Practitioner*, 9:32-36.

Vaid, R.K.; Kumar, A.; Barua S.; Rana, R. and Vihan V.S. (2004). Abortion in goats- A storms of Zoonotic diseases. *Intas Polivet*, 5 (II): 302-308.

Vaid, R.K.; Rana R.; Kumar .A. and Vihan V.S. (2003). Isolation of *E.coli* of serogroups indicating probable presence of VTEC from diarrheal kids. *Indian Journal of Animal Sciences*, 73: 1197-1199.

Vander Schalie, J.; Bradway, D. S.; Besser, T. E. and Evermann, J. F. (1994). Evaluation of a kinetic enzyme-linked immunosorbent assay for detection of caprine arthritis-encephalitis virus specific antibodies. *J. Vet. Diagn. Investig.*, 6:30-33.

Ved Prakash; Chandra, R.; Rao, V.D.P. and Garg ,S.K. (1994). Cultivation of goat pox virus in established cell line. *Indian Journal of Virology*, 10: 60–63.

Venkteswarulu, K.; Rao, D.S.T.; Reddy, K.S.; Ram Rao and Gaffar, A.A. (1994). Clinico biochemical findings in sub clinical ketosis in cross bred cows. *Indian J. Vet. Med.*, 14: 6-8.

Vihan, V.S. (1980). Gangrenous mastitis in crossbred ewes- A report on two clinical cases. *Indian Vet. Journal*, 57: 257-258.

Vihan, V.S. (1981). A note on evaluation of cerebrospinal fluid in health and diseases in sheep. *Indian Veterinary Journal*, 58: 938-940.

Vihan, V.S. (1982). Studies on metabolic derangements with special reference to pregnancy toxemia in sheep and goats. Ph.D. thesis submitted to (SAU, Kanpur).

Vihan, V.S. (1986). Sheep and goat immunoglobulin and their affect on neonatal survivability and performance. *World Review of Animal Production*, 22 (4): 65-68.

Vihan, V.S. (1987). Hexachlorethane toxicity in goats. *Indian Vet. Journal*, 64: 180-181.

Vihan, V.S. (1988). Immunoglobulin levels and their affect on neonatal survival in sheep and goats. *Small Ruminants Research*, 1: 135-148.

Vihan, V.S. (1989). Determination of NA Gase activity in milk for diagnosis of sub clinical caprine mastitis. *Small Ruminant Research*, 2: 359-366.

Vihan, V.S. (1989). Glutaraldehyde coagulation test for detection of Hypo-Gammaglobulinaemia in neonatal kids. *Indian Vet. Journal*, 66: 101-105.

Vihan, V.S. (1989). Use of improved media for detection of K99 antigen in kid enterotoxigenic strains of *E. coli*. *Indian Journal Vet. Med.*, 9: 36-37.

Vihan, V.S. (1990). Improved method for primary cultivation of *Mycobacterium paratuberculosis* from goat faeces. *Indian J. Vet. Med.*, 10: 96-100.

Vihan, V.S. (1991). Goat breeds of India, their physiological and reproductive norms. *Vet.*, 3: 26-30.

Vihan, V.S. (1991). N-Acetyl-B-D-Glucosaminidase activity, somatic cells, percent fat and protein in sub-clinical mastitis. Indian *Journal Vet. Med.*, 10 (1): 84-86.

Vihan, V.S. (1992). Epidemiological investigation on neonatal kid mortality due to enteropathogenic colibacillosis. *Preventive Veterinary Medicine*, 13: 179-183.

Vihan, V.S. (1992). Review on important helminthic diseases of goats in India. *Indian J. Anim. Prod. Mgmt.*, 8: 111-116.

Vihan, V.S. (1993). Review on contagious caprine pleuropneumonia. *Vet.*, 4 (1): 5-7.

Vihan, V.S. (1993). Use of *Escherichia coli* vaccine for passive protection against neonatal colibacillosis in goats. *Small Ruminant Research*, 11: 179-185.

Vihan, V.S. (1994). Distribution of *Escherichia coli* serotypes in diarrhoeic kids amongst the goat flocks. *Indian Vet. J.*, 71: 633-638.

Vihan, V.S. (1994). Rehydration therapy of neonatal kids affected with colibacillosis. *Indian Vet. J.*, 71: 51-55.

Vihan, V.S. (2002). Effect of salinomycin on oocyst and live weight gain in experimental caprine coccidiosis. *Indian Journal of Veterinary Medicine*, 22 (2):2100-102.

Vihan, V.S. (2002). Evaluation of ELISA using wild type species-specific sonicated antigen for diagnosis of *Mycobacterium paratuberculosis* in goats. *Indian Journal of Veterinary Medicine*, 22 (2):21-24.

Vihan, V.S. and Kumar, S. (1999). Disease incidence in goats: Economic implications and options. *Vet.*, 6 : 23-27.

Vihan, V.S. and Panwar, H.S. (1987). A note on toxicity of *Acacia Leucopholea* (Ramja) in sheep. *Indian Journal of Animal Research*, 21 (1): 53-55.

Vihan, V.S. and Rai, P. (1984). Clinical,haematological and therapeutic studies on metabolic disorders in sheep and goat. *Indian Vet. J.*, 61 (4): 300-303.

Vihan, V.S. and Rai, P. (1985). Experimental pregnancy toxemia in sheep and goats. *Indian Vet. Journal*, 62: 958-963.

Vihan, V.S. and Rai, P. (1985). Studies on blood serum changes in experimental hypocalcaemia in sheep. *Indian Veterinary Journal*, 62: 573-576.

Vihan, V.S. and Rai. P. (1987). Certain haematological and bio-chemical attributes during pregnancy and parturition and post-parturition in sheep and goats. *Indian Journal of Animal Sciences*, 57 (11) 1205-1210.

Vihan, V.S. and Sahni, K.L. (1979) Efficacy of four anthelmintics in growing lambs under semi-arid condition. *Indian Journal of Animal Research*, 13 (1) 15-18.

Vihan, V.S. and Sahni, K.L. (1980). A note on the efficacy of different therapeutic measures adopted in clinical cases of colibacillosis in lambs and kids. *Veterinary Res. Journal*, 3 (1): 47-49.

Vihan, V.S. and Sahni, K.L. (1981). Note on the quantitative changes occurring in the gamma-globulin level during pregnancy and lactation. *Indian Journal Anim. Sci.*, 51: 992-993.

Vihan, V.S. and Sahni, K.L. (1981). Note on the seasonal changes in body temperature, pulse and respiration rates of Jamunapari goats in semi-arid conditions. *Indian Journal Anim. Sci*, 52 (1) : 115-117.

Vihan, V.S. and Sahni, K.L. (1982). Note on the relationship between serum gamma-globulin concentration and resistance to early postnatal diseases in lambs. *Indian Journal of Anim. Sci.*, 52 (10): 979-980.

Vihan, V.S. and Sahni, K.L. (1982). Studies on various factors effecting on lamb mortality. *Indian Veterinary Journal*, 59: 138-142.

Vihan, V.S. and Sahni, K.L. (1985). Note on the effect of gamma-globulin therapy on the survivability and performance of young lambs. *Indian Veterinary Journal*, 62: 306-308.

Vihan, V.S. and Sahni, K.L. (1987). Observation on efficacy of various indirect diagnostic tests for detection of sub-clinical mastitis in goats. *Indian Vet. Journal*, 64: 715-716.

Vihan, V.S. and Saini, A.L. (1987). Effect of some health practices on the growth rate and survival of Jamunapari goats. *Indian Journal of Animal Production Management*, 3 (4): 219-222.

Vihan, V.S. and Saini, A.L. (1987). Influence of body temperature of new born lambs and kids on their survivability- A note. *Indian Journal Anim. Prod.*, 3 (3): 144-146.

Vihan, V.S. and Singh, N. (1989). Studies on the strain of *Escherichia coli* serotypes isolated from diarrhea kids and their resistance to drugs. *Indian Journal of Anim. Sci.*, 59 (5): 655-656.

Vihan, V.S.; Nem Singh and Singh, S.V. (1988). Prevalence of clinical coccidiosis in kids under semi-arid conditions. *Indian Journal of Anim. Sci.*, 58 (10): 1178-1180.

Vihan, V.S.; Sahni, K.L.; Sinha, N.K. and Joshi, J.D. (1981). Effect of serum gamma-globulin levels on the survivability and growth in lambs and kids. *Indian Journal Anim. Sci.,* 51: 1059-1062.

Vihan, V.S.; Saini, A.L. and Khan, B.U. (1986). Note on preliminary investigation on mortality in Jamunapari kids. *Indian Vet. Med. Journal,* 10: 166-169.

Vihan, V.S.; Saini, A.L.; Singh, B. and Prakash, B. (1986). Factors affecting kid mortality of Jamunapari goats in semi arid conditions. *Indian Journal of Veterinary Medicine,* 6 (1): 8-10.

Vihan, V.S.; Singh, N. and Singh, S.V. (1989). Preliminary observations on faecal culture test for the detection of *M. paratuberculosis* in goats. *Indian Journal of Anim. Sci.,* 59 (12): 1522-1524.

Vihan, V.S.; Singh, N. and Singh, S.V. (1994). Passive immunization of kids against neonatal colibacillosis by vaccinating of goats with whole cell bacterin. *World Review of Animal Production,* 29 (1) 18-23.

Vihan, V.S.; Singh, N. and Singh, S.V. (1997). Evaluation of combined therapeutic regimes in the treatment of natural form of paratuberculosis in goats. *Indian J. of Vet. Med.,* 17: 108-110.

Vihan, V.S.; Singh, S.V. and Singh, N. (1990). Prevalence, pathogenicity and serotypes of *E. coli* associated diarrhoea in newborn kids. *Indian Journal Anim. Sci.,* 60: 793-795.

Vihan, V.S.; Sinha, N.K. and Sahni, K.L. (1981). Seasonal changes in body temperature, pulse and respiration rate in different genetic groups of sheep. *Indian Veterinary Journal,* 58: 717-721.

Vihan, V.S.; Wani, G.M. and Sahni, K.L. (1979). A note on the therapeutics of experimental rumen acidosis in goats. Veterinary Research Bulletin, 2:84-85.

Vihan, V.S.; Wani, G.M. and Sahni, K.L. (1982). Observation on changes in blood serum in experimental rumen acidosis in goats. *Indian Veterinary Journal,* 59: 998-1000.

Vihan,V.S. and Nem Singh (1988). Epidemiological and clinicopathological studies on colibacillosis in kids. *Indian Journal of Anim. Sci.,* 58 (2): 233-236.

Vihan,V.S. and Rai, P. (1983). Metabolic profiles at different physiological stages in sheep and goats. *Indian Journal of Veterinary Medicine,* 3: 1-8.

Vihan,V.S. and Rai, P.(1984).Studies on biochemical and haematological changes in metabolic derangements of sheep and goats. *Indian Journal of Veterinary Medicine,* 4 (1): 9-12.

Vihan,V.S.and Rai, P. (1984). Studies on haematological and biochemical changes in experimental pregnancy toxemia induced by fasting in sheep and goat. *Indian Journal of Veterinary Medicine,* 4 (1): 32-36.

Vihan. V.S. (2002). Impact of *Haemonchus contortus* on live body weight gain in experimentally infected yearling goats. *Indian Journal of Veterinary Medicine*, 22(2): 94- 95.

Vinod K Gupta; Kumar, A. and Vihan, V.S. (2005). Clinical parturient paresis (Milk fever) in goat -A case report. *Veterinary Practitioner*, 6 (1): 17.

Vinod K Gupta; Kumar, A.; Arora, N., Vihan, V.S. (2003). Efficacy of doramectin in ecto and endo parasitic infestation in goats. *Indian Vet. Med. J.*, 27: 81-82.

Vinod K. Gupta ; Kumar, A.; Rana, R. ; Vihan, V.S. and Barua, S. (2005). Therapeutic effect of different antimicrobial combinations on survivability of kids suffering from collibacillosis. *Vet. Pract.*, 6 (2) 112-123.

Vinod K. Gupta; Kumar, A.; Vihan, V.S. and Sharma S.D. (2008). Studies on Haemogram in sub-clinical ketosis in goats and sheep in organized farming system. *Indian Journal of small Ruminants*, 14 (1): 114-117.

Vinod K. Gupta; Kumar, A.; Vihan, V.S. and Sharma, S.D. (2008).Alteration in biochemical parameters in sub-clinical ketosis in goats under organized farming system. *Indian Vet. Journal*, 85:1234-1236

Vinod K. Gupta; Kumar, A.; Vihan,V. S. and Sharma, S.D. (2007). Prevalence of ketosis in goats maintained under organized farming system. *Indian Vet. J.*, 84 (11):1169-1172

Vinod K. Gupta; Sharma, S.D.; Vihan, V.S. and Kumar, A. (2008). Prevalence and changes in haemogram in sub-clinical ketosis in sheep reared under organized farming system. *Indian Journal of Animal Sciences*, 78 (5): 453-456.

Vinod K. Gupta; Sharma,S.D.; Vihan, V.S. and Kumar, A. (2007).Influence of kidding seasons on prevalence of sub-clinical ketosis in goats under organized farm conditions. *Indian Journal of Small Ruminants*, 13 (2) 202-204

Vinod Kumar Gupta; Vihan, V. S. and Sharma, S. D. (2006). Effect on milk yield in sub-clinical ketotic goats reared under semi intensive system management. *Veterinary Practitioner*, 7:110-111.

Wadha, D. R.; Prasad ,B.; Verma ,S. and Sharma, A. K. (2002). Occurrence of PPR like outbreak among migratory goat and sheep in Himachal Pradesh. *Ind. J. Vet. Med.*, 22(2):105-106.

Wang ,S.H. and Jiang, H.X. (1988). Studies on attenuated goat pox virus vaccine. The manufacture of freeze-dried tissue culture goat pox modified virus vaccine. *Acta Veterinaria et Zootechnica Sinica*, 19: 279–284.

Webbs, G.; Jennings, D.M.; Redding, A.J. and Mellor, P.S. (1980). Sheep and goat pox, transmission of capripox viruses by various flies indicated the need for a reassessment of the methods of controlling this disease. In: Annual Report, Institute for Animal Health, Pirbright, UK.

West, H.J. (1996). Maternal under nutrition during late pregnancy in sheep, its relationship to maternal condition, gestation length, hepatic physiology and glucose metabolism. *Br. J. Nutr.*, 4 (75) :593-605.

Winter, P.; Fuchs K.; Walshe, K. and Colditz, I.G. (2003). Serum amyloid A in the serum and milk of ewes with mastitis induced experimentally with *Staphylococcus epidermidis. Vet. Rec.*, 152 (2003): 558–562.

Young, J. D.; and Griffith, J. W. (1985). Spontaneous Pasteurella pneumonia in adult laboratory goats complicated by super infection with *Corynebacterium pseudotuberculosis* and *Muellerius capillaris. Laboratory Animal Science*, 35(4): 409-411.

Zanoni, R. G.; Vogt, H. R.; Pohl, B.; Bottcher, J.; Bommeli, W. and Peterhans, E. (1994). An ELISA based on whole virus for the detection of antibodies to small-ruminant lentiviruses. *Zentbl Veterinarmed.*, 41:662-669.

□□□

Index

B

D

E

F

G

H

I

J

K

L

M

P

Q

R

S

T

U

V

W

Y

Z

□□□

About the Author

Dr. V.S.Vihan is currently Principal Scientist and Head, Animal Health Division at Central Institute for Research on Goats, Makhdoom, Mathura (U.P.). He has over 37 years of research experience in the field of diseases of Small Ruminants. He is the pioneer worker who has long experience working in the field of goat and sheep diseases in country. His work on metabolic disorders, development of vaccine and control of neonatal mortality caused due to *E.coli* infection, development of diagnostic kits against brucellosis and Johne's disease and diagnostic test of sub-clinical caprine mastitis and Herbal drug development are accredited to his accomplishment. His research work has been cited in number of text books of Veterinary Medicine and Goat Medicine. Dr Vihan has undergone advanced training in the field of Animal Health Research at National Institute of Animal Health, Tsukuba, Japan and All-Russian Research Institute for Veterinary Virology and Microbiology, Pokrov, Russia. He has published more than 250 research papers in National and International Journals. Besides that, he has published 3 books, 30 popular articles, 25 research bulletins, 150 papers presented in National and International Seminars, Symposiums. He has recipients of many awards Viz. D.C. Blood Gold Medal award for outstanding research in Veterinary Medicine. He is member of many scientific societies of India and also was the member of Editorial Board of International Journal of goat association of Small Ruminant Research.